KU-281-846

EXTRACELLULAR MATRIX-CELL INTERACTION
Molecules to Diseases

EXTRACELLULAR MATRIX-CELL INTERACTION
Molecules to Diseases

Edited by

Yoshifumi Ninomiya
Okayama University Medical School, Okayama, Japan

Bjorn Reino Olsen
Harvard Medical School, Boston, U.S.A.

Toshiro Ooyama
Toho University School of Medicine, Tokyo, Japan

JAPAN SCIENTIFIC SOCIETIES PRESS
Tokyo

KARGER
Basel·Freiburg·Paris·London·New York·New Delhi·
Bangkok·Singapore·Tokyo·Sydney

© Japan Scientific Societies Press, 1998

All rights reserved. No part of this publication may be reproduced or transmitted in any form or by any means, electronic or mechanical, including photocopy, recording, or any information storage and retrieval system without permission in writing from the publisher.

Published jointly by:

Japan Scientific Societies Press
2-10 Hongo, 6-chome, Bunkyo-ku, Tokyo 113, Japan
ISBN 4-7622-1893-6

and

S. Karger AG
P.O. Box, CH-4009 Basel, Switzerland
ISBN 3-8055-6741-3

Sole distribution rights outside Japan granted to S. Karger AG, Basel.

Designed by Gow Michiyoshi, Tokyo

Printed in Japan by Daishowa Printing Co. Ltd., Tokyo

Preface

Twenty years ago when we asked medical students what extracellular matrix was they answered, "It is abundant and exists on the outside of cells. It is inert, does not change much but builds up a complex architecture". If medical students were asked the same question today, they would answer "Extracellular matrix plays an important role in development, cell differentiation, cancer metastasis, and pathogenesis of various disorders". As they train to become medical doctors or basic research scientists, they now realize that extracellular matrix is essential for understanding normal physiology as well as pathogenesis of many diseases.

A large quantity of different macromolecules is present in extracellular matrix. Several major types of molecules are represented by families of proteins whose members exhibit tissue or cell specific expression. Interactions between these molecules result in the formation of a complex structure, and the interaction of extracellular matrix with cells is important for cell differentiation and maintenance of tissue architecture.

The basis for many acquired and genetic disorders that involve extracellular matrix has been defined by studies utilizing molecular genetics and transgenic/knockout mouse technique. This book contains current information on the interaction between some representative extracellular molecules and cells. The chapters in the first half deal with functional roles of extracellular molecules or their receptors, including collagenase, fibrillins, laminin, aggrecan, tenascin-C, elastin, collagen, and PDGF receptors. The balance of the chapters discuss regulation of gene expression and disease-related analyses, including collagen genes and B-myb, type I collagen, type IV collagen, extracellular matrix gene expression and function, structure of vitreous humor, chondrogenic differentiation, type IIA procollagen, cartilage collagens, cartilage hypertrophy, and collagen-induced arthritis. It is our hope that this information will provide readers with a basis for understanding the roles of extracellular molecules, their interactions with cells and pathogenesis of diseases that involve extracellular matrix.

Publication of this volume was originally planned on the occasion of the retirement of Professor Yutaka Nagai from Tokyo Medical and Dental University in March, 1994. Dr. Nagai spent two years (1961–1963) in the laboratory of Dr. Jerry Gross at Massachusetts General Hospital/Harvard Medical School in Boston where he showed that the products of vertebrate collagenase digestion of collagen formed segment-long-spacing (SLS) aggregates. The following year he was in the laboratory of Dr. Karl Piez at the National Institutes of Health in Bethesda, where he determined the biochemical features of the collagenase products, TC^A and TC^B. This book is dedicated to Dr. Nagai for his work on the biochemistry of mammalian metalloproteinases, particularly collagenase.

The editors would like to thank the authors of individual chapters for their contributions, without which this volume would not have been possible. They are also indebted to Mr. Tugio Miyazaki, Editor of Japan Scientific Societies Press, who guided them throughout the preparation procedures. They are grateful to Mr. Haruo Naito, President of Eisai Co., Ltd. for generous financial support, Mr. Masayuki Aso of Eisai Co., Ltd. for his various comments and to Dr. Teruo Nishida, Yamaguchi University, and Dr. Charles Boyd, University of Hawaii, for valuable suggestions on the publication.

May 1998   Yoshifumi NINOMIYA
                              Bjorn Reino OLSEN
                              Toshiro OOYAMA

Contents

Contributors

Andrikopoulos, K. Brookdale Center for Developmental and Molecular Biology, Mount Sinai School of Medicine, New York, New York 10029, U.S.A.

Brand, D. Department of Internal Medicine, University of Tennessee, Research Service of the Veterans Affairs Medical Center, Memphis, Tennessee 38163, U.S.A.

Charbonneau, N.L. Shriner's Hospital for Crippled Children, Department of Biochemistry and Molecular Biology, Oregon Health Sciences University, Portland, Oregon 97201-3095, U.S.A.

Cremer, M.A. Department of Internal Medicine, University of Tennessee, Research Service of the Veterans Affairs Medical Center, Memphis, Tennessee 38163, U.S.A.

de Crombrugghe, B. Department of Molecular Genetics, M.D. Anderson Cancer Center, Houston, Texas 77030, U.S.A.

Denton, C.P. Department of Molecular Genetics, M.D. Anderson Cancer Center, Houston, Texas 77030, U.S.A.

Di Liberto, M. Cellular Biochemistry and Biophysic Program, Memorial Sloan-Kettering Cancer Center, New York, New York 10021, U.S.A.

Dzamba, B.J. Shriner's Hospital for Crippled Children, Department of Biochemistry and Molecular Biology, Oregon Health Sciences University, Portland, Oregon 97201-3095, U.S.A.

Greenwel, P. Brookdale Center for Developmental and Molecular Biology, Mount Sinai School of Medicine, New York, New York 10029, U.S.A.

Hasegawa, M. Department of Clinical Physiology, Toho University School of Medicine, Ota-ku, Tokyo 143-0015, Japan

Hayashi, T. Department of Life Sciences, Graduate School of Arts and Sciences, The University of Toyo, Meguro-ku, Tokyo 153-0041, Japan

Hiraki, Y. Department of Biochemistry, Osaka University Faculty of Dentistry, Suita, Osaka 565-0871, Japan

Inagaki, Y. Department of Internal Medicine, National Kanazawa Hospital, Kanazawa 920-8650, Japan

Ishizeki, K. Department of Oral Anatomy, Iwate Medical University School of Dentistry, Morioka, Iwate 020-0023, Japan

Kang, A.H.   Department of Internal Medicine, University of Tennessee, Research Service of the Veterans Affairs Medical Center, Memphis, Tennessee 38163, U.S.A.

Katsura, N.   Department of Oral Biochemistry, School of Dentistry, Nagasaki University, Sakamoto, Nagasaki 852-8102, Japan

Keene, D.R.   Shriner's Hospital for Crippled Children, Department of Biochemistry and Molecular Biology, Oregon Health Sciences University, Portland, Oregon 97201-3095, U.S.A.

Kimura, T.   Department of Orthopaedic Surgery, Osaka University Medical School, Suita, Osaka 565-0871, Japan

Kitagawa, Y.   Nagoya University BioScience Center and Graduate School of Agricultural Sciences, Nagoya University, Nagoya 464-0814, Japan

Krane, S.M.   Department of Medicine, Harvard Medicial School and Medical Services (Arthritis Unit) Massachusetts General Hospital, Boston, Massachusetts 02114, U.S.A.

Kumagai, C.   Graduate School of Agricultural Sciences, Nagoya University, Nagoya 464-0814, Japan

Kusakabe, M.   Division of Experimental Animal Research, RIKEN, Tsukuba, Ibaraki 305-0074, Japan

Kypreos, K.E.   Department of Biochemistry, Boston University School of Medicine, Bosotn, Massachusetts 02118, U.S.A.

Linsenmayer, T.F.   Department of Anatomy and Cellular Biology, Tufts University Medical School, Boston, Massachusetts 02111, U.S.A.

Lu, R.

Department of Physiological Optics, School of Optometry, University of Alabama at Birmingham, Birmingham, Alabama 35294, U.S.A.

Mayne, R.

Department of Cell Biology, University of Alabama at Birmingham, Birmingham, Alabama 35294, U.S.A.

Mori, S.

Second Department of Internal Medicine, Chiba University School of Medicine, Chiba 260-0856, Japan

Myers, L.K.

Department of Pediatrics, University of Tennessee, Research Service of the Veterans Affairs Medical Center, Memphis, Tennessee 38163, U.S.A.

Naito, I.

Division of Ultrastructural Biology, Shigei Medical Research Institute, Okayama 701-0202, Japan

Niimi, T.

Nagoya University BioScience Center, Nagoya University, Nagoya 464-0814, Japan

Ninomiya, Y.

Department of Molecular Biology and Biochemistry, Okayama University Medical School, Shikata, Okayama 700-8558, Japan

Nurminskaya, M.

Department of Anatomy and Cellular Biology, Tufts University Medical School, Boston, Massachusetts 02111, U.S.A.

Oganesian, A.

Department of Pathology, University of Washington, Seattle, Washington 98195, U.S.A.

Olsen, B.R.

Department of Cell Biology, Harvard Medical School and Harvard-Forsyth Department of Oral Biology, Harvard School of Dental Medicine, Boston, Massachusetts 02115, U.S.A.

Ono, T.   Department of Oral Biochemistry, School of Dentistry, Nagasaki University, Sakamoto, Nagasaki 852-8102, Japan

Ooyama, T.   Department of Clinical Physiology, Toho University School of Medicine, Ota-ku, Tokyo 143-0015, Japan

Rabe, C.   Department of Molecular Genetics, M.D. Anderson Cancer Center, Houston, Texas 77030, U.S.A.

Ramirez, F.   Brookdale Center for Developmental and Molecular Biology, Mount Sinai School of Medicine, New York, New York 10029, U.S.A.

Ren, Z.-X.   Department of Cell Biology, University of Alabama at Birmingham, Birmingham, Alabama 35294, U.S.A.

Rosloniec, E.F.   Department of Internal Medicine, University of Tennessee, Research Service of the Veterans Affairs Medical Center, Memphis, Tennessee 38163, U.S.A.

Sado, Y.   Division of Immunology, Shigei Medical Research Institute, Okayama 701-0202, Japan

Saito, Y.   Second Department of Internal Medicine, Chiba University School of Medicine, Chiba 260-0856, Japan

Sakai, L.Y.   Shriner's Hospital for Crippled Children, Department of Biochemistry and Molecular Biology, Oregon Health Sciences University, Portland, Oregon 97201-3095, U.S.A.

Sakakura, T.   Department of Pathology, Mie University School of Medicine, Tsu, Mie 514-0001, Japan

Sandell, L.J.   Department of Orthopaedic Surgery, Washington University School of Medicine, St. Louis, Missouri 63110, U.S.A.

Seyer, J.M.   Department of Internal Medicine, University of Tennessee, Research Service of the Veterans Affairs Medical Center, Memphis, Tennessee 38163, U.S.A.

Shukunami, C.   Department of Biochemistry, Osaka University Faculty of Dentistry, Suita, Osaka 565-0871, Japan

Sonenshein, G.E.   Department of Biochemistry, Boston University School of Medicine, Boston, Massachusetts 02118, U.S.A.

Stuart, J.M.   Department of Internal Medicine, University of Tennessee, Research Service of the Veterans Affairs Medical Center, Memphis, Tennessee 38163, U.S.A.

Tajima, S.   Department of Dermatology, National Defense Medical College, Tokorozawa, Saitama 359-0042, Japan

Takarada, Y.-I.   Department of Clinical Physiology, Toho University School of Medicine, Ota-ku, Tokyo 143-0015, Japan

Tanaka, S.   Brookdale Center for Developmental and Molecular Biology, Mount Sinai School of Medicine, New York, New York 10029, U.S.A.

Terato, K.   Department of Internal Medicine, University of Tennessee, Research Service of the Veterans Affairs Medical Center, Memphis, Tennessee 38163, U.S.A.

Watanabe, H.   Craniofacial Developmental Biology and Regeneration Branch, National Institute of Dental Research, National Institutes of Health, Bethesda, Maryland 20892-4370, U.S.A.

Yamada, Y.   Craniofacial Developmental Biology and
                    Regeneration Branch, National Institute
                    of Dental Research, National Institutes of
                    Health, Bethesda, Maryland 20892-4370,
                    U.S.A.
Yamato, M.   Institute of Biomedical Engineering,
                    Tokyo Women's Medical University,
                    Shinjuku-ku, Tokyo 162-0054, Japan
Yokote, K.   Second Department of Internal Medicine,
                    Chiba University School of Medicine,
                    Chiba 260-0856, Japan
Zhao, W.   Department of Medicine, Harvard Medi-
                    cal School and Medical Services (Arthritis
                    Unit) Massachusetts General Hospital,
                    Boston, Massachusetts 02114, U.S.A.
Zhu, Y.   Department of Orthopaedic Surgry, Wa-
                    shington University School of Medicine,
                    St. Louis, Missouri 63110, U.S.A.

Extracellular Matrix-Cellular Interaction: Molecules to Diseases (Y. Ninomiya et al., eds.), pp. 1–21,
Japan Sci. Soc. Press, Tokyo/S. Karger, Basel (1998)

The Role of Collagenases in Remodeling of the Extracellular Matrix: Lessons from a Targeted Mutation in Col1a1 That Encodes Resistance to Collagenase Cleavage

STEPHEN M. KRANE AND
WEIGUANG ZHAO

Department of Medicine, Harvard Medical School and Medical Services (Arthritis Unit) Massachusetts General Hospital, Boston, Massachusetts 02114, U.S.A.

Under physiological conditions, when bone and other connective tissues are remodeled, components of the extracellular matrix (ECM) are degraded and removed and new components synthesized and deposited, although the rates of connective tissue resorption and deposition vary in different tissues. The remodeling events are regulated in each tissue by specific cells that respond to signals from other cells and the ECM itself. In most adult tissues other than bone, such as skin and visceral organs, the activity of cells that function in ECM remodeling is low, and therefore the rates of "turnover" of the ECM are low. The "turnover" of *collagen*, the most abundant component of the ECM in adult tissues, estimated by measurements of urinary excretion of markers of collagen breakdown indicate that bone is the major source of these markers (*1, 2*).

1

I. MATRIX METALLOPROTEINASES

Three main classes of enzymes have been implicated in the degradation of collagen. The first includes members of the matrix metalloproteinase (MMP) family (*3-9*). The second includes lysosomal cysteine proteinases such as cathepsins B, L, and K which cleave collagens in the telopeptide domain (*10-13*) but can cleave gelatins into smaller fragments. The third class includes serine proteinases such as plasmin, generated from plasminogen activator (*14, 15*). Plasmin and cathepsin B as well as active MMPs themselves probably have a role in the activation of the MMP zymogens (*16, 17*). After activation, certain MMPs have the capacity to degrade different types of interstitial collagen substrates during embryonic development and physiological postnatal remodeling and in pathological processes such as local invasion by malignant tumors, resorption of periodontal structures in periodontal disease, or the destruction of joints in rheumatoid arthritis. The latter has been a subject of major interest in our laboratory (*6, 17-19*).

The MMPs are members of a large subfamily of proteinases, also termed matrixins, that contain tightly bound zinc. Each of these enzymes contains a catalytic zinc-binding domain that include a sequence of motif HEXXH in which the Glu (E) acts as a catalytic base (*8*). There have been more than 20 different MMP gene products described, of which 15 have been identified in humans. Included in the MMP subfamily (*4-9*) are genes in humans that encode at least three collagenases, three stromelysins, several gelatinases, and cell-bound forms of MMPs that possess a transmembrane domain (*20*). Although there is considerable conservation of amino acid sequences and sequence motifs among the human MMPs, only the collagenases can cleave native, undenatured, "interstitial" collagens (*e.g.*, types I, II, III, X) within the uninterrupted triple helical domain at neutral pH. Among these collagenases are "fibroblast" collagenase (MMP-1 or collagenase-1), "neutrophil" collagenase

(MMP-8 or collagenase-2) and the rodent-type interstitial collagenases (*21–23*) (homologous to human collagenase-3 or
MMP-13 (*24, 25*)) and a *Xenopus laevis* collagenase). Another
collagenase so far identified only in *X. laevis* is called
collagenase-4 or MMP-18 (*26*). According to one report, the
72kDa gelatinase has collagenolytic activity if purified free of the
Tissue Inhibitor of Metallo Proteinases (TIMPs) and by this
criterion could also be considered as a collagenase (*27*). One of
the transmembrane (MT) MMPs, MT-MMP-1, when expressed
as a soluble protein using a construct in which the transmembrane sequence was deleted also had collagenase activity (*28*).
Nevertheless, only a single interstitial collagenase has been
identified so far in the mouse, the homologue of human
collagenase-3.

The collagenases have the unique capacity to initiate the
process of collagen degradation by making a single proteolytic
cleavage across the three chains of the collagen molecular trimer
at a locus 3/4 of the distance from the amino-terminus. In type
I collagen, collagenase cleavage occurs between Gly775/Ile776
in the $\alpha 1$(I) chain and Gly775/Leu776 in the $\alpha 2$(I) chain,
yielding a larger (A) and a smaller (B) helical fragment (*29*). In
other interstitial collagens, such as type II collagen, the cleavage
site is at a corresponding locus between a Gly and Leu. There
are many Gly/Ile and Gly/Leu sequences distributed throughout the component chains of type I collagen, yet only the bonds
between residues 775 and 776 are cleaved by collagenases.
Valuable information about collagenase specificity had been
obtained through studies of small synthetic peptides that could
be cleaved by collagenase but had a relatively high K_M (>1 mM)
(*7, 30, 31*). Nevertheless, the results of the latter studies did not
explain how collagenases cleave the natural triple helical substrates with a much lower K_M and cleave denatured collagens
more slowly than native collagens.

II. EXPRESSION OF COLLAGENASE-RESISTANT TYPE I COL-LAGEN

1) In Vitro

Our initial approach, in collaboration with Drs. Hong Wu, Xin Liu, and Rudolf Jaenisch at the Whitehead Institute at Massachusetts Institute of Technology, Cambridge, MA, to understanding how collagenases cleave native collagens was to produce mutated $\alpha 1(I)$ chains of type I collagen in cell culture that were resistant to cleavage by collagenases in the helical domain. We made several collagenase-resistant mutations in the mouse *Col1a1* gene and tested them in cultured Mov13 cells (*32, 33*) which do not express the endogenous *Col1a1* gene due to a retroviral insertion. We found, with substitution of Pro for

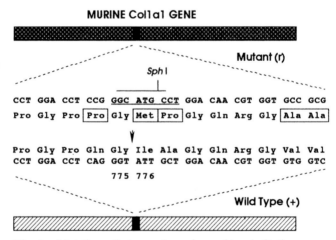

Fig. 1. Mutations that encode amino acid substitutions around the collagenase cleavage site in the $\alpha 1(I)$ chain of type I collagen. The mutations in the mutant (**r**) *Col1a1* gene are shown on top and the wild type (+) gene on the bottom. The mutated amino acids are boxed. This particular mutation included a new *Sph*I site for genotyping.

Ile776 at the P_1' position, that the collagen was totally resistant to collagenase cleavage. Another collagen encoded by a mutation (Mut IV) (Fig. 1) that has a double substitution of Gln774Pro and Ala777Pro at P_2 and P_2', Ile776Met and a double substitution of Val782Ala, Val783Ala, was also not cleaved. Of additional interest was our observation that cleavage of the wild type $\alpha 2(I)$ chains in the type I heterotrimer is dependent on the presence of cleavable sequences in the $\alpha 1(I)$ chains, suggestive of a dominant negative effect. We could thus replace a relatively small number of amino acids in the $\alpha 1(I)$ chain and have cultured cells secrete a type I collagen totally resistant to collagenase cleavage. Although there was strong circumstantial evidence suggesting a role for collagenases in resorption of type I collagen in different tissues, particularly under pathological conditions, we reasoned that expression of this mutation *in vivo* could be used to obtain more rigorous proof for such a role and reveal new or unanticipated functions of these proteinases.

2) In Vivo

We therefore introduced one of the mutations that resulted in resistance to collagenase (Mut IV above) into the endogenous *Col1a1* gene (*34*, *35*) via homologous recombination in embryonic stem (ES) cells using the "hit-and-run" strategy of Hasty *et al.* (*36*) as shown in Fig. 2. (We chose to target the substrate rather than the enzyme(s) because of the potential redundancy in collagenase genes in the mouse analogous to collagenase redundancy in humans, although at the present time only one mouse collagenase gene has been identified.) The ES cells were implanted in blastocysts and the mice carrying the mutation were bred to homozygosity. The mice, termed Col1a1[tm1 Jae], that carried the mutation (**r**) on both alleles (**r/r**) had normal embryonic development and no abnormalities as young adults, yet the type I collagen extracted from the skin and tendons of homozygotes was not cleaved (as predicted) either by rat interstitial collagenase or human fibroblast collagenase at what was

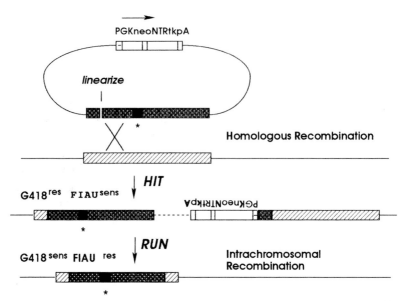

Fig. 2. Schematic representation of the gene targeting strategy used in the experiments described (*34, 35*) modified from the procedure of Hasty *et al.* (*36*). The black insert indicates the sequences shown in Fig. 2 with the (*) indicating the mutation shown in Fig. 2.

previously considered to be the single site at Gly775/Ile776 in the helical domain in the $\alpha 1(I)$ chain (*35*). We had based the design of the targeted mutation on observations that collagenases are able to attack triple helical collagen because the region around the cleavage site is relatively poor in Pro (hydroxyproline (Hyp)) in the (-Gly-X-Y-) triplet -Y- position and since Pro (*i.e.* Hyp) in the -Y- position stabilizes the triple helix the area is susceptible to unwinding (*37*). When we had obtained collagen from the r/r mice sufficient for analysis by circular dichroism (*38, 39*), we found (in preliminary experiments with Dr. H.P. Bächinger, Shriners Hospital, Portland, OR) that the T_M of the r/r collagen was $\sim 1°C$ higher than $+/+$ collagen; it is of interest that adding Pro to only 2/333 triplets would have such a profound effect on structure.

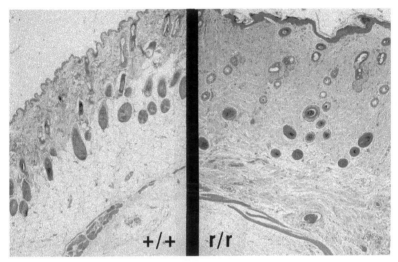

Fig. 3. Skin phenotype of **r/r** mice. (left) Gross appearance of 6-month old male $+/+$ and **r/r** mice. Note the mottled hair loss over the dorsum of the **r/r** mice. (right) Histological sections of skin stained with H & E from $+/+$ and **r/r** mice. Note the dense collagen ECM extending from the epidermis to the deep muscular layer as described by Liu *et al.* (*35*).

III. DEVELOPMENT OF A PHENOTYPE IN THE COLLAGEN-ASE-RESISTANT (Colla1^{tm1 Jae}) MICE

Although embryonic development of the **r/r** mice was normal, we observed that young adult **r/r** mice, beginning at approximately 4 weeks of age, displayed abnormalities such as thickened skin with patchy hair loss (Fig. 3 (left)), joint contractures and bony deformities (*35*). Skin from **r/r** mice, examined histologically with H & E and trichrome staining showed a remarkable increase in collagen extending through to the deep dermis (Fig. 3 (right)). Extraction of the skin collagen and analysis by SDS-PAGE and delayed reduction showed that the collagen was >95% type I. The uterus of previously pregnant r/r females revealed nodules of collagen (again, type I by SDS-PAGE) beneath the endometrium, in some areas extending to the serosal surface, either diffuse or as nodular masses. On examination of homozygous mice as young as 4 weeks of age, dorsiflexion contractures were regularly observed at the ankle joints (40). We thus demonstrated that interference with collagenase cleavage within the helical domain had no effect on embryonic development but that fibrosis of selected tissues began to be evident within weeks after birth.

The presence of fibrosis in the dermis and uterine stroma of the r/r mice suggests that there could be a dynamic relationship in tissues such as dermis and uterus between collagen synthesis, on one hand, and collagen degradation on the other. There are data indicating that collagenase-1 (MMP-1) is expressed in *human* keratinocytes in inflammation and wounding (41-43). Data from several laboratories indicate that epithelial cells such as keratinocytes as well as various mesenchymal cells bind to type I collagen matrices using $\alpha 2\beta 1$ integrins on the cell surfaces (44-46). In wounded epidermis, migrating keratinocytes continue to express these collagen-binding $\alpha 2\beta 1$ integrins but then redistribute them such that they are concentrated on the fronto-basal portions of the cell (45). Ligation via this integrin is associated with induction of expression of collagenase in human keratinocytes, MMP-1 (44-45); antibodies to the $\alpha 2\beta 1$ integrin blocks the collagenase induction. We have collaborated in experiments involving the potential role of collagenase in adhesion and migration of human keratinocytes on type I collagen (45) with Drs. William Parks and Howard Welgus (Washington University, St. Louis, MO). HaCat cells, a human keratinocyte line, that do not express collagenase-1 in response to collagen do not migrate on collagen but do migrate on gelatin. If collagenase-1 is induced with a ligand such as epidermal growth factor (EGF), however, HaCat cells migrate on type I collagen; the migration and collagenase activity were inhibited by a synthetic peptide-hydroxamate compound. The HaCat cells also did not migrate on collagen from r/r mice although collagenase-

1 was induced by the **r/r** as well as +/+ collagen. We concluded that collagenase-1 activity is required for efficient human keratinocyte migration on its type I collagen substrate and that this could play a role in skin wound healing. It is possible that trauma and subsequent inflammation in mice induce collagenase-expression in the basal layer of the epidermis; the collagenase might then normally participate in the removal of collagen deposited under such stimuli. In the case of the **r/r** mice, even though collagenase is induced and presumably secreted and activated, it is ineffective in removing the collagen. By analogy, cleavage at the helical site must be important in the prevention of dermal fibrosis in mice, in view of the abnormal phenotype that we observed. Interestingly, the human keratinocytes require active collagenase for migration and although collagen from the **r/r** mice can induce collagenase gene expression, the collagen is not cleaved and the cells do not migrate (*45*). In other systems, *e.g.*, those utilizing melanoma cells, binding to type I collagen also induces collagenase gene expression; cleavage (? partial) of the collagen then "reveals" sites for binding by the $\alpha v \beta 3$ integrin, ligation of which results in sustained viability of the cells through inhibition of programmed cell death (apoptosis) (*46*).

In the report of our observations on the phenotype of the **r/r** mice we stated (*35*), "Preliminary observations suggest that, while overall bone development was normal in Col1a1[tm1 Jae] mice, deformities of the tibia and increased deposition of trabecular and cortical bone in femurs and tibias were frequent after the age of 6 months, suggesting that bone remodeling may also be affected". We are currently in the process of examining the skeleton in greater detail. Although our studies have not yet been completed, we have been impressed by the presence of anterior bowing deformities in the tibias and suggestive evidence of old fractures. The volume of bone in the metaphysis appeared to be normal in young mice, inconsistent with osteopetrosis. In older animals, however, the volume of bone appears to be excessive; some, but not all of this excessive bone can be accounted for by accumulation of woven bone from incomplete fracture healing. A similar increase in the volume of bone in the medullary canal of the femurs was also seen, but this was mostly woven bone and could also be due to fractures.

In addition, we have begun to measure the responses of the bones from the collagenase-resistant mice to parathyroid hormone (PTH) (47). PTH is thought to stimulate bone resorption through a direct action on osteoblasts and stromal cells and then through an indirect action to increase the differentiation and function of osteoclasts. PTH acting on osteoblasts and stromal cells increases collagenase gene transcription and synthesis. In order to measure responses to PTH, we have injected vehicle or PTH subcutaneously for three days over a hemi-calvarium according to the protocol of Boyce and co-workers (48, 49). Bone resorption was quantitated by measurements of (1) resorptive bone area/total bone area and (2) osteoclast number. Osteoclasts were identified by their morphology, staining for tartrate-resistant acid phosphatase (TRAP) and in situ hybridization using a 92 kDa gelatinase riboprobe. The responses of the bones to PTH was also assessed by in situ hybridization with collagenase and 92 kDa gelatinase (MMP-9) riboprobes. We observed that calvarial bone resorption in response to PTH was markedly reduced in the r/r mice resistant to collagenase cleavage at the helical site in type I collagen. Nevertheless, PTH still induced expression of collagenase and 92 kDa gelatinase in periosteal cells in r/r as well as +/+ mice.

Thus, these preliminary experiments suggest that cleavage of collagen at the helical site by collagenase may be necessary for induction of osteoclastic bone resorption by PTH. Several possible mechanisms might explain how collagenase could be involved in the bone resorptive actions of PTH. For example, Chambers and colleagues (50) have proposed that the action of collagenase produced by osteoblasts or stromal cells might be required for denuding of collagen on bone surfaces in order for osteoclasts to attach. We have also considered that the degradation of this surface layer of collagen might release stored matrix-bound growth factors or that cleavage products resulting from

the action of collagenase on type I collagen might be biologically active on osteoclasts. The latter possibility is supported by the
findings in a recent report indicating that collagen fragments
produced by interstitial collagenase can activate osteoclastic
bone resorption (*51*). Finally, based on the findings described
earlier regarding keratinocyte migration (*45*) and observations
on melanoma cells (*46*), collagenase cleavage of type I collagen
could expose cryptic integrin binding sites, *e.g.*, for $\alpha v \beta 3$,
involved in osteoclast migration or prevention of osteoclast
apoptosis. Failure of collagenase to cleave at the helical locus
would interrupt this chain of events.

IV. SOME COLLAGENASES CLEAVE TYPE I COLLAGEN AT AN AMINOTELOPEPTIDE SITE

Using collagen extracted from two young heterozygous $r/+$
mice we observed that only $\sim$25% was cleaved by collagenase
although there was $\sim$equal expression of the **r** and $+$ alleles
determined by analyzing the cyanogen bromide (CNBr) cleavage
fragments (*35*). (In the engineered mutation the insertion of Met
for Ile776 resulted in the formation of an additional CNBr
cleavage in $\alpha 1(I)CB7$ (*32*).) These results are consistent with a
dominant negative effect of the mutation. Mechanisms proposed
to account for the exquisite specificity of the cleavage of triple
helical collagens by collagenases must take this dominant negative effect into consideration. Our data on the effects of collagenases on cleavage of triple helical collagen that contains the
targeted mutations in the $\alpha 1(I)$ chains around the helical cleavage site indicate that all three chains in the type I heterotrimer
must have the proper cleavable sequences (*32, 33, 35*). Even a
single mutated $\alpha 1(I)$ chain prevents the cleavage of the other
two $+/+$ chains (*i.e.*, another $\alpha 1(I)$ chain and an $\alpha 2(I)$ chain).
These findings could account for our observations that even old
mice heterozygous for the Col1a1[tm1 Jae] mutation develop dermal
fibrosis. We were nevertheless perplexed as to what mechanism
could have accounted for normal remodeling of type I collagen

during embryonic development if collagenase could not work on the mutated collagen.

In the course of analysis of the effects on the mutant type I collagen substrate of murine collagenases, we observed that even though the products of helical cutting were not detected, there was a diminution in the amount of crosslinked β and γ components. Subsequently, we observed that highly purified rat interstitial collagenase (*21, 35*) (97% amino acid sequence identity with the mouse (*23*)) made an additional cleavage between Gly/Val in the N-telopeptide that begins with the putative crosslinking Lys, *i.e.,* LysSerAlaGlyValSerValPro. The cleavage is C-terminal to the crosslinking Lys, eight residues before the start of the major collagen helix containing the triplet GlyProMet. We determined that the capacity to cleave type I collagen in the N-telopeptide is a property of both the mouse and the rat interstitial collagenases which have ∼97% amino acid sequence identity with each other but only ∼50% amino acid sequence identity with human MMP-1 (*24*). Human MMP-1 and an homologous sheep MMP-1 do not make this N-telopeptide cleavage. Our studies of expressed recombinant chimeric molecules of the mouse (MMP-13) and human (MMP-1) interstitial collagenases (*52*) indicate that this property is determined by amino acid sequences in the distal two thirds of the catalytic domain of the mouse collagenase extending to the zinc-binding domain as shown diagrammatically in Fig. 4. It is not known whether cleavage of collagens in the N-telopeptide also depends on the helical conformation of downstream sequences. Based on these observations, we postulated that this newly recognized N-telopeptidase activity might suffice for collagen degradation during embryonic development in the **r/r** mice, but it is apparent from examination of the older mice that helical cleavage of type I collagen is essential for normal remodeling events.

A homologue of the rat and mouse fibroblast/osteoblast interstitial collagenases, structurally different from the human neutrophil and fibroblast collagenases (52–53% amino acid sequence identity), now called collagenase-3 or MMP-13, has

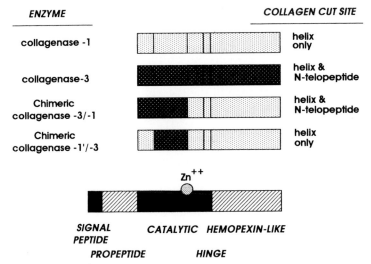

Fig. 4. Schematic representation of the domain structure (7) of the collagenases. The cDNAs for human collagenase-1 (MMP-1) and the mouse interstitial collagenase (collagenase-3 or MMP-13) or chimeric molecules as shown were expressed in *Escherichia coli* and cleavage of $+/+$ or $\mathbf{r}/\mathbf{r}$ collagen was assayed to determine helical and N-telopeptide cleavage activity as described by Krane *et al.* (52).

been cloned from a human breast carcinoma cDNA library (24). Human collagenase-3 has ∼86% sequence identity to the rat and mouse collagenases and clearly cleaves at the N-telopeptide of type I collagen. Analysis of recombinant hybrid human collagenase-3/collagenase-1 enzymes indicates that the capacity to cleave at the N-telopeptide is determined by sequences in the N-terminus of collagenase-3 (52). Collagenase-3 is particularly potent in proteolysis of type II collagen at the same site cleaved by MMP-1 (53, 54). In addition, following the initial cleavage, collagenase-3 cleaves the 1/4(B) fragment at Gly778/Gln779. Collagenase-3 also behaves as a more potent gelatinase than collagenase-1 (MMP-1) (54). Thus, the collagenase-3 enzymes that have a unique pattern of substrate specificity could have different roles in collagen degradation *in vivo*.

It is helpful to consider the specificity of certain collagenases to cleave at the helical and N-telopeptide loci with respect
to their protein structures. There are some indications that
determinants of specificity for collagen substrates are located in
the C-terminal hemopexin-like portion of the collagenases (55).
For example, exchange of the C-terminal domain of stromelysin-
1, which does not cleave collagens with uninterrupted helices
(e.g., "interstitial" types I, II, III) for that of the human fibroblast collagenase (MMP-1) which does cleave these collagens,
results in a chimeric proteinase that also does not cleave interstitial collagens even though the chimeric enzyme retains the
capacity to bind collagen (55). Further analysis of chimeric
proteinases indicates that the ability to cleave in the triple helix
results from a composite of elements derived from both halves of
the collagenase molecule (56). Other evidence supporting an
important role for the C-terminal domain is derived from studies
of recombinant neutrophil collagenases in which the catalytic
domains are preserved but contain engineered mutations encoding C-terminal truncations and still retain non-specific proteinase activity but do not attack helical collagens (57). The X-ray
crystal structures of the catalytic domain of the human neutrophil collagenase (MMP-8) (58), human MMP-1 (59), and the
full-length porcine MMP-1 (60) have been reported. In the latter
structure, the catalytic domain is connected to the hemopexin-
like domain by an exposed proline-rich flexible linker. The
hemopexin-like domain contains four, four-stranded antiparallel
β-sheets stabilized on its fourfold axis by a cation, probably
calcium, and thus constitutes a four-bladed β-propeller structure. Bode (61) suggests that the hemopexin-like domain with
the linker fully stretched could be positioned over the active site
of the catalytic domain and "(sandwich) a triple-helical collagen
substrate bound to its active site like a waffle in a waffle iron".
The mouse interstitial collagenase, with the hemopexin domain
deleted, retains the capacity to cleave at the N-telopeptide (62).

V. PERSPECTIVE

Collagenase-3 is highly expressed in the distal hypertrophic chondrocytes of the mouse (*63–65*) and human (*66, 67*) growth plate and is localized to cells expressing type X collagen. This MMP is also expressed in osteoblasts and other mesenchymal cells in the diaphysis. Parathyroid hormone related peptide (PTHrP) is a major regulator of events at the growth plate (*68*) and we have preliminary evidence (*69*) that there is decreased expression of collagenase-3 in the growth plate and bone shaft in mouse embryos in which the PTH/PTHrP receptor has been knocked out. Collagenase-3 could thus have a major role in skeletal remodeling events not only during embryogenesis but also in the adult skeleton. We (*47*) and others (*51*) have recently obtained evidence that collagen cleavage by osteoblast/stromal cell collagenase may be critical for osteoclast recruitment and function in response to resorptive stimuli during endochondral growth. Collagenase-3 is also expressed in chondrocytes in osteoarthritic cartilage (*54, 70, 71*) and is probably responsible for the excessive degradation of the cartilage extracellular matrix in that disease. We therefore suggest that collagenase-3 might function during embryonic development and remodeling during early postnatal life through its action to cleave at the N-telopeptide site. Cleavage at the N-telopeptide may be sufficient for remodeling of type I collagen—as well as type II collagen—containing extracellular matrices during this period, particularly in skeletal tissues where the enzyme is predominantly expressed (*63–67*). Under other circumstances, in organs where there is a requirement for rapid collagen degradation during a brief window of time, *e.g.*, in the uterus immediately postpartum, degradation at the helical site becomes critical. It remains to be demonstrated, however, whether N-telopeptide cleavage actually occurs *in vivo* and whether N-telopeptidase activity is necessary for other remodeling events.

We have shown in our studies of the **r/r** mice that embryonic development appears to be normal and we do not detect a phenotype until ∼4 weeks of age. As animals normally grow older, there is increased formation of stable intermolecular crosslinks in collagen (*72, 73*). Increasing collagen crosslinking *in vitro* has been shown to decrease both the rate and extent of degradation by collagenases (*74*). We therefore speculate that solubilization of the more mature collagen and further degradation would require proteolysis at both the N-telopeptide and helical sites (Fig. 5). We have presented the hypothesis (see above and (*35, 52*)) that N-telopeptidase activity could suffice for interstitial collagen remodeling during development and early postnatal life prior to the formation of mature covalent crosslinks. Once these mature crosslinks are formed, helical cleavage would be essential for remodeling. Cleavage at the helical sites would result in unwinding beginning at the cleavage ends and progressive uncoiling (denaturation), decreasing the number and strength of non-covalent forces laterally aligning and holding the molecules in the collagen fibrils. These effects of collagenase would also be important for critical interactions (migration, invasion, and prevention of apoptosis) of cells in these tissues with their extracellular matrix (*3, 41–46, 51*). The

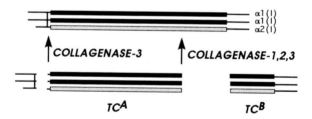

Fig. 5. Scheme for cleavage of type I collagen by different collagenase gene products. The three different collagenase genes and their products have so far been identified only in humans. The mouse interstitial collagenase, the homologue of human collagenase-3, is the only product so far identified in mice.

results of the experiments proposed here would be interpreted in light of our observations in the Col1a1$^{tm1\ Jae}$ **r/r** mice, where introduction of mutations encoding amino acid substitutions around the collagenase cleavage site in type I collagen produces resistance to collagenase cleavage and results in various disturbances of cellular function. Excessive action of collagenases leading to breakdown of type II as well as type I collagens in the specific ECM is a major cause of morbidity in several human diseases, including destructive forms of joint and bone disease. Understanding mechanisms by which the collagenases work should help in the design of new therapeutic approaches to these diseases.

SUMMARY

Degradation of type I collagen takes place through the action of collagenases at a highly conserved helical cleavage site between Gly775 and Ile776 of the α1(I) chain and an homologous site in the α2(I) chain. Mutations at or around this site render type I collagen resistant to collagenase digestion *in vitro*. We have introduced a mutation around the collagenase cleavage site into the endogenous *Col1a1* gene by gene targeting that renders the type I collagen resistant to cleavage at this site. Mice carrying the targeted mutation had normal embryonic development but beginning at ~4 weeks of age fibrosis of the dermis became evident in addition to joint contractures and skeletal deformities. Postpartum involution of the uterus in the mutant mice was also impaired, with persistence of collagenous nodules in the uterine wall. Although type I collagen from the homozygous mutant mice was resistant to cleavage by human MMP-1 or rat fibroblast collagenases at the helical site, only rat and mouse interstitial collagenase and their human homologue (collagenase-3, MMP-13) cleaved collagen trimers at an additional, novel site in the nonhelical N-telopeptide domain. Our results suggest that cleavage by collagenase-3 at the N-telopeptide site could account for resorption of type I collagen during connective tissue remodeling in embryonic and early adult life. During intense collagen resorption, however, such as in the uterus at the time of parturition, in bone in response to PTH and in the dermis later in life, cleavage at the helical site by collagenases is essential for normal collagen turnover and avoidance of fibrosis.

Acknowledgments

Original research described here was supported by USPHS Grants AR-03564, AR-44855, and AR-07258 to S. M. Krane and HL-41484 to R. Jaenisch.

REFERENCES

1. Kivirikko, I.K. (1970) *Int. Rev. Connect. Tiss. Res.* **5**, 93–163.
2. Krane, S.M. (1987) *Adv. Meat Res.* **4**, 325–331.
3. Alexander, C.M. and Werb, Z. (1989) *Curr. Opin. Cell. Biol.* **1**, 974–982.
4. Woessner, J.F.J. (1991) *FASEB J.* **5**, 2145–2154.
5. Murphy, G. and Docherty, A.J. (1992) *Am. J. Respir. Cell. Mol. Biol.* **7**, 120–125.
6. Krane, S.M. (1994) *Ann. N.Y. Acad. Sci.* **732**, 1–10.
7. Birkedal-Hansen, H., Moore, W.G.I., Bodden, M.K. *et al.* (1993) *Crit. Rev. Oral Biol. Med.* **4**, 197–250.
8. Bode, W., Gomis-Rüth, F.X., and Stöcker, W. (1993) *FEBS Lett.* **331**, 134–140.
9. Sang, Q.A. and Douglas, D.A. (1996) *J. Protein Chem.* **15**, 137–160.
10. Burleigh, M.C. (1977) In *Proteinases in Mammalian Cells and Tissues* (Barrett, A.J., ed.), pp. 285–309, Elsevier, Amsterdam.
11. Eeckhout, Y. and Vaes, G. (1977) *Biochem. J.* **166**, 21–31.
12. Drake, F.H., Dodds, R.A., James, I.E. *et al.* (1996) *J. Biol. Chem.* **271**, 12511–12516.
13. Bossard, M.J., Tomaszek, T.A., Thompson, S.K. *et al.* (1996) *J. Biol. Chem.* **271**, 12517–12524.
14. Werb, Z., Mainardi, C.L., Vater, C.A., and Harris, E.D. Jr. (1977) *N. Engl. J. Med.* **296**, 1017–1023.
15. Werb, Z. and Aggeler, J. (1978) *Proc. Natl. Acad. Sci. USA* **75**, 1839–1843.
16. Harris, E.D. Jr. and Vater, C.A. (1982) *Methods Enzymol.* **82**, 423–452.
17. Harris, E.D. Jr., Welgus, H.G., and Krane, S.M. (1984) *Collagen Rel. Res.* **4**, 493–512.
18. Krane, S.M. (1993) In *Arthritis and Allied Conditions. A Textbook of Rheumatology* (McCarty, D.J. and Koopman, W.J., eds.), 12th Ed., pp. 763–

779, Lea & Febiger, Philadelphia.
19. Krane, S.M., Conca, W., Stephenson, M.L., Amento, E.P., and Goldring, M.B. (1990) *Ann. N.Y. Acad. Sci.* **580**, 350–354.
20. Sato, H., Takino, T., Okada, Y. *et al*. (1994) *Nature* **370**, 61–65.
21. Roswit, W.T., Halme, J., and Jeffrey, J.J. (1983) *Arch. Biochem. Biophys.* **225**, 285–295.
22. Quinn, C.O., Scott, D.K., Brinckerhoff, C.E., Matrisian, L.M., Jeffrey, J.J., and Partridge, N.C. (1990) *J. Biol. Chem.* **265**, 22342–22347.
23. Henriet, P., Rousseau, G.G., and Eeckhout, Y. (1992) *FEBS Lett.* **310**, 175–178.
24. Freije, J.M.P., Díez-Itza, I., Balbín, M. *et al*. (1994) *J. Biol. Chem.* **269**, 16766–16773.
25. Billinghurst, R.C., Dahlberg, L., Ionescu, M. *et al*. (1997) *J. Clin. Invest.* **99**, 1534–1545.
26. Stolow, M.A., Bauzon, D.D., Li, J. *et al*. (1996) *Mol. Biol. Cell* **7**, 1471–1483.
27. Aimes, R.T. and Quigley, J.P. (1995) *J. Biol. Chem.* **270**, 5872–5876.
28. Ohuchi, I., Imai, K., Fujii, Y., Sato, H., Seiki, M., and Okada, Y. (1997) *J. Biol. Chem.* **272**, 2446–2451.
29. Gross, J. (1981) In *Cell Biology and Extracellular Matrix* (Hay, E.D., ed.), pp. 217–258, Plenum, New York.
30. Fields, G.B., Van Wart, H.E., and Birkedal-Hansen, H. (1987) *J. Biol. Chem.* **262**, 6221–6226.
31. Sottrup-Jensen, L. and Birkedal-Hansen, H. (1989) *J. Biol. Chem.* **264**, 393–401.
32. Wu, H., Byrne, M.H., Stacey, A. *et al*. (1990) *Proc. Natl. Acad. Sci. USA* **87**, 5888–5892.
33. Hasty, K.A., Wu, H., Byrne, M. *et al*. (1993) *Matrix* **13**, 181–186.
34. Wu, H., Liu, X., and Jaenisch, R. (1994) *Proc. Natl. Acad. Sci. USA* **91**, 2819–2823.
35. Liu, X., Wu, H., Byrne, M., Jeffrey, J., Krane, S., and Jaenisch, R. (1995) *J. Cell Biol.* **130**, 227–237.
36. Hasty, P., Ramirez-Solis, R., Krumlauf, R., and Bradley, A. (1991) *Nature* **350**, 243–246.
37. Brown, R.A., Hukins, D.W., Weiss, J.B., and Twose, T.M. (1977) *Biochem. Biophys. Res. Commun.* **74**, 1102–1108.
38. Bätge, B., Notbohm, H., Diebold, J. *et al*. (1990) *Eur. J. Biochem.* **192**, 153–159.
39. Bächinger, H.P., Morris, N.P., and Davis, J.M. (1993) *Am. J. Med. Genet.* **45**, 152–162.
40. Zhao, W., Byrne, M.H., Tsay, A., and Krane, S.M. (1996) *J. Bone Mineral Res.* **11**, S107.
41. Saarialho-Kere, U.K., Kovas, S.O., Pentland, A.P., Olerud, J., Welgus,

H.G., and Parks, W.C. (1993) *J. Clin. Invest.* **92**, 2858–2866.

42. Saarialho-Kere, U.K., Vaalamo, K., Airola, K., Niemi, K.-M., Oikarinen, A.I., and Parks, W.C. (1995) *Invest. Dermatol.* **104**, 982–988.
43. Inoue, M., Kratz, G., Haegerstrand, A., and Stähle-Bäkdahl, M. (1995) *Invest. Dermatol.* **104**, 479–483.
44. Vihinen, P., Riikonen, T., Laine, A., and Heino, J. (1996) *Cell Growth Diff.* **7**, 439–447.
45. Pilcher, B.K., Dumin, J.A., Sudbeck, B.D., Krane, S.M., Welgus, H.G., and Parks, W.C. (1997) *J. Cell Biol.* **137**, 1445–1457.
46. Montgomery, A.M.P., Reisfeld, R.A., and Cheresh, D.A. (1994) *Proc. Natl. Acad. Sci. USA* **91**, 8856–8860.
47. Zhao, W., Byrne, M.H., Boyce, B.F., and Krane, S.M. (1997) *J. Bone Mineral Res.* **12** (Suppl.), S110.
48. Yates, A.J.P., Gutierrez, G.E., Smolens, P. *et al.* (1988) *J. Clin. Invest.* **81**, 932–938.
49. Boyce, B.F., Aufdemorte, T.B., Garrett, I.R., Yates, A.J.P., and Mundy, G.R. (1989) *Endocrinology* **125**, 1142–1150.
50. Chambers, T.J., Darby, J.A., and Fuller, K. (1985) *Cell. Tiss. Res.* **421**, 671–675.
51. Holliday, L.S., Welgus, H.G., Fliszar, C.J., Veith, G.M., Jeffrey, J.J., and Gluck, S.L. (1997) *J. Biol. Chem.* **272**, 22053–22058.
52. Krane, S.M., Byrne, M.H., Lemaître, V. *et al.* (1996) *J. Biol. Chem.* **271**, 28509–28515.
53. Knäuper, V., López-Otín, C., Smith, B., Knight, G., and Murphy, G. (1996) *J. Biol. Chem.* **271**, 1544–1550.
54. Mitchell, P.G., Magna, H.A., Reeves, L.M. *et al.* (1996) *J. Clin. Invest.* **9**, 761–768.
55. Murphy, G., Allan, J.A., Willenbrock, F., Cockett, M.I., O'Connel, J.P., and Docherty, A.J.P. (1992) *J. Biol. Chem.* **267**, 9612–9618.
56. Sanchez-Lopez, R., Alexander, C.M., Behrendtsen, O., Breathnach, R., and Werb, Z. (1993) *J. Biol. Chem.* **268**, 7238–7247.
57. Hirose, T., Patterson, C., Pourmotabbed, T., Mainardi, C.L., and Hasty, K.A. (1993) *Proc. Natl. Acad. Sci USA* **90**, 2569–2573.
58. Bode, W., Reinemer, P., Huber, R., Kleine, T., Schnierer, S., and Tschesche, H. (1994) *EMBO J.* **13**, 1263–1269.
59. Lovejoy, B., Cleasby, A., Hassell, A.M. *et al.* (1994) *Science* **263**, 375–377.
60. Li, J., Brick, P., O'Hare, M.C. *et al.* (1995) *Structure* **3**, 541–549.
61. Bode, W. (1995) *Structure* **3**, 527–530.
62. Lemaitre, V., Jungbluth, A., and Eeckhout, Y. (1997) *Biochem. Biophys. Res. Commun.* **230**, 202–205.
63. Mattot, V., Raes, M.B., Henriet, P. *et al.* (1995) *J. Cell. Sci.* **108**, 529–535.
64. Fuller, K. and Chambers, T.J. (1995) *J. Cell. Sci.* **108**, 2221–2230.
65. Gack, S., Vallon, R., Schmidt, J. *et al.* (1995) *Cell Growth Diff.* **6**, 759–767.

66. Johansson, N., Saarialho-Kere, U., Airola, K. *et al*. (1997) *Dev. Dynamics* **208**, 387–397.
67. Stähle-Bäckdahl, M., Sandstedt, B., Bruce, K. *et al*. (1997) *Lab. Invest.* **76**, 717–728.
68. Lanske, B., Karaplis, C.A.C., Kaechong, L. *et al*. (1996) *Science* **273**, 663–666.
69. Kovacs, C.S., Lanske, B., Byrne, M., Krane, S.M., and Kronenberg, H.M. (1997) *J. Bone Mineral Res.* **12**, S116.
70. Borden, P., Solymar, D., Sucharczuk, A., Lindman, B., Cannon, P., and Heller, R.A. (1996) *J. Biol. Chem.* **271**, 23577–23581.
71. Reboul, P., Pelletier, J-P., Tardiff, G., Clouter, J.-M., and Martel-Pelletier, J. (1996) *J. Clin. Invest.* **97**, 2011–2019.
72. Light, N.D. and Bailey, A.J. (1980) *Biochem. J.* **189**, 111–124.
73. Bailey, A.J., Sims, T.J., Avery, N.C., and Halligan, E.P. (1995) *Biochem. J.* **305**, 385–390.
74. Vater, C.A., Harris, E.D. Jr., and Siegel, R.C. (1979) *Biochem. J.* **181**, 639–645.

Extracellular Matrix-Cellular Interaction: Molecules to Diseases (Y. Ninomiya et al., eds.), pp. 23-40, *Japan Sci. Soc. Press, Tokyo/S. Karger, Basel (1998)*

Two Fibrillins, Multiple Fibrillinopathies, How Many Functions?

BETTE J. DZAMBA, NOE L. CHARBONNEAU, DOUGLAS R. KEENE, AND LYNN Y. SAKAI

Shriner's Hospital for Crippled Children, Department of Biochemistry and Molecular Biology, Oregon Health Sciences University, Portland, Oregon 97201-3095, U.S.A.

Fibrillin-1 and fibrillin-2 are highly homologous gene products with the same overall arrangement of unique and repetitive domains (Fig. 1). Each of the major types of repetitive domains is homologous between the two fibrillins at the level of primary structure (*1–5*) and is thought to be an independently-folding unit (*6, 7*). Most of the remaining primary structure, including the unique amino and carboxyl terminal ends and the "hybrid" domains (*2*), are also highly homologous between the two fibrillins. The only region of nonhomologous sequence occurs between the first 8-cys domain and the 4th generic epidermal growth factor (EGF)-like repeat; it is proline-rich in fibrillin-1 and glycine-rich in fibrillin-2.

Genetic analyses in humans (*8*) and mice (*9*) have demonstrated the importance of fibrillins to the extracellular matrix. Mutations in FBN1 have been identified in individuals with the Marfan syndrome (*10*), the MASS phenotype (*11*), familial ectopia lentis (*12*), severe "neonatal" Marfan syndrome (*12–14*), Shprintzen-Goldberg syndrome (*15*), and aortic aneurysm (*16*,

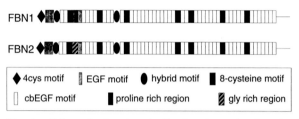

Fig. 1.   Schematic drawings indicating the major domains in fibrillins.

17). Connective tissues primarily affected in these diseases include skeletal, cardiovascular, and ocular tissues. Mutations in FBN2 have been identified in families with congenital contractural arachnodactyly (CCA) (*18*), who normally present symptoms in the skeletal but not the cardiovascular or ocular systems, and in individuals with severe neonatal forms of CCA (*19*), where cardiovascular complications are also manifested.

Targeting a deletion of exons 19–24 in mouse *fbn1* results in death from cardiovascular complications before three weeks of age in mice homozygous for the deletion (*9*). Heterozygous mice showed no obvious phenotype, presumably due to the low level of expression of the mutant allele (*9*), consistent with the observation that a low level of expression of a mutant allele results in milder human disease (*11*). These results demonstrated that fibrillin-1 is required for maintenance of cardiovascular connective tissues, and is not required for morphogenesis or prenatal growth.

Since fibrillins are structural components of microfibrils (*5, 20–22*), it is widely assumed that mutations in fibrillins must adversely affect either the assembly of fibrillin molecules into microfibrils or the stability of microfibrils over time. One possibility is that mutant molecules are assembled into microfibrils, resulting in all microfibrils being composed of mutant and wild-type molecules (Fig. 2A). If the mutation renders the mutant molecule more susceptible to proteolysis, then over time microfibrils may be weakened or degraded and the integrity of the

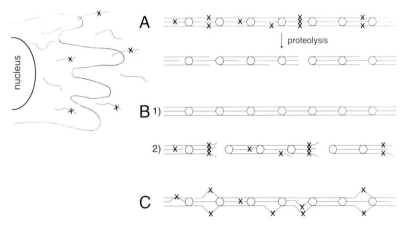

Fig. 2.   Cartoon depicting three possible effects of mutations in fibrillins on microfibrils.

connective tissue will be compromised (Fig. 2A). Consistent with this possibility are these observations: most mutations identified in individuals with the Marfan syndrome are missense mutations predicted to disrupt calcium-binding in a single EGF-like domain (*10*); calcium stabilizes fibrillin against proteolysis (*23*). Another possibility is that mutant molecules adversely affect assembly of wild-type proteins, leading to two populations of microfibrils: 1) a greatly diminished number of completely wild type microfibrils and 2) shorter or more frag-mented microfibrils (Fig. 2B). A third possibility is that both mutant and wild-type molecules are assembled into microfibrils, resulting in structurally and functionally abnormal microfibrils (Fig. 2C). This latter possibility has been suggested by rotary shadowing of microfibrils from different patient samples (*24, 25*).

In order to understand the molecular mechanisms under-lying the dominant negative effects of mutations in fibrillins in humans, more basic information is required. First, the distribu-tion of fibrillins in cells and tissues as well as in individual microfibrils must be defined. Second, the cellular and molecular steps in assembling the microfibril must be identified, and finally, model cell culture or organismal systems must be established to test the effect of mutations on these steps in secretion and assembly.

I. DISTRIBUTION OF FIBRILLINS

In the Marfan syndrome and other fibrillinopathies, the major connective tissues which are affected include skeletal, cardiovascular, and ocular tissues. In contrast, mutations in FBN2 seem to affect primarily skeletal tissues, leading to joint contractures which tend to resolve with time. In rare cases (19), mutations in FBN2 also lead to cardiovascular abnormalities. Do these phenotypic manifestations of disease reflect the tissue distribution patterns of the fibrillins?

Fibrillin-1 has been found in every connective tissue examined (20). However, fibrillin-1-containing microfibrils are particularly abundant in the major tissues affected in the Marfan syndrome. Fibrillin-1 is present in ocular tissues (26), particularly in the ciliary zonule where microfibrils form the only apparent structural elements (8). In large blood vessels, elastic fibers composed of elastin and fibrillin microfibrils are abundant and uniquely arranged in concentric lamellar sheets around the lumen of the vessel (8). Fibrillin-1 is found in long fibers in the perichondrium/periosteum of all skeletal elements (20, 27), is abundant in tendons, ligaments, and joint capsules, and is present in bone (28), and in cartilage (27). It is abundant in the dura mater (29), consistent with another major manifestation of Marfan syndrome, dural ectasia. Fibrillin-1 is also present in tissues like skin, lung, kidney, peripheral nerves, and skeletal muscle, which are not primarily affected in any of the fibrillinopathies.

Less is known about the tissue distribution of fibrillin-2, particularly in human. *In situ* hybridization studies of developing mouse (30) have demonstrated that *fbn2* appears earlier than *fbn1* in most tissues examined (lung, skeletal tissues). The heart and aorta were exceptional in that in these tissues *fbn1* was expressed very early in development (*30, 31*). In addition, these studies (*30*) suggested that, based upon expression in elastic fiber rich tissues, the function of fibrillin-2 may be to mediate elastic fiber assembly. In contrast, because of the later expression of fibrillin-1, it was proposed that its function may be to maintain connective tissue homeostasis. Expression of fibrillin-2 in adult or postnatal tissues has not been demonstrated.

Therefore, in order to determine the tissue distribution of fibrillin-2 in both developing and postnatal tissues as well as to investigate whether microfibrils are heteropolymers of fibrillins-1 and -2 or different populations of homopolymers of fibrillin-1 and homopolymers of fibrillin-2, we have produced monoclonal antibodies to a recombinant peptide of human fibrillin-2 (rF37, described in ref. *27*). We have selected two monoclonal antibodies which have been mapped to disulfide-linked epitopes which are present either in the first 8-cys domain or in the 4th generic EGF-like domain, flanking the glycine-rich sequence, in fibrillin-2. Neither of these monoclonal antibodies crossreacts with recombinant peptides of human fibrillin-1 nor with authentic human fibrillin-1 from cell culture media. Both of the antibodies recognize authentic human fibrillin-2 from cell culture media, and both also crossreact with bovine fibrillin-2.

Using these monoclonal antibodies, we have found that fibrillin-2 is present in all fetal human tissues (15–19 weeks of gestation) in an identical distribution as fibrillin-1. We did not find any preferential distribution to elastic fiber rich tissues. In fact, fibrillin-2 as well as fibrillin-1 was present in tissues like hyaline cartilage, cornea, and the ciliary zonule, which are known to contain no elastin. Control sections of adjoining tissues like the perichondrium and sclera were clearly positive for elastin at these time points. We did, however, determine that fibrillin-1 staining in fetal tissues was limited to the loose connective tissue surrounding skeletal muscle and tendon at 10–

11 weeks of gestation (*27*), while at these time points fibrillin-2 staining was ubiquitous. These results at the protein level confirmed conclusions from earlier *in situ* hybridization studies that *fbn2* expression precedes *fbn1* expression in many tissues. Fibrillin-1 staining was equally ubiquitous by 15–16 weeks of gestation (*27*).

In postnatal and adult tissues, fibrillin-1 continues to be widely distributed throughout the connective tissue. In contrast, fibrillin-2 staining disappears from most tissues rather abruptly. Neonatal skin, for example, contains very little fibrillin-2, which is limited to the dermal-epidermal junction and to some of the thick elastic fibers; in contrast, fibrillin-1 forms arboreal patterns of immunostaining in the papillary dermis that connect to the thick parallel elastic fibers of the reticular dermis. Fibrillin-2 staining was abundant in 9-year-old tendon but not in skeletal muscle; fibrillin-1 staining was abundant in both of these tissues. In adult aorta, fibrillin-1 is present in the intima, media, and adventitia, whereas only very small amounts of fibrillin-2 can be found in the media, and none in the adventitia or intima. Smaller blood vessels apparently do not contain fibrillin-2 very soon after birth; fibrillin-1 is abundant in all vasculature after birth.

In summary, the tissue distribution of fibrillin-1 is consistent with the Marfan phenotype. Similarly, the postnatal distribution of fibrillin-2 (which was found to be most abundant in tendon, peripheral nerve, and perichondrium) is consistent with the phenotype of CCA. However, it is unclear from distribution alone why disease manifestations are not also found in other tissues containing fibrillin-1, or during gestation when most fetal tissues contain abundant quantities of fibrillin-2. Clearly, the abundance of fibrillin containing microfibrils compared to other structural elements as well as the physiological functions and stresses on the connective tissue must play roles in why disease is manifested in one tissue but not in another. In addition, there may be specific molecular interactions which are more important in one connective tissue than in another, and these molecular interactions may be disturbed by mutations in FBN1 or FBN2. More knowledge than tissue distribution is required to understand why dominant negative effects are manifested in the cardinal systems affected in the fibrillinopathies.

II. ASSEMBLY OF FIBRILLINS

Investigations of assembly of fibrillins into microfibrils have been complicated by the existence of two highly homologous fibrillins. The structures of the two fibrillins do not suggest whether they might assemble as homopolymers or heteropolymers. The two proteins cannot be easily resolved by biochemical methods, and until now, suitable immunochemical reagents have not be available to distinguish one from the other. We have initially approached these investigations by 1) performing immunolocalization with fibrillin-1 and fibrillin-2 specific antibodies, 2) identifying which fibrillins are synthesized by cultured cells, and 3) determining high molecular weight forms of fibrillins in cell and organ culture matrices.

1. Immunolocalization of Fibrillins

Using human fetal skin, we have labeled microfibrils with a polyclonal antiserum (pAb9543) which reacts with fibrillin-1 but not with fibrillin-2 (9) in combination with monoclonal antibodies specific for fibrillin-2. Different gold sizes were used to distinguish the rabbit antiserum from the mouse monoclonal antibodies. Results revealed gold labeling of both sizes on all bundles of microfibrils. Moreover, individual microfibrils were well labeled with both sizes of gold, demonstrating that both fibrillin-1 and fibrillin-2 were present in the same individual microfibril (Fig. 3). When microfibrils were extracted from fetal skin, immunolabeled with fibrillin-1 pAb9543 and monoclonal antibodies to fibrillin-2, and visualized by negative stain and electron microscopy, individual beaded microfibrils were also labeled with antibodies to both fibrillin-1 and fibrillin-2 (Fig. 4). Based upon these results, together with our observations

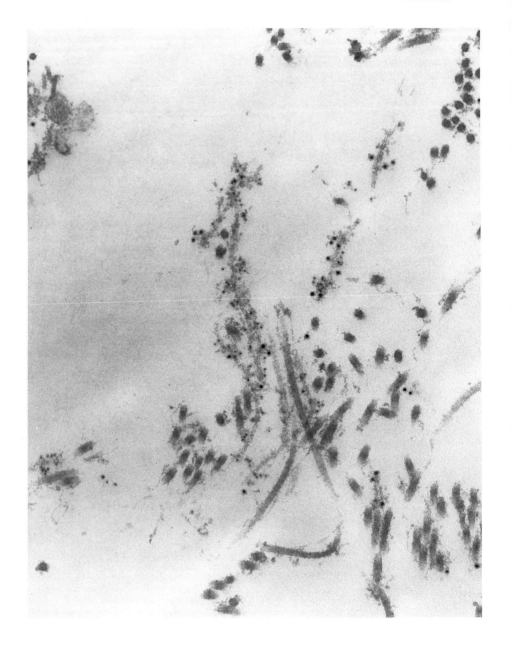

finding that fibrillins-1 and -2 are codistributed in all fetal
tissues, we conclude that fetal microfibrils are heteropolymers of
the two fibrillins.

Do homopolymeric microfibrils also exist? In most postnatal tissues where little if any fibrillin-2 staining is found,
microfibrils are probably homopolymers of fibrillin-1. Are there
differences between heteropolymeric and homopolymeric microfibrils? Electron microscopic visualization of fibrillin-1/fibrillin-
2 microfibrils compared to fibrillin-1 microfibrils demonstrates
no obvious structural differences. However, immunolocalization
of fetal microfibrils, using single monoclonal antibodies specific
for either fibrillin-1 or fibrillin-2, does not yield long stretches of
the periodic antibody-induced banding patterns typically seen
when postnatal microfibrils are labeled with fibrillin-1 antibodies (see refs. *22, 32*). This could reflect structural differences
in the heteropolymer or it might represent differences between
fetal and postnatal extracellular matrices. In postnatal tissues
which retain expression of fibrillin-2, heteropolymers and/or
homopolymers of both fibrillins may be present.

2.   Expression and Assembly of Fibrillins by Cells in Culture

In order to investigate assembly of microfibrils *in vitro*, we
screened various cell lines by immunofluorescence of the extracellular matrix and Western blotting of proteins secreted into the
medium. Our goals were to identify cell lines which produce
only fibrillin-1 microfibrils, cell lines which produce only fibril-
lin-2 microfibrils, and cell lines which incorporate both fibrillin-
1 and fibrillin-2 into fibrils. Western blotting of cell culture
media using enhanced chemiluminescence detection methods
revealed only one cell line, A204 (a rhabdomyosarcoma from
American Type Culture Collection), which secretes fibrillin-1
but not fibrillin-2, and one cell line, JAR (a choriocarcinoma

---

←Fig. 3.   Immunolocalization of fibrillins-1 and -2 in fetal human skin,
demonstrated by two sizes of gold.

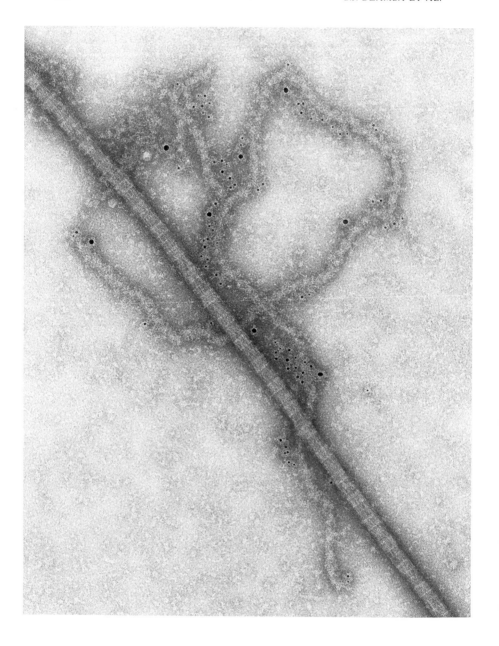

from ATCC), which secretes fibrillin-2, but not fibrillin-1. These results at the protein level were confirmed by Northern analyses of JAR cells. However, neither of these cell lines assembles fibrillin fibrils in the extracellular matrix. All other cell lines examined (postnatal and fetal skin fibroblasts, fetal lung fibroblasts, embryo, postnatal tendon fibroblasts, postnatal ligament fibroblasts, aortic smooth muscle cells, and established cell lines, SW1353 (chondrosarcoma), HT1080 (fibrosarcoma), MG63 (osteosarcoma), U2OS (osteosarcoma), WISH (transformed amniotic epithelial cells), U251MG (glioblastoma), and HaCaT (transformed keratinocytes, provided by Dr. Norbert Fusenig) secreted both fibrillin-1 and fibrillin-2 into the media. However, only some of these cell lines assemble both fibrillins into fibrils in the extracellular matrix.

Based upon these studies, our conclusions are that assembly of fibrillins into fibrils in the extracellular matrix is dependent upon 1) cell type and 2) concentration of fibrillins present. Skin, tendon, and ligament fibroblasts, vascular smooth muscle cells as well as vascular endothelial cells, and various sarcoma cell lines secrete fibrillins into the media and assemble them into fibrils in the extracellular matrix. In contrast, epithelial cells like HaCaT (transformed keratinocytes), as well as normal keratinocytes, and WISH cells secrete fibrillins differentially into the media and incorporate fibrillins into nonfibrillar forms in the extracellular matrix.

Immunofluorescence of HaCaT cells yielded only diffuse pericellular staining, while staining of WISH cells appeared punctate. Immunoblotting demonstrated that WISH cells secrete both fibrillins into the medium. In contrast, immunoblotting of HaCaT cells and of normal keratinocytes identified only fibrillin-2 secreted by these cells into the culture media; fibrillin-1 appeared to be directionally deposited into the extracellular

---

←Fig. 4.   Immunolocalization of fibrillins-1 and -2 in negatively stained extracts of fetal human skin, demonstrated by two sizes of gold.

matrix and was not found in the media. In addition, a dimer-sized nonreducibly crosslinked species containing fibrillin-1, but not fibrillin-2, was identified by immunoblotting of HaCaT, keratinocyte, and WISH cell matrix extracts. This high molecular weight species of fibrillin-1 was not identified in cell matrix extracts of fibroblasts. To test whether this species is a transient intermediate, we analyzed HaCaT cells grown for long periods of time (weeks) during which microfibrils would be expected to be assembled by fibroblasts. After extraction of the extracellular matrix and molecular sieve chromatography, rotary shadowing electron microscopy failed to identify beaded microfibrils. Taken together, these data suggest that keratinocytes, polarized cells which normally sit on basement membranes, selectively deposit fibrillin-1 into a basement membrane-like extracellular matrix, and assemble fibrillin-1 into a nonfibrillar high molecular weight species.

This fibrillin-1 "dimer" may be present in the lamina densa of the dermal-epidermal junction. Using "section-surface" labeling techniques (33, 34), we have recently immunolocalized fibrillin-1 to the lamina densa of the dermal-epidermal junction. Labeling was found all along the lamina densa, not only in areas where microfibrils were observed to intersect the lamina densa. While we cannot exclude the possibility that fibrillin-1 is present as "microfibrils" within the lamina densa, we suggest that fibrillin-1 may be specially crosslinked by keratinocytes and deposited into the lamina densa to stabilize protein-protein interactions and/or to target and concentrate specific growth factors required in the lamina densa, close to the basal surface of the keratinocytes.

3. The Effects of Concentration on Assembly of Fibrillins

We have been aware of major differences between how fibroblasts assemble fibrillin and how they assemble fibronectin since 1990, when a well-utilized cell culture assay was devised for analyzing immunofluorescent fibrillin fibrils in fibroblast samples from individuals with the Marfan syndrome (35). When plated at high concentrations $(2.5 \times 10^5$ cells/ml), wild-type fibroblasts will assemble a striking immunofluorescent fibrillar network of fibrillin after 3–4 days in culture. This is in dramatic contrast with the assembly of fibronectin, which becomes apparent within hours rather than days, on cells plated at the lowest concentrations. Fibronectin assembly seems to take place even on isolated cells and has been described as an integrin-mediated process occurring on the cell surface (*36*).

When fibroblasts are examined at different time points prior to the demonstration of a fibrillin fibril network, fibrillin-1 can be easily identified as monomers both in the medium and in extracts of the cell layer. These results suggest that fibrillin assembly is a protein-protein concentration-dependent process: fibrillin is secreted as single monomers both into the medium and into the extracellular matrix (where it cannot be detected as bright fibrils); in the extracellular matrix, when sufficient concentrations of fibrillin (or associated microfibrillar proteins) are achieved, fibrils may spontaneously assemble. This hypothesis would predict that high concentrations of pure fibrillin (and any required associated microfibrillar components) will assemble in solution. Efforts to test this hypothesis have been hampered by difficulties in purifying sufficient quantities of either authentic fibrillins from cell culture medium and in expressing recombinant full-length fibrillins. Furthermore, it is clear from our recent data that preparations of authentic fibrillin from fibroblast medium were probably composed of fibrillin-1 as well as small quantities of fibrillin-2. Since we have now identified cell culture sources of pure fibrillin-1 (A204) and pure fibrillin-2 (JAR), we are currently attempting to quantitate a critical concentration of fibrillin-1 and fibrillin-2 required for fibril formation.

Results from the cell culture assay of fibroblasts from individuals with the Marfan syndrome demonstrated a gross reduction of immunofluorescent fibrillin fibrils in 90% of these samples (*35*). Studies of fibrillin biosynthesis (*37, 38*) indicated that in some cases mutations in fibrillin-1 may lead to defective incorporation of fibrillin in the extracellular matrix. Rotary shadowing of microfibrils isolated from fibroblast cell cultures suggested that microfibrils are structurally abnormal in many cases (*24, 25*). Biosynthetic studies are normally conducted over a period of 24 hr or less, when fibrillin immunofluorescence is negative. Cell culture immunofluorescence assays are carried out after cells are cultured for 3–4 days, when microfibrils cannot be visualized by rotary shadowing and electron microscopy. Rotary shadowing analyses are typically performed on microfibrils which have been allowed to assemble in culture over a 2–3 week period. By all of these approaches, the data indicate that at least some mutations result in defects in fibrillin assembly in the extracellular matrix. However, further investigations are required to establish which of several possible different mechanisms (Fig. 2) result in disease and whether any of these different mechanisms can be correlated with clinical phenotypes. Systematic studies of individual mutations, using all of these approaches as well as modeling in an animal model, would be extremely useful to help bring together the current available data. In addition, basic information, which is currently lacking, is a prerequisite for further understanding.

4.   Disulfide-bond Formation and Assembly of Fibrillins

In extraction and pulse-chase studies using organ cultures of chick aortae, fibrillins were shown to form intermolecular disulfide crosslinks (*39*). Fibrillins are now known to be proteins composed of multiple repeated domains which are thought to be stabilized by intrachain disulfide bonds. Recently, intrachain disulfide bonds have been confirmed in selected domains (*32*), and folding patterns of selected calcium-binding EGF-like domains (*6*) and an 8-cys domain (*7*) have been determined. There are 177 predicted intrachain disulfide bonds stabilizing 47 EGF-like repeats, 7 8-cys domains, and 2 hybrid domains.

Additional cysteine residues are present in the amino and carboxyl terminal domains, and there is one extra cysteine in the first hybrid domain of both fibrillin-1 and fibrillin-2. The amino terminal domain has four cysteines, and the carboxyl terminal domain has two cysteines. We are currently determining whether any of these cysteines are paired or free for intermolecular interactions.

Interestingly, we have found that the amino terminal domains of fibrillin-1 and fibrillin-2 are highly homologous to sequences at the amino terminus of latent transforming growth factor-β1 binding protein (LTBP) (40). We refer to this domain as the 4-cys domain (Fig. 1). LTBP also contains tandemly repeated calcium-binding EGF-like domains, several 8-cys domains, and a hybrid domain. However, LTBP does not contain a carboxyl terminal domain homologous to the fibrillins; in fact, carboxyl terminal sequences of LTBP are rather short. In addition, LTBP does not contain an extra cysteine in its hybrid domain, and its total size is smaller (and shorter) than the fibrillins. If initial assembly of fibrillins occurs through intermolecular disulfide bond formation between the amino terminal end (the region between the amino terminus and the first hybrid domain) and the carboxyl terminal end, then LTBPs would not be expected to participate in this assembly process.

5. Other Microfibrillar Proteins

Many other proteins, including the LTBPs, have been suggested to be associated with microfibrils. The major evidence supporting these conclusions has been immunolocalization. Additional evidence, provided by binding studies, isolation of crosslinks, investigations of biosynthesis and assembly, and genetic analyses, has not been forthcoming. All of the putative microfibrillar proteins are candidate genes for heritable disorders of connective tissue similar to the Marfan syndrome. Demonstration of mutations in any of these genes which result in similar phenotypes would be strong evidence for participation in microfibril structure.

SUMMARY

Dissecting the functions of the fibrillins has been aided by genetic analyses in humans and mice. It is clear from these studies that the fibrillins perform important roles in vascular and skeletal growth and maintenance of connective tissue integrity. However, the mechanisms behind these roles remain obscure. Structural information provided by analyses of the specific types of domains (calcium-binding EGF-like domains; 8-cys domains; hybrid domains; amino and carboxyl termini; pro/gly-rich regions) will provide the basis for understanding important molecular interactions (ligand interactions; binding to growth factors; assembly of microfibrils). Systematic investigations of these functions, using mutational analyses and *in vitro* and organismal models, are now required.

Acknowledgments
We gratefully acknowledge the expert technical assistance of Robert N. Ono, Catherine C. Ridgway, and Sara F. Tufa and funding from the Shriners Hospital for Children (to LYS and DRK).

REFERENCES

1. Maslen, C.L., Corson, G.M., Maddox, B.K., Glanville, R.W., and Sakai, L.Y. (1991) *Nature* **352**, 334–337.
2. Corson, G.M., Chalberg, S.C., Dietz, H.C., Charbonneau, N.L., and Sakai, L.Y. (1993) *Genomics* **17**, 476–484.
3. Lee, B., Godfrey, M., Vitale, E. *et al.* (1991) *Nature* **352**, 330–334.
4. Pereira, L., D'Alessio, M., Ramirez, F. *et al.* (1993) *Hum. Mol. Genet.* **2**, 961–968.
5. Zhang, H., Apfelroth, S.D., Hu, W. *et al.* (1994) *J. Cell Biol.* **124**, 855–863.
6. Downing, A.K., Knott, V., Werner, J.M., Cardy, C.M., Campbell, I.D., and Handford, P.A. (1996) *Cell* **85**, 597–605.
7. Yuan, X., Downing, A.K., Knott, V., and Handford, P.A. (1997) *EMBO J.* **16**, 6659–6666.
8. Dietz, H.C., Ramirez, F., and Sakai, L.Y. (1994) In *Advances in Human*

Genetics (Harris, H. and Hirschhorn, K., eds.), vol. 22, pp. 153–186, Plenum Publishing Corp., New York.

9. Pereira, L., Andrikopoulos, K., Tian, J. *et al.* (1997) *Nature Genet.* **17**, 218–222.
10. Dietz, H.C. and Pyeritz, R.E. (1995) *Hum. Mol. Genet.* **4**, 1799–1809.
11. Dietz, H.C., McIntosh, I., Sakai, L.Y. *et al.* (1993) *Genomics* **17**, 468–475.
12. Kainulainen, K., Karttunen, L., Puhakka, L., Sakai, L., and Peltonen, L. (1994) *Nature Genet.* **6**, 64–69.
13. Wang, M., Price, C.E., Han, J. *et al.* (1995) *Hum. Mol. Genet.* **4**, 607–613.
14. Putnam, E.A., Cho, M., Zinn, A.B., Towbin, J.A., Byers, P.H., and Milewicz, D.M. (1996) *Am. J. Med. Genet.* **62**, 233–242.
15. Sood, S., Eldadah, Z.A., Krause, W.L., McIntosh, I., and Dietz, H.C. (1996) *Nature Genet.* **12**, 209–211.
16. Francke, U., Berg, M.A., Tynan, K. *et al.* (1995) *Am. J. Hum. Genet.* **56**, 1287–1296.
17. Milewicz, D.M., Michael, K., Fisher, N., Coselli, J.S., Markello, T., and Biddinger, A. (1996) *Circulation* **94**, 2708–2711.
18. Putnam, E.A., Zhang, H., Ramirez, F., and Milewicz, D.M. (1995) *Nature Genet.* **11**, 456–458.
19. Wang, M., Clericuzio, C.L., and Godfrey, M. (1996) *Am. J. Hum. Genet.* **59**, 1027–1034.
20. Sakai, L.Y., Keene, D.R., and Engvall, E. (1986) *J. Cell Biol.* **103**, 2499–2509.
21. Maddox, B.K., Sakai, L.Y., Keene, D.R., and Glanville, R.W. (1989) *J. Biol. Chem.* **264**, 21381–21385.
22. Sakai, L.Y., Keene, D.R., Glanville, R.W., and Bächinger, H.P. (1991) *J. Biol. Chem.* **266**, 14763–14770.
23. Reinhardt, D.P., Ono, R.N., and Sakai, L.Y. (1997) *J. Biol. Chem.* **272**, 1231–1236.
24. Kielty, C.M. and Shuttleworth, C.A. (1994) *J. Cell Biol.* **124**, 997–1004.
25. Kielty, C.M., Rantamäki, T., Child, A.H., Shuttleworth, C.A., and Peltonen, L. (1995) *J. Cell. Sci.* **108**, 1317–1323.
26. Wheatley, H.M, Traboulsi, E.I., Flowers, B.E. *et al.* (1995) *Arch. Ophthalmol.* **113**, 103–109.
27. Keene, D.R., Jordan, C.D., Reinhardt, D.P. *et al.* (1997) *J. Histochem. Cytochem.* **45**, 1069–1082.
28. Keene, D.R., Sakai, L.Y., and Burgeson, R.E. (1991) *J. Histochem. Cytochem.* **39**, 59–69.
29. Kapoor, R., Sakai, L.Y., Funk, S., Roux, E., Bornstein, P., and Sage, E.H. (1988) *J. Cell Biol.* **107**, 721–730.
30. Zhang, H., Hu, W., and Ramirez, F. (1995) *J. Cell Biol.* **129**, 1165–1176.
31. Yin, W., Smiley, E., Germiller, J. *et al.* (1995) *J. Biol. Chem.* **270**, 1798–1806.

32. Reinhardt, D.P., Keene, D.R., Corson, G.M. *et al.* (1996) *J. Mol. Biol.* **258**, 104–116.

33. Schlötzer-Schrehardt, U., von der Mark, K., Sakai, L.Y., and Naumann, G.O.H. (1997) *Invest. Ophthalmol. Visual Sci.* **38**, 970–984.

34. Maddox, B.K., Keene, D.R., Sakai, L.Y. *et al.* (1997) *J. Biol. Chem.* **272**, 30993–30997.

35. Hollister, D.W., Godfrey, M., Sakai, L.Y., and Pyeritz, R.E. (1990) *N. Engl. J. Med.* **323**, 152–159.

36. Wu, C., Keivens, V.M., O'Toole, T.E., McDonald, J.A., and Ginsberg, M.H. (1995) *Cell* **83**, 715–724.

37. Milewicz, D.M., Pyeritz, R.E., Crawford, E.S., and Byers, P.H. (1992) *J. Clin. Invest.* **89**, 79–86.

38. Aoyama, T., Francke, U., Dietz, H.C., and Furthmayr, H. (1994) *J. Clin. Invest.* **94**, 130–137.

39. Sakai, L.Y. (1990) In *Elastin: Chemical and Biological Aspects* (Tamburro, A.M. and Davidson, J.M., eds.), pp. 215–227, Congedo Editore, Galatina, Italy.

40. Olofsson, A., Ichijo, H., Moren, A., ten Dijke, P., Miyazono, K., and Heldin, C.H. (1995) *J. Biol. Chem.* **270**, 31294–31297.

Extracellular Matrix-Cellular Interaction: Molecules to Diseases (Y. Ninomiya et al., eds.), pp. 41–69, Japan Sci. Soc. Press, Tokyo/S. Karger, Basel (1998)

Role of Interchain Disulfide-bonding in Dimer-trimer Switch during Laminin Heterotrimer Assembly

YASUO KITAGAWA,[*1,*2] TOMOAKI NIIMI,[*1] AND CHINO KUMAGAI[*2]

*Nagoya University BioScience Center[*1] and Graduate School of Agricultural Sciences,[*2] Nagoya University, Nagoya 464-8601, Japan*

Laminins constitute a family of basement membrane glycoproteins which affect tissue morphogenesis by their effects on proliferation, migration, and differentiation of various types of cells (*1–4*). The best studied Engelbreth Holm Swarm (EHS) tumor isoform (laminin-1) is composed of $\alpha1$ (400 kDa), $\beta1$ (220 kDa), and $\gamma1$ (210 kDa) chains, assembled and disulfide-bonded in a cross-shaped structure with three short arms and one rod-like long arm (*5–7*; Fig. 1A). The long arm is the site of chain assembly (*5–19*). It has many repeats of a heptad motif which form a hydrophobic surface along the α-helix with two charged edges at each side (*20–22*). Interchain hydrophobic interactions at this surface drive the chain assembly, and ionic interactions at the edges determine the chain selectivity (Fig. 1B). Since only $\alpha\beta\gamma$ trimers with the same N- to C-terminus orientation are selectively formed, the intracellular assembly of laminin chains is a highly controlled process.

Many variants of laminin chains have been cloned (Fig. 2A): from mouse, $\alpha1$ (*19*), $\alpha2$ (*23*), $\alpha3$ (*24*), $\alpha4$ (*25*), $\alpha5$ (*26*),

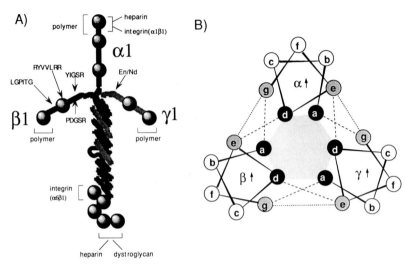

Fig. 1.   A) Structure of mouse laminin-1. Sites for the interaction with entactin/nidogen (En/Nd), heparin, integrins, dystroglycan, and self-polymerization (polymer) are indicated. Biologically active partial amino acid sequences are indicated by a single letter. B) Coiled-coil association of laminin chain long arms supported by the heptad repeats of $(abcdefg)_n$. Hydrophobic amino acids often appear at a and d and charged amino acids appear at e and g. Shadowed area indicates interchain hydrophobic interaction and dotted line indicates interchain ionic interaction.

$\beta 1$ (*18*), $\beta 2$ (*27*), $\beta 3$ (*28*), $\gamma 1$ (*17*), and $\gamma 2$ (*29*); and from human, $\alpha 1$ (*30*), $\alpha 2$ (*31, 32*), $\alpha 3$ (*33, 34*), $\alpha 4$ (*35, 36*), $\beta 1$ (*37*), $\beta 2$ (*38, 39*), $\beta 3$ (*40*), $\gamma 1$ (*41*), and $\gamma 2$ (*42, 43*). Thus, mammalian has five α variants, three β variants, and two γ variants. In $\alpha 3$, $\alpha 4$, $\beta 3$, and $\gamma 2$, the domains forming the short arms are abbreviated (Fig. 2A). Miner *et al.* (*44*) recently found that mouse $\alpha 3$ has two mRNAs corresponding to the large and the petite forms produced by alternative splicing of a single gene (Fig. 2A).

Random assembly of one of five α chains, one of three β chains, and one of two γ chains might produce thirty different $\alpha\beta\gamma$ heterotrimers. As summarized in Fig. 1B, however, only

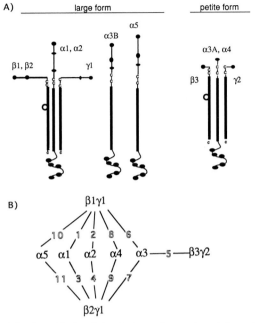

Fig. 2. A) Laminin chain variants cloned from human and mouse. B) Combination of laminin chain variants forming heterotrimers of laminin-1 through laminin-11.

eleven laminin isoforms have been found in mammalian (5), showing that random assembly is limited. Laminin-1 ($\alpha 1\beta 1\gamma 1$), laminin-2 ($\alpha 2\beta 1\gamma 1$), laminin-6 ($\alpha 3\beta 1\gamma 1$), laminin-8 ($\alpha 4\beta 1\gamma 1$), and laminin-10 ($\alpha 5\beta 1\gamma 1$) have $\beta 1\gamma 1$ dimer as the common component while laminin-3 ($\alpha 1\beta 2\gamma 1$), laminin-4 ($\alpha 2\beta 2\gamma 1$) and laminin-7 ($\alpha 3\beta 2\gamma 1$), laminin-9 ($\alpha 4\beta 2\gamma 1$) and laminin-11 ($\alpha 5\beta 2\gamma 1$) have $\beta 2\gamma 1$ dimer as the common precursor. This shows that $\beta \gamma$ dimer acts as a core for heterotrimer assembly and that $\beta 1\gamma 1$ and $\beta 2\gamma 1$ dimers can associate with any α chains. On the other hand, $\beta 3\gamma 2$ dimer can associate only with $\alpha 3$ chain to form laminin-5 ($\alpha 3\beta 3\gamma 2$). Since combinations like $\beta 3\gamma 1$, $\beta 1\gamma 2$, and $\beta 2\gamma 2$ are not known, $\beta 3$ and $\gamma 2$ have high selectivity for $\beta \gamma$ dimer formation.

The intracellular process of heterotrimer assembly explains why the $\beta\gamma$ dimer acts as the core structure. Our analyses on various cell lines (45–47) showed that β and γ chains first form the disulfide-bonded dimer and then α is disulfide-bonded to $\beta\gamma$ dimer to form $\alpha\beta\gamma$ trimer. Assembly in bovine aortic endothelial cells summarized in Fig. 3 (47) substantiated the replaceable association of either α1 or αx to β1γ1 dimer to form α1β1γ1 or αxβ1γ1. This explains the mechanism of laminin-1, laminin-2, laminin-8 , laminin-6, and laminin-10 formation with β1γ1 dimer as the core structure.

By developing chain-specific antibodies against *Drosophila* laminin chains and taking advantage of the recombinant DNA technique, we have dissected the *in vivo* process of heterotrimer assembly. By expressing mouse α, β, or γ chain in monkey COS1 cells, we show that a common mechanism is conserved among different animal species to form hybrid heterotrimers. We also show that C- and N-termini of the long arm have a distinct role in the trimer assembly. Chain specific antisera show that the

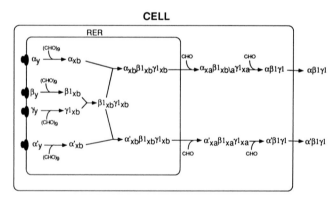

Fig. 3. Intracellular assembly of laminin chains in bovine aortic endothelial cells forming two heterotrimers of $\alpha\beta$1γ1 and $\alpha'\beta$1γ1. Subscripts xa and xb indicate stable intermediates at different steps of glycan processing. y indicates unglycosylated form. $(CHO)_9$ shows high mannose oligosaccharide chain transferred *en bloc* and CHO shows terminally transferred monosaccharide.

disulfide-bonding between *Drosophila* β and γ chains has a crucial role in the heterotrimer assembly. These observations prompted us to propose a "dimer-trimer switch model" supported by the interchain disulfide-bonding during the selective and replaceable assembly of laminin heterotrimers.

I. EXPRESSION OF LONG ARM SEQUENCE OF MOUSE LAMININ α1, β1, OR γ1 CHAINS IN COS1 CELLS AND ASSEMBLY OF MONKEY-MOUSE HYBRID LAMININ

By denaturation and renaturation of proteolytic fragments of laminin-1, Engel and coworkers showed *in vitro* that E8 fragments of three chains, which correspond to the C-terminal end of the long arm, mimic normal intracellular assembly (*48, 49*). By preparing recombinant or synthetic peptides comprised of mouse α1, β1, and γ1 partial sequences, Yamada and coworkers (*50–53*) showed that chain selection is controlled by defined sites at the C-terminal end of the long arm. However, we do not know whether these *in vitro* data reflect the microenvironment of intracellular chain assembly.

To address this question, we expressed mouse α1, β1, or γ1 sequence covering various regions of the long arm in monkey kidney cells (COS1 cells) and detected monkey-mouse hybrid laminin with correct chain selection. Despite this selective assembly, most of the overexpressed recombinant laminin chains preferentially formed disulfide-bonded homopolymers. Since such homopolymers have not been observed in the natural process, we show that the long arm is not sufficient for the correct assembly.

Figure 4A summarizes the plasmids constructed for the expression of various long arm sequences of mouse α1, β1, and γ1 in COS1 cells. To determine the critical region of β1 long arm for the assembly, we prepared the pEFβ1ΔN series covering β1 sequence which extended from the C-terminal E8 region to the N-terminal end of the short arm. Since *in vitro* experiments showed the importance of the C-terminal end, pEFβ1ΔNC1

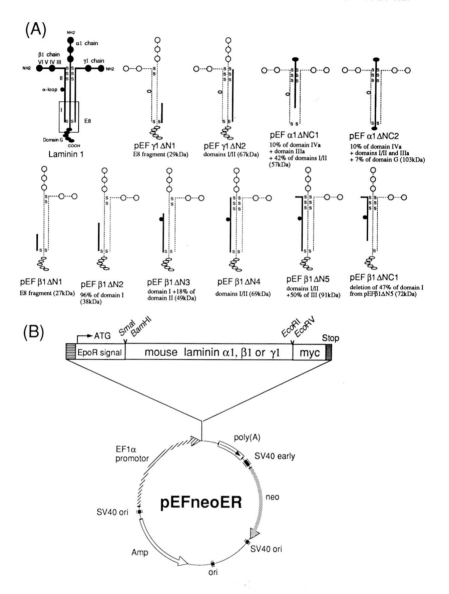

(A)

Laminin 1

pEF γ1 ΔN1
E8 fragment (29kDa)

pEF γ1 ΔN2
domains I/II (67kDa)

pEF α1 ΔNC1
10% of domain IVa
+ domain IIIa
+ 42% of domains I/II
(57kDa)

pEF α1 ΔNC2
10% of domain IVa
+ domains I/II and IIIa
+ 7% of domain G (103kDa)

pEF β1 ΔN1
E8 fragment (27kDa)

pEF β1 ΔN2
96% of domain I
(38kDa)

pEF β1 ΔN3
domain I +18% of
domain II (49kDa)

pEF β1 ΔN4
domains I/II (69kDa)

pEF β1 ΔN5
domains I/II
+50% of III (91kDa)

pEF β1 ΔNC1
deletion of 47% of domain I
from pEFβ1ΔN5 (72kDa)

(B)

ATG   SmaI BamHI   EcoRI EcoRV
                                                              Stop
EpoR signal | mouse laminin α1, β1 or γ1 | myc

poly(A)

EF1α promotor

SV40 early

pEFneoER

neo

SV40 ori

SV40 ori

Amp

ori

was prepared in which the E8 region was truncated. pEFγ1ΔN1 and pEFγ1ΔN2 cover the C-terminal end and the whole of mouse γ1 long arm, respectively, while pEFα1ΔNC1 and pEFα1ΔNC2 cover the N-terminus and the entire mouse α1 long arm, respectively. β1 and γ1 are disulfide-bonded at N- and C-termini while α1 is disulfide-bonded to β1 and γ1 only at N-termini of the long arm. Since we followed the process of chain assembly by non-reducing and reducing sodium dodecyl sulfate (SDS) gel electrophoresis, all constructs are designed to contain at least one cysteine residue for interchain disulfide-bonding.

To detect the transient expression of mouse α1, β1, and γ1 chains in COS1 cells and to analyze their assembly with endogenous monkey laminin chains, the cells were labeled with [^{35}S]-methionine/cysteine and the cell lysate was immunoprecipitated with anti-EHS tumor laminin antibody two days after transfection with various plasmids. The precipitate was separated by SDS gel electrophoresis under non-reducing condition where the inter-chain disulfide-bondings are preserved (Fig. 5). The chain composition of the complex was confirmed by two-dimensional SDS gel electrophoresis in Fig. 6, in which non-reducing electrophoresis in the first dimension (left to right) is followed by reducing electrophoresis in the second dimension (top to bottom). In this electrophoresis, monomeric proteins come to the diagonal while proteins disulfide-bonded to each other migrate to the area below the diagonal and form a vertical line for each disulfide-bonded complex.

←Fig. 4.   Plasmids encoding long arm regions of mouse α1, β1, and γ1 chains. (A) Simplified mouse laminin structure, its major domains and cysteine residues for interchain disulfide-bonding (S) are shown at top left of the panel. The long arm sequence encoded by each plasmid is shown by a thick line and other parts by dotted lines. Under the name of each plasmid encoded sequences are briefly explained together with calculated molecular mass of the products. (B) Map of pEFneoER and the strategy of plasmid construction.

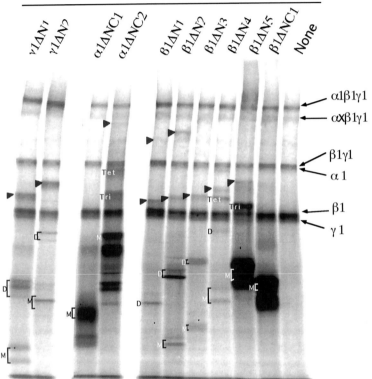

Fig. 5. Non-reducing SDS gel electrophoresis of immunoprecipitates with anti-EHS tumor laminin antibody from COS1 cells expressing mouse laminin chains. COS1 cells transfected with indicated plasmids were labeled with [^{35}S]methionine/cysteine, cell lysate was immunoprecipitated with anti-EHS tumor laminin antibody and precipitates were separated by SDS electrophoresis under non-reducing condition. Mock transfected cells were also labeled and immunoprecipitated with the same antibody (None). Arrowheads indicate the migration positions of monkey-mouse hybrids. M, D, Tri, and Tet indicate the migration positions of the monomer, homodimer, homotrimer, and homotetramer of mouse chains, respectively.

The lane None in Fig. 5, to which the cell lysate from mock transfected COS1 cells was applied, shows two laminin trimers in addition to $\beta1\gamma1$ dimer, monomeric $\alpha1$, $\beta1$, and $\gamma1$. Two dimensional electrophoresis in Fig. 6 confirms that one trimer was composed of $\alpha1$, $\beta1$, and $\gamma1$ chains while the other had a smaller α variant chain (tentatively referred to as αx) instead of the $\alpha1$ chain. Given this endogenous laminin expression, we studied the behavior of recombinant mouse laminin chains by focusing on the bands which appeared specifically in transfectants.

The lanes pEF$\gamma1\Delta$N1 or pEF$\gamma1\Delta$N2 in Fig. 5 show a transfection-dependent band (indicated by arrowheads) which migrated between the endogenous $\beta1/\gamma1$ monomer and the $\beta1\gamma1$ dimer. When the immunoprecipitates from pEF$\gamma1\Delta$N2 transfected cells were separated by two-dimensional SDS gel electrophoresis (Fig. 6A), the band was separated into two spots aligned in the vertical line indicated as $\beta1\Delta\gamma1$ at the bottom. Since the migration distance in the second dimension confirmed that these spots are the monkey $\beta1$ chain and the mouse recombinant $\gamma1$ chains, we conclude that the long arm of mouse $\gamma1$ can selectively interact with monkey $\beta1$ to form disulfide-bonded hybrid $\beta1\Delta\gamma1$ dimer. The $\beta1\Delta\gamma$ band produced by pEF$\gamma1\Delta$N1 (Fig. 5) gave essentially the same pattern of two-dimensional electrophoresis. Thus, the C-terminal half of domain I (Fig. 4A) is sufficient to allow mouse $\gamma1$ chain to assemble with monkey $\beta1$ chain. Despite the formation of the hybrid $\beta1\Delta\gamma1$ dimer, further assembly of monkey $\alpha1$ chain to form the disulfide-bonded $\alpha1\beta1\Delta\gamma1$ trimer was not detected even for the construct covering the entire long arm (pEF$\gamma1\Delta$N2). This suggests that the homology between monkey and mouse $\gamma1$ sequence is not enough to support this interaction, or that the short arm of $\gamma1$ is needed for the $\alpha\beta\gamma$ trimer assembly.

In addition to $\beta1\Delta\gamma$ dimer and monomeric $\Delta\gamma$ (M), formation of disulfide-bonded homodimer of $\Delta\gamma$ (indicated by "D" in the pEF$\gamma1\Delta$N1 and pEF$\gamma1\Delta$N2 lanes in Fig. 5) was observed. Two-dimensional electrophoresis of a larger amount of radio-

active immunoprecipitates from pEFγ1ΔN2 transfected cells
(Fig. 6A) showed the formation of homotrimer as well. A less
glycosylated product of pEFγ1ΔN2 tended to form even more
complicated disulfide-bonded complexes aligned in a horizontal
line (Fig. 6A). SDS electrophoresis of the radiolabeled media
showed that such inappropriate laminin complexes were secreted, probably through the default pathway.

The long arm of mouse α1 chain was less active in complex
formation with monkey β1 and γ1 chains. Most of the pEFα1-
ΔNC1 product, in which the C-terminal end of the long arm was
truncated, remained as monomers (the pEFα1ΔNC1 lane in Fig.
5). When the entire long arm was expressed by pEFα1ΔNC2,
most of the product formed homopolymers and a very small
amount was recruited for Δα1β1γ1 hybrid formation (the
pEFα1ΔNC2 lane in Fig. 5 and Fig. 6B). Increased reactivity of
α chain by the addition of the C-terminal end of the long arm
is consistent with the *in vitro* experiment by Utani *et al*. (*50,
51*), but our *in vivo* experiment shows that the self-association is
also activated.

To identify the functional domains of the mouse β1 long
arm for chain assembly, we transfected COS1 cells with pEFβ1-
ΔN plasmids which encode different lengths of the C-terminus
of the mouse β1 long arm (Fig. 4A). As indicated by arrowheads in pEFβ1ΔN1 lane in Fig. 5, the mouse β1E8 fragment
formed disulfide-bonded Δβ1γ1 dimer and α1Δβ1γ1 trimer.
This ability was increased in the product of pEFβ1ΔN2 in
which the sequence extended to the α-loop. Two-dimensional
electrophoresis of precipitates from the pEFβ1ΔN2 transfected
cells in Fig. 6C suggested that two forms of mouse Δβ1 chain at
different stages of glycosylation selectively interacted with
monkey γ1 chain for the formation of Δβ1γ1 dimer, but the
amount of mouse Δβ1 chain recruited for α1Δβ1γ1 trimer
formation was too small to be detected. When the mouse β1
sequence was extended beyond the α-loop, on the other hand,
the α1Δβ1γ1 trimer formation was lost and the Δβ1γ1 dimer
was gradually diminished (pEFβ1ΔN3, pEFβ1ΔN4, and

pEFβ1ΔN4 lanes in Fig. 5). This apparently conflicting result suggests a crucial role of the α-loop.

α-Loop is a domain found only in laminin βs (see Figs. 1A and 4A) where the repeats of heptad motif are interrupted and the sequence is looped out by intrachain disulfide-bondings. Electron microscopic observation of laminin-1 shows a bending of the long arm at this position (5–7). By scoring the interchain ionic interactions among laminin chains through their heptad motifs in the long arm, Beck et al. suggested that an anti-clock-wise chain arrangement of α1-β1-γ1 when viewed from the N-termini is most probable based on the reported C-terminal sequences up to the α-loop. However, the probability of the N-terminal half was increased when this chain order was altered around the α-loop (20). The result of diminished hybrid trimer formation in the pEFβ1ΔN3 and pEFβ1ΔN4 transfectants also predicts this altered chain arrangement.

When the E8 region was truncated from the mouse β1 long arm sequence (pEFβ1ΔNC1 lane in Fig. 5), complex formation was completely lost and the product remained as a monomer. Recombinant mouse β1 chains also formed homopolymers as they did for γ1 and α1 chains. Two-dimensional electrophoresis of immunoprecipitates from pEFβ1ΔN5 transfected cells (Fig. 6D) showed the formation of disulfide-bonded homopolymers.

Since a transient expression system was employed in the above experiments, there is no doubt but that the population of COS1 cells expressing recombinant products was limited. This implies that a significant part of monkey laminin chains did not have a chance to interact with mouse counterparts. Nevertheless, we could detect chain-selective disulfide-bonding of correct pairs such as mouse β1 and monkey γ1 chain, mouse γ1 and monkey β1 chain, mouse α1 chain and monkey β1γ1 dimer, and mouse β1-monkey γ1 dimer and monkey α1 chain. No pairing between α1s, β1s, and γ1s of different origins was detected. Mouse chains assembled with monkey chains in the right order in which the $\beta\gamma$ dimer formed first and the α chain came next to

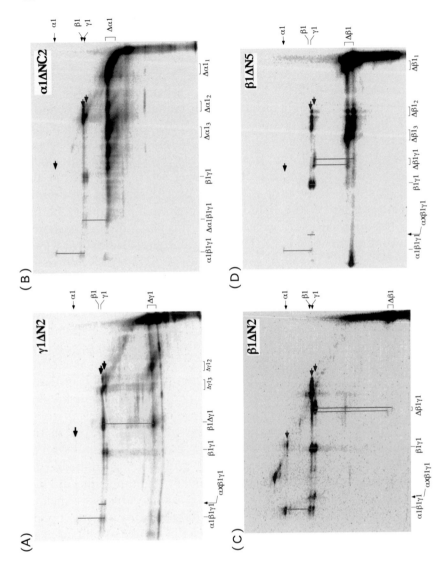

form the $\alpha\beta\gamma$ trimer (45-47). Neither $\alpha 1$ chain interacted directly with $\beta 1$s and $\gamma 1$s of either origin.

Despite the chain selective hybrid formation, a large fraction of ectopic laminin chains were disulfide-bonded to form homopolymers. Since such homopolymers have not been found among natural laminins, the process observed in COS1 cells was obviously out of control. This could be due to the overexpression of a single type of laminin chain in a limited sub-population of COS1 cells. However, compared to the high level of laminin synthesized in embryonal carcinoma F9 cells (54), for example, the expression in COS1 cells could not be considered overexpressed. If we assume any supporting machinery such as molecular chaperones which support the laminin chain assembly, however, relative overexpression of a single type of recombinant chain in excess of the "capacity" of poor-laminin producing cells like COS1 could lead to uncontrolled chain assembly.

II. DISTINCT ROLES OF $\beta 1$ LONG ARM DOMAINS FOR HETEROTRIMER FORMATION

Probably due to the expression of a single type of recombinant chain in a limited population of COS1 cells in excess of their capacity of laminin chain processing, the transient gene expression system resulted in an uncontrolled chain assembly.

←Fig. 6. Two-dimensional electrophoresis of immunoprecipitates with anti-EHS tumor laminin antibody from COS1 cells expressing mouse laminin chains. Radiolabeled lysate from COS1 cells was immunoprecipitated with anti-EHS tumor laminin antibody and precipitates were separated by SDS electrophoresis under non-reducing condition in the first dimension (from left to right) followed by reducing condition in the second dimension (from top to bottom). Migration positions of disulfide-bonded complexes in the first dimension and monomers in the second dimension are indicated at the bottom and right margins of each panel, respectively. $\Delta\alpha 1$, $\Delta\beta 1$, and $\Delta\gamma 1$ indicate truncated mouse laminin chains. Vertical lines are drawn to connect spots which co-migrated in the first dimension due to disulfide-bonding to each other.

To avoid such problems, we employed mouse embryonal carcinoma F9, the expert of the laminin production, and analyzed
the assembly after isolating stable transfectants. For this, F9 cells
were transfected with pEFβ1ΔN1 or pEFβ1ΔN5 shown in Fig.
4A (referred to as pEFβ1ΔS or pEFβ1ΔL hereafter) and *neo*-
resistant clones were selected. The cells were metabolically
labeled with [³⁵S]methionine/cysteine, radioactive cell lysate or
medium was immunoprecipitated with anti-EHS tumor laminin
antibody and the precipitate was separated by SDS gel electro-

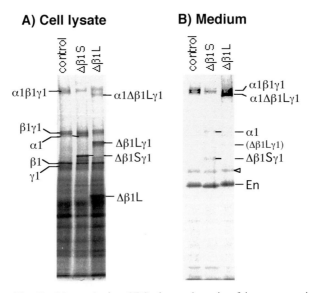

Fig. 7.   Non-reducing SDS electrophoresis of immunoprecipitates with
anti-EHS tumor laminin antibody from F9 cells expressing partial mouse
laminin β1 chain. F9 cells expressing none (control), Δβ1S or Δβ1L were
labeled with [³⁵S]methionine/cysteine, the cell lysate and medium were
immunoprecipitated with anti-EHS tumor laminin antibody. Immunoprecipitates were separated by SDS electrophoresis under non-reducing
condition. "En" indicates entactin/nidogen (150 kDa) which has strong
affinity to the short arm of γ1 in α1β1γ1 trimer. The band indicated by
open arrowhead is an unidentified protein immunoprecipitated even from
the control cells.

phoresis under non-reducing condition (Fig. 7). Chain composition of the trimer and dimer was confirmed by two-dimensional SDS electrophoresis (Fig. 8).

In the cells expressing $\Delta\beta 1S$ or $\Delta\beta 1L$, exogenous $\beta 1$ was recruited to the hybrid dimer formation and the band corresponding to $\Delta\beta 1S\gamma 1$ or $\Delta\beta 1L\gamma 1$ dimer was detected (Fig. 7A, lanes of $\Delta\beta 1S$ and $\Delta\beta 1L$). Two-dimensional gel electrophoresis in Fig. 8 confirms that endogenous $\gamma 1$ was selected by exogenous $\beta 1$ for the formation of disulfide-bonded hybrid dimers of $\Delta\beta 1S\gamma 1$ or $\Delta\beta 1L\gamma 1$. In the cells expressing $\Delta\beta 1L$, the exogenous $\beta 1$ was further recruited to the $\alpha 1\Delta\beta 1L\gamma 1$ trimer formation (Fig. 7 and Fig. 8), showing that the $\beta 1$ sequence covering the whole long arm can support the association of endogenous $\alpha 1$ and its disulfide-bonding to $\Delta\beta 1L\gamma 1$ dimer. As endogenous $\alpha 1\beta 1\gamma 1$ trimer, the $\alpha 1\Delta\beta 1L\gamma 1$ trimer was actively secreted into the medium (Fig. 7B, $\Delta\beta 1L$ lane). In contrast, C-terminal of the long arm could not support the disulfide-bonding of $\alpha 1$ to $\Delta\beta 1S\gamma 1$ dimer and no band of hybrid trimer was detected in the cells expressing $\Delta\beta 1S$ (Figs. 7 and 8). This suggests that the pair of cysteines for interchain disulfide-bonding between $\alpha 1$ and $\gamma 1$ is not well oriented without the extension of coiled-coil formation between $\beta 1$ and $\gamma 1$ to N-terminal end of the long arm.

Despite the failure of disulfide-bond formation, Figs. 7 and 8 suggest that $\Delta\beta 1S\gamma 1$ dimer can associate with $\alpha 1$ to form $\alpha 1\Delta\beta 1S\gamma 1$ trimer by the following observations. First, the level of endogenous $\alpha 1\beta 1\gamma 1$ trimer is reduced both in the cell lysate and the medium of $\Delta\beta 1S$ expressing cells when compared with the level in the control cells (Fig. 7). This suggests that one component of $\alpha 1\beta 1\gamma 1$ trimer is depleted due to the expression of $\Delta\beta 1S$. Second, the level of monomeric $\alpha 1$ in the cell lysate is increased in $\Delta\beta 1S$ expressing cells (Fig. 7A). This is not clear in Fig. 7 but the two-dimensional electrophoresis in Fig. 8 clearly shows that monomeric $\alpha 1$ is specifically increased relative to other spots. This result is consistent with the model that a part of $\alpha 1$ is assembled to $\alpha 1\Delta\beta 1S\gamma 1$ trimer without disulfide-bonding to $\Delta\beta 1S\gamma 1$ dimer, but migrated as monomer in the SDS

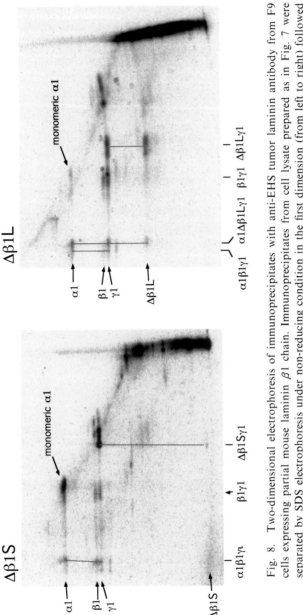

Fig. 8. Two-dimensional electrophoresis of immunoprecipitates with anti-EHS tumor laminin antibody from F9 cells expressing partial mouse laminin β1 chain. Immunoprecipitates from cell lysate prepared as in Fig. 7 were separated by SDS electrophoresis under non-reducing condition in the first dimension (from left to right) followed by reducing condition in the second dimension (from top to bottom). Migration positions of disulfide-bonded complexes in the first dimension and monomers in the second dimension are indicated at the bottom and left margins of each panel. Vertical dotted lines show proteins which are assumed to be disulfide-bonded to each other.

electrophoresis. Third, a small amount of monomeric α1 was secreted to the medium by $\Delta\beta$1S expressing cells together with $\Delta\beta$1Sγ1 dimer (Fig. 7B). Our previous studies suggest that there is a selection mechanism which allows only the $\alpha\beta\gamma$ trimers to leave the endoplasmic reticulum (45–47). Since monomeric α1 together with $\Delta\beta$1Sγ1 dimer is secreted by $\Delta\beta$1S expressing cells, the C-terminal sequence of β1 corresponding to the β1 segment of E8 fragment is enough to assemble the trimer acceptable to this selection mechanism.

The above results clearly showed that the mouse β1 long arm sequence has distinct domains for the trimer assembly. Consistent with the results of *in vitro* experiments, the C-terminal end of β1 is adequate for the chain selective assembly of α1β1γ1 trimer. Our *in vivo* experiment made it possible to analyze the function of the entire long arm and demonstrated that extension of the coiled-coil formation between β1 and γ1 to N-terminal end is essential for α1 to orient itself for the interchain disulfide-bonding to β1 and γ1. Since α1, β1, and γ1 short arms are combined by three pairs of cysteines at N-terminal end of the long arm, and correct orientation of these short arms is important for the interaction of laminin with cells and other basement membrane components (Fig. 1A), we suggest that N-terminal domain of β1 long arm is important for the organization of the functional laminin structure.

III.   DISULFIDE-BONDING BETWEEN *DROSOPHILA* LAMININ β AND γ CHAINS IS ESSENTIAL FOR α CHAIN TO FORM $\alpha\beta\gamma$ TRIMER

Drosophila laminin was first isolated by Fessler *et al.* and characterized to have α (400 kDa), β (220 kDa), and γ (200 kDa) chains (55). The cDNA clones for α (56), β (57, 58), and γ (59) have been isolated. *Drosophila* α has the domain organization similar to mammalian α5 with the sequence homology of 35% (26). *Drosophila* β and γ have 41 and 30% sequence homology to mouse β1 and γ1, respectively (58). *Drosophila*

laminin has the conserved domains found in mammalian laminins such as epidermal growth factor (EGF)-like repeats and
globular domains in the short arms, the heptad repeats to form
α-helix coiled-coil structure in the long arm and the domain G
at the C-terminal end of α chain.

Taking advantage of a hemocyte cell line of Kc 167 (60),
which produces a large amount of basement membrane proteins,
we prepared chain specific polyclonal antibodies directed
against *Drosophila* α, β, and γ. Analysis with the antibodies
revealed that β and γ form stable dimer before they are disulfide-bonded to each other. In contrast, α associates with neither
monomeric β, monomeric γ, nor $\beta\gamma$ dimer without disulfide-
bonding but only with disulfide-bonded $\beta\gamma$ dimer to form $\alpha\beta\gamma$
trimers. These results demonstrated that the interchain disulfide-
bonding between β and γ is essential for $\alpha\beta\gamma$ trimer formation.

When the medium from Kc 167 culture was separated by
SDS gel electrophoresis under reducing condition, laminin α, β,
and γ were detected among the major proteins stained with
Coomassie brilliant blue (CBB) (Fig. 9A and B). Assignment of
the bands corresponding to the α, β, and γ chains was confirmed by co-sedimentation of the three bands in sucrose velocity
sedimentation (Fig. 9A) (61) and their relative migration distance in SDS gel electrophoresis. We prepared chain specific
polyclonal antibodies directed against α, β, or γ by immunizing
rabbits with a grounded gel piece containing each protein.
Resulting antibodies gave strong signals at the migration positions of α, β, and γ in an immunoblot of the medium separated
by SDS gel electrophoresis under reducing condition (Fig. 9C,
D, and E, lanes under "Kc"), showing the immunochemical
specificity. When the extract from stage 17 *Drosophila* embryos
was immunoblotted, weak bands of various sizes were detected
in addition to the main bands of intact α, β, and γ (Fig. 9C, D,
and E, lanes under "emb"). The anti-α antibody detected bands
migrating faster than intact α, suggesting that α is subjected to
a processing. In contrast, anti-β and -γ antibodies detected
bands migrating more slowly than intact β and γ (Fig. 9D and

E, lanes under "emb"). This may indicate that a part of β and γ is cross-linked to some components in the embryos through covalent bonding resistant to the reduction with 2-mercaptoethanol.

To study the assembly of laminin in Kc 167 cells, cells were metabolically labeled with [^{35}S]methionine/cysteine and the radiolabeled cell lysate and medium were immunoprecipitated by chain specific antibodies. The immunoprecipitates were separated by two-dimensional SDS gel electrophoresis (Fig. 10).

When radiolabeled medium was immunoprecipitated either with anti-α, -β, or -γ antibody, disulfide-bonded $\alpha\beta\gamma$ trimer, disulfide-bonded $\beta\gamma$ dimer, and α monomer were commonly precipitated in addition to a series of disulfide-bonded trimers containing β, γ, and processed fragments of α (Fig. 10A, C, and E). Distinct from mammalian cells, Kc 167 cells secreted monomeric α into the medium together with disulfide-bonded $\beta\gamma$ dimer. However, since α is precipitated either with anti-β or -γ antibody (Fig. 10C and E) and disulfide-bonded $\beta\gamma$ dimer is precipitated with anti-α antibody (Fig. 10A), it is clear that *Drosophila* α forms a stable $\alpha\beta\gamma$ trimer with disulfide-bonded $\beta\gamma$ dimer. This indicates that *Drosophila* α is secreted after forming $\alpha\beta\gamma$ trimer even if it is not disulfide-bonded to disulfide-bonded $\beta\gamma$ dimer. Absence of monomeric β and γ in the medium immunoprecipitated either with anti-β or -γ shows that β and γ are not secreted before disulfide-bonding.

Immunoprecipitation of radiolabeled cell lysate either with anti-α, -β, or -γ antibody (Fig. 10B, D, and F) confirmed that *Drosophila* laminin chains are assembled in the same order as mammalian laminin chains. Since $\beta\gamma$ dimer is the only disulfide-bonded dimer detected, we can conclude that β and γ first form disulfide-bonded $\beta\gamma$ dimer and then α associates with this dimer to form disulfide-bonded $\alpha\beta\gamma$ trimer. The advantage of employing the chain-specific antibodies provided the following novel information. First, anti-β (anti-γ) antibody immunoprecipitates monomeric $\gamma(\beta)$ and monomeric α in addition to monomeric $\beta(\gamma)$, disulfide-bonded $\beta\gamma$ dimer, disulfide-bonded $\alpha\beta\gamma$ trimer,

A) Sucrose velocity sedimentation

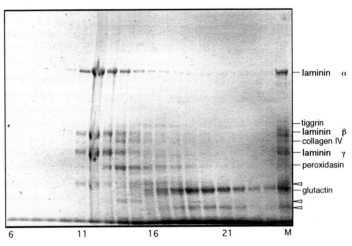

laminin α

tiggrin
laminin β
collagen IV
laminin γ
peroxidasin

glutactin

Fraction number

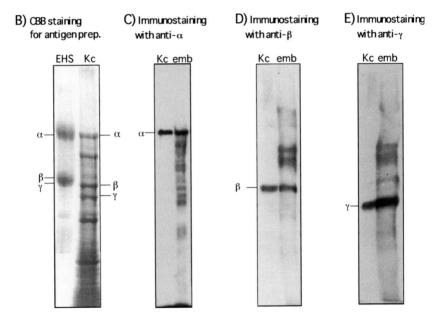

B) CBB staining
for antigen prep.

C) Immunostaining
with anti-α

D) Immunostaining
with anti-β

E) Immunostaining
with anti-γ

and a series of disulfide-bonded trimers containing β, γ, and processed fragments of α (Fig. 10D or F). This result shows that $\beta(\gamma)$ forms a stable complex with $\gamma(\beta)$ before it is disulfide-bonded to $\gamma(\beta)$. Immunoprecipitation of monomeric α with anti-$\beta(\gamma)$ antibody might suggest that $\beta(\gamma)$ also forms a stable complex with α. Second, since anti-α antibody immunoprecipitates neither monomeric β nor γ (Fig. 10B), however, it is likely that monomeric α is immunoprecipitated with anti-$\beta(\gamma)$ antibody due to its association with disulfide-bonded $\beta\gamma$ dimer. At the same time, this result implies that α does not have affinity to the stable $\beta\gamma$ dimer without disulfide-bonding. Third, anti-α antibody precipitates disulfide-bonded $\beta\gamma$ dimer in addition to monomeric α, disulfide-bonded $\alpha\beta\gamma$ trimer and a series of disulfide-bonded trimers containing β, γ and processed fragments of α (Fig. 10B). These results taken together show that α associates with neither monomeric β, monomeric γ, nor $\beta\gamma$ dimer without disulfide-bonding but only with disulfide-bonded $\beta\gamma$ dimer.

The chain specific antibodies thus showed the order of laminin chain assembly summarized in Fig. 11. β and γ first recognize each other and form a stable $\beta\gamma$ dimer. They are then disulfide-bonded probably through the paired cysteines at C-termini of *Drosophila* β and γ. α associates only with this disul-

---

←Fig. 9. Chain specific antibodies directed against *Drosophila* α, β, and γ. A) The conditioned medium of Kc 167 cells was separated by velocity sedimentation and fractionated (fraction 1 to be the bottom). CBB staining of the reduced SDS electrophoresis of the fractions is shown. The original conditioned medium was separated in lane M. Assignment of each band is indicated at the right margin. B) Mouse EHS tumor laminin (EHS) and the conditioned medium of Kc 167 cell culture (Kc) after concentration were separated by 4% SDS-polyacrylamide gel under reducing condition and stained with CBB. C), D), and E) The medium from Kc 167 cell (Kc) and the extract from *Drosophila* at stage 17 (emb) were separated under reducing condition and immunoblotted with the affinity purified anti-α antibody (panel C), anti-β antiserum (panel D), or anti-γ antiserum (panel E).

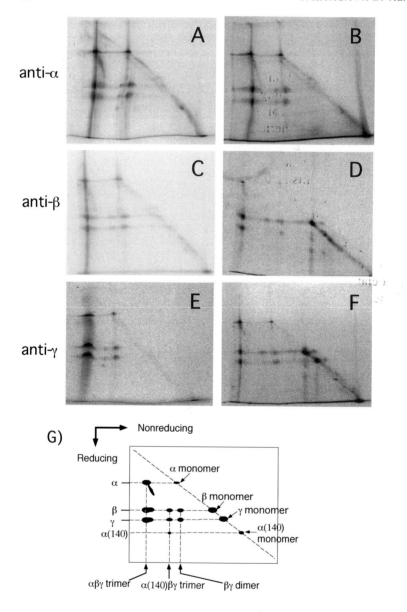

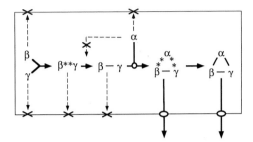

Fig. 11.   *Drosophila* laminin assembly in Kc 167 cells. The symbols * * and – indicate the chain association without and with interchain disulfide-bonding.

fide-bonded $\beta\gamma$ dimer. $\alpha\beta\gamma$ trimer is transported to the secretion pathway with α either disulfide-bonded or not -bonded to disulfide-bonded $\beta\gamma$ dimer. Neither monomeric β, γ nor $\beta\gamma$ dimer is suggested to be secreted from the cells.

IV.   ROLE OF INTERCHAIN DISULFIDE-BONDING

Laminin chains have many cysteines for intrachain disulfide-bonding in short arms, in domain G of $\alpha 1$ and in α-loop of $\beta 1$ but they are depleted in the long arms (*17–19*). Through a limited number of cysteine residues in the long arm, $\beta 1$ and $\gamma 1$ are disulfide-bonded at N- and C-termini while $\alpha 1$ is disulfide-bonded to $\beta 1$ and $\gamma 1$ only at N-termini of the long arm. The role of these interchain disulfide-bondings in the heterotrimer

---

←Fig. 10.   Two-dimensional SDS electrophoresis of radiolabeled medium and cell lysate from Kc 167 cell culture. Kc 167 cells were radiolabeled with [^{35}S]methionine and the medium (panels A, C, and E) or the cell lysate (panels B, D, and F) was immunoprecipitated with affinity purified anti-α antibody (panels A and B), anti-β antiserum (panels C and D), or anti-γ antiserum (panels E and F). Immunoprecipitates were separated by two-dimensional electrophoresis. Panel G is a schematic representation of the gels showing migration positions of laminin chains.

assembly has not been elucidated in previous studies. The homopolymer formation of the ectopically expressed mouse long arms in COS1 cells confirmed that the disulfide-bonding is a rapid reaction which fixes even unfavorable pairs before the chains can dissociate for another round of association. This implies that protein disulfide isomerase in the endoplasmic reticulum produces an environment different from an *in vitro* system and that the interchain disulfide-bonding needs to be well controlled to avoid mismatching.

Two lines of evidence showed that the interchain disulfide-bonding of α chain to $\beta\gamma$ dimer is not essential, at least for selective transport of $\alpha\beta\gamma$ heterotrimer to the secretion pathway. $\alpha 1\Delta\beta S\gamma 1$ trimer formed in F9 cells was secreted into the medium without disulfide-bonding of $\alpha 1$ to $\Delta\beta S\gamma 1$ dimer. Kc 167 cells showed that *Drosophila* $\alpha\beta\gamma$ heterotrimer can be secreted either with or without disulfide-bonding of α chain to disulfide-bonded $\beta\gamma$ dimer. These suggest that the interchain disulfide-bonding of α chain is important only for the organization of functional laminin structure after secretion.

In contrast, analysis of *Drosophila* laminin assembly employing the chain-specific antibodies uncovered a crucial role of the interchain disulfide-bonding between β and γ chains. Precipitation of monomeric β (or γ) chain with anti-γ (or -β) antibody showed that β and γ chains form stable dimer before they are disulfide-bonded to each other, while α chain associates with neither monomeric β, monomeric γ nor $\beta\gamma$ dimer without disulfide-bonding but only with disulfide-bonded $\beta\gamma$ dimer to form $\alpha\beta\gamma$ trimers. These clearly show that the interchain disulfide-bonding between β and γ chains is important for their orientation in the structure of $\beta\gamma$ dimer favorable to accept α chain. Based on this observation, we would like to propose a "dimer-trimer switch model" during the laminin heterotrimer assembly (Fig. 12).

For α to associate with a $\beta\gamma$ dimer *in vivo* environment, the orientation of β and γ chains in the $\beta\gamma$ dimer should be shifted from that of dimer formation (Step 1) to that of trimer formation

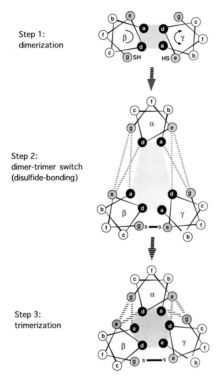

Step 1:
dimerization

Step 2:
dimer-trimer switch
(disulfide-bonding)

Step 3:
trimerization

Fig. 12. Role of interchain disulfide-bonding in dimer-trimer switch during laminin heterotrimer assembly. See Fig. 1B for keys and the text for details.

(Step 3). We assume that the disulfide-bonding between β and γ (Step 2) is important for this shift. According to the model proposed by Beck *et al.* (*20*), dimerization of β and γ driven by the interchain ionic interaction at two charged edges of α-helixes of β and γ should be specific enough to avoid mismatching but, at the same time, should be flexible enough to change the interacting partners from between β and γ to between $\beta(\gamma)$ and α. The α-helixes of β and γ may be rotating relative to each other in Step 1. With a high concentration of laminin chains *in vitro*, the α-helixes of α may find a chance to get into either side of the dimerized $\beta\gamma$ α-helixes to form new interchain ionic interactions for the trimerization. In the *in vivo* environment where the relative concentration of laminin chains is limited, however, the dimerized $\beta\gamma$ helixes need the disulfide-bonding to open the space for α-helix of α to get in (Step 2). In other words, the interchain disulfide-bonding between β and γ functions as a switch which changes the orientation of β and γ chains in $\beta\gamma$ dimer from dimerization state (Step 1) to trimerization state (Step 3).

SUMMARY

Laminins are heterotrimers with α, β, and γ chains held together by a triple-stranded α-helical coiled-coil structure. This structure is highly conserved among laminins from mammalians and invertebrates. Various isoforms of laminin chains have been found and this diversity raised the question how laminin chains are selected for the heterotrimer assembly. We here studied *in vivo* the process of the heterotrimer assembly using three experimental systems. First, we expressed mouse laminin α1, β1, or γ1 sequence covering various regions of the long arm in monkey COS1 cells and found that a common mechanism is shared by laminin chains of different animal origins. We also found that the sequences in E8 domain at C-terminal end of the long arm are crucial for the assembly. Second, we expressed mouse laminin β1 covering either C-terminal end or the whole long arm in mouse embryonal carcinoma F9 cells. Both fragments were disulfide-bonded to endogenous γ1 and formed trimers with α1, which was actively secreted into the medium. However, β1 sequence covering C-terminal end could not support the disulfide-bonding of α1 at N-terminus of the long arm. Third, we developed chain specific antibodies directed against *Drosophila* laminin α, β, and γ and found that the disulfide-bonding between *Drosophila* β and γ is essential for α to form $\alpha\beta\gamma$ heterotrimer. Based on these observations, we proposed a

32 kDa including 82 amino acids derived from the out-of-frame sequence (Fig. 1). Since this truncated protein shows impaired transcriptional elongation, the message may be unstable. Reverse transcriptase PCR (RT-PCR) showed a reduced level of aggre- can mRNA in homozygous *cmd* mice while normal levels of mRNA for type II collagen and link protein (*26*). Even if translated, this polypeptide, without the carboxyl-terminal do- main important for intracellular trafficking (*27*), may not be secreted from the cell. Furthermore, the molecule consists of an incomplete PTR structure and will not bind to hyaluronan. Our results identify a specific molecular defect in the aggrecan gene which provides an explanation for the lack of functional ag- grecan production in *cmd* mice. Besides providing molecular genetic evidence for an inherited disorder, this finding has potential implications for the analysis of related disorders in humans.

Another aggrecan gene defect is nanomelia in chicken, which bears a phenotype similar to the *cmd* mouse. A point mutation in the aggrecan gene was found at the end of exon 12 encoding CS domain in the aggrecan gene of chick nanomelia (*27*). In nanomelia, a truncated protein was localized in the endoplasmic reticulum (*28*), suggesting that G3 is important for intracellular trafficking.

2.   *Cmd Heterozygotes (29)*

Cmd heterozygotes are apparently normal at birth. Since heterozygous *cmd* mice have only one normal allele of the aggrecan gene, some metabolic differences in aggrecan in the heterozygotes may occur and create abnormal phenotypic changes after birth.

The levels of mRNA for aggrecan and for the $\alpha 1$ chain of type II collagen (*Col2a1*) in embryonic limb cartilage of *cmd* heterozygotes and homozygotes were measured and compared with those of the wild type mice by quantitative RT-PCR. The quantitative PCR experiments showed that the levels of aggrecan mRNA in *cmd* heterozygotes and in homozygotes decreased to

81% and 41%, respectively, compared to that in the wild type mice. The levels of *Col2a1* mRNA of both the heterozygote and of the homozygote were similar to that in the wild type mice. The levels of chondroitin sulfate in spinal cartilage from 90 day-old *cmd* heterozygous mice was reduced to 87% of the wild type. The amount of chondroitin sulfate (μmoles/g, mean $\pm$ S.D.) in the spine of mice at 6 months was 2.30 ± 0.04 for the wild type and 2.00 ± 0.11, ($p<0.0001$) for the heterozygote. A similar decrease in chondroitin sulfate in the spine of day 14 mice was found.

Since aggrecan is important in cartilage development, its reduced level is likely to affect the growth of the heterozygotes. Actually, *cmd* heterozygotes show two phenotypes: a slight dwarfism and late onset spinal misalignment. Approximately 28 days after birth, the heterozygotes are noticeably smaller than the wild type mice. According to our observations, the body weights measured on day 90 after birth were significantly different ($p<0.0001$) between wild type and the heterozygote mice. The body weights (mean $\pm$ S.D.) of the heterozygotes were 27.6 ± 2.1 g for males ($n=14$) and 24.9 ± 0.9 g for females ($n=10$), while those of the wild type animals were 35.8 ± 3.0 g for males ($n=13$) and 29.8 ± 2.1 g for females ($n=10$).

The most notable phenotype of the *cmd* heterozygotes is a misalignment of the cervical and thoracic spine which develops about one year after birth. We found that the heterozygotes die after 12–15 months, while the wild type mice live for 2–2.5 years. No heterozygotes lived longer than 19 months ($n=17$) whereas all of the wild type mice ($n=20$) lived for more than two years. No gender differences in survival were found. Close observations of the heterozygotes revealed that they develop a marked lordosis of the cervical spine and kyphosis of the thoraco-lumbar spine. In mice, the cervical spine is particularly susceptible to gravitational loading because the animals have to support their head. Mice with spinal distortions suddenly developed a spastic gait disturbance and showed decreased movement; they were unable to eat and starved to death within one month following acquisition of the gait disturbance. X-ray analysis was performed on 1-year old wild type and heterozygous mice. From a total of five mice of each genotype, two heterozygotes showed misalignment between vertebrae C3 and C4, and another showed a compression fracture of C4. No wild type mice showed spine misalignment. In histological examinations, the heterozygotes showed herniation of the vertebral disc, deformation of the vertebral bodies, and degenerative changes of the cervico-thoracic spine. Compression of the spinal cord by the herniated disc was also observed in these mice, explaining their spastic gait. Disappearance of the apophysis of the vertebral bodies was also observed. In contrast, we found no specific changes in the spine of one-year old wild type mice. Alcian blue staining in the heterozygotes showed reduction of glycosaminoglycans, and toluidin blue staining also confirmed their significant reduction. Alcian blue stained tissues surrounded the chondrocytes in the disc cartilage of the heterozygotes, while diffuse staining was observed in the wild type. In spinal degeneration, pathological changes are usually found in both the intervertebral discs and the facet joints. In the *cmd* heterozygotes, pathological changes were found entirely in the intervertebral discs, while the facet joints were apparently normal. These histological findings indicate that the primary lesion lies in the disc and that degeneration characteristic of reactive bone growth did not occur. The knee joint and other cartilaginous tissues were apparently normal in the heterozygotes.

While the turnover of collagen within the disc is estimated to be very slow (>100 years), aggrecan turnover is more rapid, with the half life of 8–300 days in rabbits (*30*). Because of this relatively rapid turnover, decreased synthesis of aggrecan is expected to significantly affect its deposition in the tissue. A certain level of aggrecan deposition in the disc may be critical for maintenance of disc function, and the proteoglycan is thought to function in maintaining the collagen network. Electron microscopy of the cartilage of the *cmd* homozygotes showed abnormal collagen fibrils which displayed an increase in the diameter, the appearance of periodic banding patterns, and bundling formation (*31*). These results support a role for aggrecan in collagen fibrillogenesis. Similar changes, such as rough fiber distribution with concentric patterns, are found in the disc of the *cmd* heterozygotes, probably due to a reduced deposition of aggrecan in the disc. Chondrocytes of the heterozygotes were abnormally packed together, contained degenerative vacuoles in their cytoplasm, and rough fibers in the matrix were organized in concentric circles which surrounded the chondrocytes. Chondrocytes in the wild type mice were distributed as individual cells in a fine extracellular matrix. Since it has been reported that aggrecan levels in cartilage decrease in normal individuals with aging (*32*), the *cmd* heterozygotes with the reduced aggrecan levels may suffer onset of spinal degeneration more quickly than the wild type. It is interesting to note that the pathological changes in the tissues of the *cmd* heterozygotes were found mainly in the specific portions of the spine which are most susceptible to gravitational load.

Cmd has been classified as an autosomal recessive disorder. Autosomal recessive inheritance is defined as inheritance in which a clinical phenotype occurs only when both alleles are defective. However, heterozygotes of some recessive disorders may have subtle differences in phenotype which may be accentuated by environmental factors. The pathologies reported here for *cmd* heterozygote mice suggest that defects in a single copy of this gene can have clinical manifestations.

SUMMARY

The aggrecan gene structure provides useful information. First, a close correlation between the exon structure and functional domains allows easy manipulation of the gene for expression of recombinant domains *in vitro* and *in vivo*. Secondly, homology studies of each domain/subdomain suggests its evolution among various hyaluronan-binding molecules. Thirdly, promoter analysis will provide clues for transcription mechanisms of cartilage-specific genes.

Cmd is a useful model to study the function of aggrecan in cartilage development. Aggrecan may play a regulatory role in the matrix assembly in the cartilage, since exogenously added aggrecan has been shown to reverse the abnormal matrix formed by cultured *cmd* chondrocytes (*33*). *Cmd* mice could be rescued by introduction of the normal aggrecan gene. If so, the strategy allows us to identify the function of each domain of aggrecan by introducing defined segments of the aggrecan gene back into the *cmd* mouse. *Cmd* is the first example of the mutation of a proteoglycan gene identified in mammals. Although several genetic disorders of collagens, including osteogenesis imperfecta and chondrodysplasia (*34–36*) have been reported, human genetic diseases caused by a defect of the aggrecan gene have not yet been identified. The *cmd* heterozygous mice show a high incidence of spinal misalignment and movement problems which develop with age, primarily involving spastic paralysis of the hind limbs. This paralysis resembles spinal paralysis in humans. It is conceivable that an analogous aggrecan gene defect causes spinal disc herniation or spondylomyelopathy, which are well known diseases in older humans. The phenotype of the heterozygous *cmd* mice may provide a useful clue for linkage studies of aggrecan gene defects in suspected human patients. Our findings support aggrecan as a candidate gene predisposing individuals to spinal problems.

REFERENCES

1. Hascall, V.C., Heinegard, D.K., and Wight, T.N. (1991) In *Cell Biology of Extracellular Matrix* (Hay, E.D., ed.), pp. 149–175, Plenum Press, New York.
2. Doege, K., Sasaki, M., Horigan, E., Hassell, J.R., and Yamada, Y. (1987) *J. Biol. Chem.* **262**, 17757–17767.
3. Doege, K.J., Sasaki, M., Kimura, T., and Yamada, Y. (1991) *J. Biol. Chem.* **266**, 894–902.

4. Doege, K.J., Garrison, K., Coulter, S.N., and Yamada, Y. (1994) *J. Biol. Chem.* **269**, 29232-29240.
5. Watanabe, H., Gao, L., Sugiyama, S., Doege, K., Kimata, K., and Yamada, Y. (1995) *Biochem. J.* **308**, 433-440.
6. Li, H. and Schwartz, N.B. (1995) *J. Mol. Evol.* **41**, 878-885.
7. Neame, P.J., Christner, J.E., and Baker, J.R. (1987) *J. Biol. Chem.* **262**, 17768-17778.
8. Neame, P.J., Christner, J.E., and Baker, J.R. (1986) *J. Biol. Chem.* **261**, 3519-3535.
9. Shinomura, T., Nishida, Y., Ito, K., and Kimata, K. (1993) *J. Biol. Chem.* **268**, 14461-14469.
10. Zimmermann, D.R. and Ruoslahti, E. (1989) *EMBO J.* **8**, 2975-2981.
11. Rauch, U., Karthikeyan, L., Maurel, P., Margolis, R.U., and Margolis, R.K. (1992) *J. Biol. Chem.* **267**, 19536-19547.
12. Bartolazzi, A., Nocks, A., Aruffo, A., Spring, F., and Stamenkovic, I. (1996) *J. Cell Biol.* **132**, 1199-1208.
13. Stamenkovic, I. and Aruffo, A. (1994) *Methods Enzymol.* **245**, 195-216.
14. Hardingham, T.E. and Fosang, A.J. (1995) *J. Rheumatol.* (Suppl.) **43**, 86-90.
15. Lark, M.W., Bayne, E.K., Flanagan, J. *et al.* (1997) *J. Clin. Invest.* **100**, 93-106.
16. Sandy, J.D., Plaas, A.H., and Koob, T.J. (1995) *Acta Orthop. Scand.* (Suppl.) **266**, 26-32.
17. Barry, F.P., Gaw, J.U., Young, C.N., and Neame, P.J. (1992) *Biochem. J.* **286**, 761-769.
18. Antonsson, P., Heinegard, D., and Oldberg, A. (1989) *J. Biol. Chem.* **264**, 16170-16173.
19. Bourdon, M.A., Krusius, T., Campbell, S., Schwartz, N.B., and Ruoslahti, E. (1987) *Proc. Natl. Acad. Sci. USA* **84**, 3194-3198.
20. Baldwin, C.T., Reginato, A.M., and Prockop, D.J. (1989) *J. Biol. Chem.* **264**, 15747-15750.
21. Fülop, C., Walcz, E., Valyon, M., and Glant, T.T. (1993) *J. Biol. Chem.* **268**, 17377-17383.
22. Metsaranta, M., Toman, D., de Crombrugghe, B., and Vuorio, E. (1991) *J. Biol. Chem.* **266**, 16862-16869.
23. Rittenhouse, E., Dunn, L.C., Cookingham, J. *et al.* (1978) *J. Embryol. Exp. Morphol.* **43**, 71-84.
24. Kimata, K., Barrach, H.J., Brown, K.S., and Pennypacker, J.P. (1981) *J. Biol. Chem.* **256**, 6961-6968.
25. Kimata, K., Takeda, M., Suzuki, S., Pennypacker, J.P., Barrach, H.J., and Brown, K.S. (1983) *Arch. Biochem. Biophys.* **226**, 506-516.
26. Watanabe, H., Kimata, K., Line, S. *et al.* (1994) *Nature Genet.* **7**, 154-157.
27. Li, H., Schwartz, N.B., and Vertel, B.M. (1993) *J. Biol. Chem.* **268**, 23504-

23511.

28. Vertel, B.M., Grier, B.L., Li, H., and Schwartz, N.B. (1994) *Biochem. J.* **301**, 211–216.

29. Watanabe, H., Nakata, K., Kimata, K., Nakanishi, I., and Yamada, Y. (1997) *Proc. Natl. Acad. Sci. USA* **94**, 6943–6947.

30. Mankin, H.J. and Lippiello, L. (1969) *J. Bone Joint Surg. (Am.)* **51**, 1591–1600.

31. Kobayakawa, M., Iwata, H., Brown, K.S., and Kimata, K. (1985) *Coll. Relat. Res.* **5**, 137–147.

32. Frymoyer, J.W. and Moskowitz, R.W. (1991) In *The Adult Spine* (Frymoyer, J.W. *et al.*, eds.), pp. 611–634, Raven Press, New York.

33. Takeda, M., Iwata, H., Suzuki, S., Brown, K.S., and Kimata, K. (1986) *J. Cell Biol.* **103**, 1605–1614.

34. Li, Y. and Olsen, B.R. (1997) *Matrix Biol.* **16**, 49–52.

35. Olsen, B.R. (1995) *Curr. Opin. Cell. Biol.* **7**, 720–727.

36. Sakai, L.Y., Burgeson, R.E., Olsen, B.R., Rowe, D.W., and Gordon, S.L. (1996) *Matrix Biol.* **15**, 211–229.

Extracellular Matrix-Cellular Interaction: Molecules to Diseases (Y. Ninomiya et al., eds.), pp. 87–107, Japan Sci. Soc. Press, Tokyo/S. Karger, Basel (1998)

Extracellular Matrix in Tissue Remodeling: Tenascin-C as a Modulator in Cell-matrix Interactions

MORIAKI KUSAKABE[*1] AND TERUYO SAKAKURA[*2]

*Division of Experimental Animal Research, RIKEN, Tsukuba, Ibaraki 305-0074[*1] and Department of Pathology, Mie University School of Medicine, Tsu, Mie 514-0001,[*2] Japan*

The various tissues and organs which form the body architecture are composed of epithelial and mesenchymal components. Epithelial components are primarily in charge of tissue specific functions, while the epithelial-mesenchymal interactions (EMIs) play a pivotal role in the dynamic changes in cell society, such as organogenesis, oncogenesis, cancer invasion, cancer metastasis, and wound healing. Grobstein has reported (*1, 2*) that mesenchymal components surrounding the epithelial components play an important role in morphogenesis and cytodifferentiation. The mesenchymal function in oncogenesis has also been studied, and in 1961 Orr (*3*) proposed the "permutation theory", that a certain mesenchymal mutation occurred at the initial phase of oncogenesis which caused the carcinogenesis in epithelial cells. Although this theory was later disproved, we would propose from our experimental data that the cancer matrix is important in the carcinogenesis of epithelial cells. Biologically, it interacts to spatiotemporally control the cell

proliferation and differentiation in organogenesis, the homeostasis of tissue specific architecture and cell behavior, such as cell
attachment, cell spreading and cell detachment during cancer
invasion, cancer metastasis, and wound healing. Extracellular
matrix (ECM) proteins, their receptors, matrix metalloproteases,
several kinds of hormones, and cytokines intricately work together to conduct these biological events.

Large molecular proteins, the major member of ECM, are
involved in the cell behavior as a ligand for each of their
receptors. Each matrix molecule influences the others by dynamic reciprocity, and numerous types of ECMs have been reported and their functions examined (*4, 5*).

Unlike other ECMs, tenascin-C appears in limited areas at
specific times, such as in the mesenchyme surrounding growing
epithelia, fetal epithelia or malignant tissues (*6-10*) and in the
connective tissue of healing wounds (*11*). By virtue of its spatiotemporally restricted expression, tenascin-C has attracted the
interest of many researchers who suspect that it is an essential
molecule not only in EMIs, but also in cancer-matrix cell interactions.

I. EPITHELIAL-MESENCHYMAL INTERACTIONS IN ORGANO-GENESIS

Multicellular organisms have established highly sophisti-
cated tissue specific architecture as they develop. The processing
of differentiation is controlled partly by epigenetic factors from
their environment and partly by epithelial cell autonomy. There
are two epigenetic factors, a directive factor which is involved in
determining the presumptive fate of organs and a permissive
factor which promotes the function of committed cells. For
example, the presumptive fate of the epithelial component of the
embryonic pituitary anlage was changed by the embryonic
submandibular mesenchyme (eSM) (*12*), and the resulting tissue
became morphologically and cytologically submandibular
gland-like. Embryonic pituitary epithelia taken from day 9 to 11

embryos can respond to a mesenchymal factor from eSM (*13*), although advanced aged pituitary epithelia has never reacted to the factor. Nevertheless, these cells do require the permissive factor from embryonic mesenchyme to differentiate into pituitary cells properly (*13*). These findings suggest that once pituitary epithelia are committed to their fate by their own mesenchyme, other mesenchymal cells cannot alter this fate. Thus, mesenchymal induction may be tissue-specific at the beginning of organogenesis, especially for non-committed embryonic epithelial components.

Furthermore, when separated mammary gland epithelial cells were sandwiched between eSM and mammary gland-like mesenchyme (MM), the epithelial cells grown in the eSM developed submandibular gland-like morphology, whereas the epithelial cells in the MM developed a mammary gland morphology (*14*). When eSM was transplanted into adult mammary gland, adjacent mammary epithelial cells began to grow into the eSM, forming submandibular gland-like tissue morphologically (*15*). These experiments in mammary gland morphogenesis revealed that behavior of the epithelial components may persistently depend upon the surrounding mesenchyme, and that the mesenchymal induction of embryonic tissues is much stronger than that of adult mesenchyme.

II. TUMOR CELL AND ITS MESENCHYME INTERACTIONS IN CARCINOGENESIS

The fact that an embryonic mesenchyme can promote cell growth in adult tissue allowed us to hypothesize that mesenchymal instability may affect the behavior of epithelial cells and sensitize them to epigenetic factors derived from surrounding mesenchyme. There are several reports regarding abnormality of cancer mesenchyme in carcinogenesis.

Adenomatosis of colon and rectum (ACR) is one of them. The fibroblasts from an ACR patient are the embryonic type; their actin filaments characteristically do not polymerize. These cells can be cultured even in a medium containing a low concentration of bovine serum (1%), although normally 10–15% serum concentration is required for normal fibroblast. Over 70% of the children of ACR patients have the same type of fibroblasts (*16*). Schor and his colleagues also reported that skin fibroblast from patients with mammary cancer, melanoma, colon polyposis, or retinoblastoma can easily migrate into collagen gel even at high cell density (*17-19*). They further reported that the frequency of this migration rose in proportion to the rise in the cell density, whereas that in normal fibroblasts was in inverse proportion to cell density. If fibroblasts cannot differentiate from embryonic type to adult type genetically, then this abnormality in the matrix must be inherited. Thus, this hereditary abnormality may cause the establishment of a pedigree with a high incidence of cancer. Although the embryonic mesenchymal cells in development are very important for morphogenesis together with epithelial cell proliferation, these mesenchymal cells in adult have already differentiated and lost the abilities observed during embryogenesis. In general, the cell proliferation is restricted in adult tissues, except for hematopoietic cells, skin and digestive tract, which have a well established cell renewal system. These experiments, therefore, caused us to wonder whether or not "matrix awakening" or "embryonic recapitulation" in adult mesenchyme truly promotes the initiation of carcinogenesis. There is good experimental evidence which answers this question (*20*). When eSM taken from the embryo of C3H mouse with murine mammary tumor viruses (MMTV) was transplanted into the mammary gland of syngeneic virgin female mice, the adult mammary epithelial cells were surrounded by the transplanted eSM, began to proliferate, formed nodules (DAN, duct-alveolar nodule) like a submandibular gland, and finally became cancer. Another experiment using a carcinogen showed that the timing of the mesenchyme transplantation and administration of the carcinogen was critically important for the incidence of cancer. Briefly, when mesenchyme was transplanted prior to the administration of carcinogen, there was a great increase in the inductive effect of embryonic mesenchyme to initiate cancer.

Taken together, these experiments suggest that the continuous stimuli from embryonic mesenchyme to epithelial cells may initiate the carcinogenesis together with epithelial responses to the mesenchymal factors.

In the next section we focus on the molecular mechanisms in cancer-mesenchyme cross-talk, and discuss tenascin-C, which has been found to be involved in tumorigenesis, as a modulator in cell-matrix interactions.

III. STRUCTURE AND BIOCHEMICAL CHARACTERIZATION OF TENASCIN-C

The structure of the tenascin-C molecule was first observed by rotary shadowing electron microscopy (21). It is probably a homohexamer with six arms in vivo (22). Each arm has a terminal knob, a thick distal segment, a thin proximal segment, a T-junction where three arms are connected to make a trimer and a central globule where two trimers connect to make a hexamer. Although the hexamer is the most common oligomer, the basic piece appears by electron microscopy to be the trimer of the three arms joined at a T-junction. Oligomers with nine or even 12 arms are occasionally found (23). Biochemical studies of human glioma U-251 showed that native tenascin was 1,500 kDa, while that of chicken tenascin is 1,200 kDa, as determined by agarose gel electrophoresis in sodium dodecyl sulfate (SDS) and by sedimentation equilibrium. Tenascin-C runs as a monomer on reducing gel, which implies the involvement of disulfide bonds at the connections (24).

The cDNAs of tenascin-C were first cloned from chicken embryos (25) and cultured fibroblast (26), respectively. The cDNAs of human, mouse, porcine, and newt tenascin-C have been cloned from various materials: human, glioblastoma (27) and melanoma (28) cell lines; mouse, fibroblast (29) and mammary tumor cell lines (30); porcine, submaxillary glands (31); and newt (32). Sixty to 70% homology of the sequence was found between chicken tenascin and the other species. The comparison in Fig. 1 shows that the domain structure of the tenascin subunit of these animals is similar but there are various alternative splicing patterns in their fibronectin type III domains.

The major form of tenascin-C consists of four structural domains: the N-terminal domain involved in the hexamer formation of tenascin-C subunits, 13 1/2-14 1/2 epidermal growth factor (EGF)-like repeats (chicken, 13 1/2; human and mouse, 14 1/2; porcine, 14), 9-15 fibronectin type III repeats (chicken, 11; human, 15; mouse, 13; porcine, 9) and the fibrinogen-like domain. Minor forms of tenascin-C cDNA were also detected: three in mouse, two each in chicken and porcine, and one in human. These clones are generated by alternative splicing of their insertional fibronectin type III repeats. The molecular weights of reduced tenascin-C range from 190 kDa to 320 kDa on SDS gels. Upon treatment with neuraminidase, they are decreased by about 10 kDa, suggesting modification with glycosylation (*33*). The sequence of full-length cDNAs of chicken, human, mouse, and porcine tenascin-C demonstrates that their encoded proteins contain a total of 1,899, 2,203, 2,019, and 1,746 amino acids, respectively. In human, the tenascin-C gene was demonstrated to be located on chromosome 9, bands q32-q34, using analysis of rodent-human somatic cell hybrids as well as *in situ* hybridization of hexabrachion cDNA probes to normal human metaphase chromosomes (*34, 35*). Mouse and porcine tenascin-C has also been mapped on chromosome 4 (*36*) and chromosome 1q21.1-q21.3 (*37, 38*), respectively, by fluorescence *in situ* hybridization.

Two tenascin-C-related genes have been reported to date. One is the tenascin-like gene on human chromosome 4 in major histocompatibility complex (MHC) class III region located immediately centromeric to *CYP21* (*39-41*). This gene was first found in congenital adrenal hyperplasia and is called the human gene X (*42*). Its duplicated expression, producing a large tenascin XB and a severely truncated XA protein has been demon-

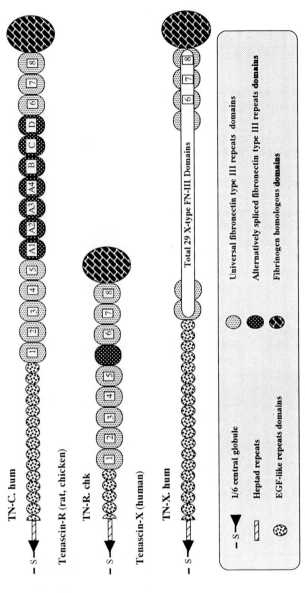

Fig. 1. Domain structure of the tenascin family. The molecules in the tenascin family are composed of heptad repeats, EGF-like repeats, fibronectin type III repeats, and fibrinogen homologous domain. The fibronectin type III repeats domain has alternatively spliced fibronectin type III repeats which are involved in the formation of the two different isoforms.

strated (*43*). The function of this X protein is still unknown, and the expression pattern shows a limited overlap. The other gene is restrictin, a chick ECM protein expressed only in the nervous system (*44*). The deduced amino acid sequence indicates that it contains 4 structural motifs: a cystein-rich segment of about 140 amino acids at the N-terminus, 4.5 EGF-like repeats, 9 fibronectin type III-like repeats, and a fibrinogen-like domain at the C-terminus. Restrictin binds to the axonal surface recognition protein F11 (*44–46*), and shows neural cell attachment activity *in vitro*. Neither the biological function of these two molecules and tenascin-C nor the relation between each molecule and tenascin-C has been established, but they probably belong to a new multigene family of ECM. These tenascin family proteins are now called cytotactin, and other tenascins, GMEM (glioma-mesenchymal extracellular matrix antigen), hexabrachion, myotendinous antigen, and J1 are called tenascin-C, restrictin is called tenascin-R (TN-R), and gene X and MHC tenascin are called tenascin-X (TN-X). The comparative structures of these molecules are summarized in Fig. 1.

IV. TENASCIN-C, A UNIQUE ECM MOLECULE IN CANCEROUS TISSUES

The most reliable pathway of signal transmission between the epithelium and mesenchyme is through the ECM (*2*). Therefore, when the cancer cell proliferates, migrates, and makes a secondary nest, new cross-talk between these cancerous cells and the surrounding matrix cells must coordinate their local contacts. For that purpose, the molecular composition of the ECM must dynamically change at each site during tumorigenesis. The mechanism of how such molecules express this specific information is still unclear. In the extracellular space, matrix molecules meet each other to make a dense meshwork in which various other substances such as cytokines work and finally form a three-dimensional scaffold. As a result, some functional domains may be formed or exposed. The further analysis of the crystal structure of these domains may offer a new aspect of research.

The ECM is probably a local regulator which serves to stabilize the structure and function of tissues. If this is true, many of the ECM components may well be redundant and functionally overlapping. In fact, the functions of collagen (47), laminin (48), fibronectin (49), and tenascin-C (50) are similar and without notable differences. They all play important roles in cell behavior such as cell adhesion, migration during morphogenesis, and carcinogenesis, and, in general, they exist throughout the body. Among them, tenascin-C is the most intriguing molecule. It is not ubiquitous, but is expressed in limited areas at specific times, such as in the dense stroma immediately surrounding growing epithelium, and in the connective tissue during an inflammatory reaction (reviewed in 23, 50, 51). Detailed observations in mouse mammary gland development and mammary cancer revealed this molecule to be oncofetal (7–9).

Tissues from a variety of human cancers as well as other diseased tissues have been examined by anti-tenascin-C antibody immunohistochemistry, and showed an enhanced expression of tenascin-C in malignant tumors and in inflammation. Histological and biochemical analyses of tenascin-C in various human tumors have indicated that this molecule is expressed in various cancer stroma and is increased in the serum (52–54), strengthening with advancement of the malignancy. Of interest is that the prognostic analysis of breast and colon cancers revealed favorable survival and no lymphogenous metastasis in patients whose cancer expressed tenascin-C strongly (55, 56). Chiquet-Ehrismann (57) and Sage and Bornstein (58) proposed that tenascin-C functions as an anti-adhesion molecule like SPARC (osteonectin) and thrombospondin.

V. COULD TENASCIN-C BE REDUNDANT IN CANCER DEVEL-
OPMENT?

During the last two decades, hints of the existence of redun-
dancies in animal development have been found in *Drosophila*
and *Calnorhabditis elegans* genetics (*59*). It was shown that
animals with single and even double or triple genomic mutations often show no phenotype, although these genes code for
highly conserved enzymes. In mice, targeted mutagenesis of
potential genes does not necessarily result in the loss of an
apparent phenotype (*60*). For example, homozygous mice that
lack the En-2 homeobox gene (*61*), the p53 tumor suppressor
gene (*62*), and the tenascin-C gene (*63*) are viable, and show no
serious abnormality in their development. The molecular basis
of genetic redundancy can probably be explained by two concepts. One is genetic overlapping. As reviewed by Tautz (*59*), the
expression of *Krupper*, a segmentation gene in *Drosophila*, is
regulated by four other genes, being promoted by low levels of
bicoid and *hunchback* and suppressed by high levels of them,
and affected by *giant* and *tailless* gene products. Thus, all four
genes are multifunctional, each having activities in addition to
its own primary function. The *Kruppel* gene can also be activated by *bicoid* and *hunchback* genes binding at the enhancer
sites. Thus, genes can interact with each other and/or DNAs and
polypeptides. The other possible concept is that of glyco-
biological overlapping. If the sugar moiety of a gene product
participates in an essential function, it is possible to consider
that other sugar molecules substituted for the lack of the gene
product. Thus, during embryonic development, redundant
genetic pathways are probably indispensable for normal morphogenesis and organostatic cell differentiation. Therefore, we
must reconsider the function of tenascin-C *in vivo*. Since
tenascin-C shows different activities, depending on the type of
producing cells and environmental condition of its expression,

its role appears to be related to tissue homeostasis. Tenascin-C acts to enable a tissue to function normally.

The publication of an earlier paper on normal development of tenascin-C gene knockout mice (*63*) did not meet with much interest, possibly because it was considered something that may happen to every gene; indeed, there are many known paradigms. A large number of genes coding for highly conserved enzymes, neurodevelopmental molecules and cancer-related molecules have been shown to be redundant in *Drosophila* and mice, because there is no apparent phenotype in the homozygous null mutant (reviewed in *59, 60, 64*). However, people intending to explore the biological function of tenascin-C were deeply disappointed at that result. Consequently, they had to conceive two simple answers. Either tenascin-C has no function, simply acting as a sort of stuffing in the extracellular spaces, or the tenascin-C function is compensated by other molecule(s). In fact, it is difficult to demonstrate any active biological function of purified tenascin-C.

In order to assess the gene function in gene disrupted animals correctly, we must get rid of any influence of the differences in mouse genetic background. Therefore, we have newly established tenascin-C knockout congenic strains which have a higher percentage of GRS/A, C3H/HeN, C57BL/6N, A/ J, and BALB/cA genetic background, respectively. In brain development, Steindler and his colleagues (*65*) examined the somatosensory cortical barrel field formation, and cerebral cortex injury in knockout mice. Their experiments demonstrated normal barrel field formation in wildtype, heterozygous, and tenascin-C knockout homozygous mice in the first postnatal week. Immunofluorescent staining of the barrel fields showed no detectable tenascin-C in the homozygous knockout mice. This study also looked for a possible compensatory expression of TN-R in the barrel field boundaries. TN-R was prominently expressed in the subcortical white matter as expected, but was not expressed in the barrel field boundaries of any of the genotypes. The study also examined the expression of tenascin-C in response to cerebral cortex wounds in late postnatal mice. The tenascin-C knockout mice showed no prominent difference in the pattern of scarring, except that they may have had an increased level of gliosis. Furthermore, in behavioral studies, tenascin-C knockout mice showed abnormal behavior such as hyperlocomotion and poor swimming ability. Biochemical analysis revealed that serotonin and dopamine transmission was decreased in the cerebral cortex, the hippocampus, or the striatum of tenascin-C knockout mouse brains (66).

However, Mitrovic and Schachner (67) recently reported detection of a truncated tenascin-C in the nervous system of tenascin-C knockout mice. They described finding low levels of tenascin-C mRNA by *in situ* hybridization and tenascin-C protein using immunocytochemistry and Western blot analysis. The authors concluded that a low level of a truncated tenascin-C protein is produced and suggested the possibility that "the failure to detect phenotypic abnormalities in the mutant mice could be due to the presence of the tenascin-C protein". The findings of Mitrovic and Schachner differ from experiments performed in our laboratories on the tenascin-C knockout mice. We could verify, on the basis of Western blotting and immunohistochemistry using several monoclonal and polyclonal antibodies, that our congenic mouse strains established from the mouse created by Saga *et al.* (63) should be considered a functional tenascin-C knockout (68). Sakai and his colleagues (69, 70) also examined the production of tenascin-C in fetal skin fibroblasts isolated from 16-day mouse embryos. Fibroblasts from wild type mice demonstrated prominent tenascin-C production, but cells from homozygous knockout mice showed no detectable tenascin-C. Sakai *et al.* used original hybrid tenascin-C knockout mice for their experiment. Forsberg and his colleagues (71) created an independent tenascin-C knockout mouse, and reported no detectable difference in healing of skin wounds and severed nerves in mice lacking tenascin-C. The genetic construct of this knockout was very similar to that of Saga *et al.*

(*63*). Forsberg and his colleagues also analyzed their knockout mice by Western blotting and immunofluorescence, and found no detectable tenascin-C protein or mRNA.

VI. TENASCIN-C CONDUCTS SIGNAL TRANSDUCTION BETWEEN CANCER AND MATRIX FIBROBLAST AS AN ENVIRONMENTAL FACTOR

Since these congenic mice also make it possible to perform a cancer cell transplantation experiment, we finally clearly demonstrated the deficient phenotype in tenascin-C knockout mice during cancer development and cancer metastasis (*72*). Recently, we performed a cancer cell transplantation experiment using our tenascin-C knockout congenic mouse. The purpose of this study was to examine the molecular mechanisms by which tenascin-C molecules are facilitated in the cross-talk between cancer and matrix cells. The following experiments were carried out using *in vivo* and *in vitro* techniques.

First, GLMT1 (2×10^6 cells/mice), which are tenascin-C nonproducing mouse mammary tumor cells under culture condition, were injected underneath the skin of both tenascin-C knockout GRS/A congenic and syngeneic GRS/A mice. Then, solid tumors were harvested, weighed, and examined immunohistologically. When these cells were transplanted in both knockout and wild-type males, tumor weights were the same between the two groups the first week. Later, tumors in control mice became significantly larger with time, whereas the tumor growth in mutants was delayed (Fig. 2). Immunohistological staining showed that tenascin was strongly expressed and deposited in the host dermal connective tissues and the growing tumor nest in wild type mice, while there was no tenascin-C in surrounding host connective tissues, dermis or subcutaneous tissue in the tumors grown in mutant mice (Fig. 3). It is known that GLMT1 expresses tenascin-C *in vivo* in response to a certain factor which may be derived from the surrounding host fibro-

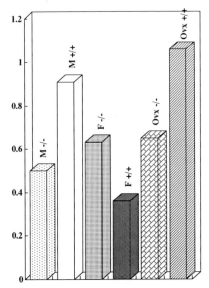

Fig. 2. Mammary tumor cell (GLMT1) tumor growth. GLMT1 (2×10^6 cells/mouse) were injected subcutaneously, allowed to grow for 3 weeks, harvested and weighed. The tumors in normal male and ovariectomized female mice grew well, whereas those in the KO mice grew poorly. Growth of tumors in normal females was much poorer than that in KO mice. M$-/-$, knockout male; M$+/+$, normal male; F$-/-$, knockout female; F$+/+$, normal female; Ovx$-/-$, ovariectomized knockout female; Ovx$+/+$, ovariectomized normal female.

blast (*73*). Therefore, we further addressed the function of fibroblasts.

Hiraiwa and her colleagues (*73*) reported that embryonic fibroblasts do produce the tenascin inducing factor, but stop producing it after birth. To know whether or not the embryonic fibroblasts of mutant mice have the ability to produce this factor, we prepared cultured embryonic fibroblasts from both mutant and control mice. The cultured supernatant from control fibroblasts stimulated the human epidermal carcinoma (A431) to produce tenascin *in vitro*, whereas the cultured supernatant from

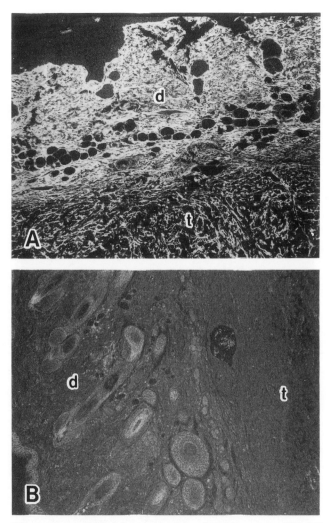

Fig. 3. Immunohistochemistry of the GLMT1 tumors. GLMT1 tumors in both normal (A) and knockout (B) mice were immunohistochemically stained by anti-tenascin antibody. A: numerous tenascin-C molecules were deposited in the dermis of the host animal and in the tumor. B: no tenascin-C molecules were detectable either in the host animal or in the tumor. d, dermis; t, tumor.

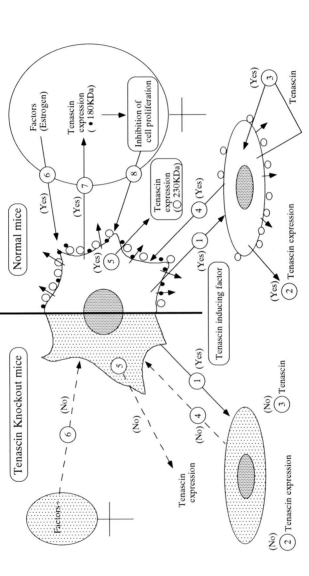

Fig. 4. Possible molecular mechanisms in cancer-mesenchyme interactions. In the normal mouse, eight cross-talks may be working in tumorigenesis. These interactions may be conducted by tenascin-C as a local modulator.

mutant fibroblasts lacked this ability. Of interest was that only tenascin-C could compensate for the disability of these mutant fibroblasts, while other ECMs, such as fibronectin, laminin, and the molecules poly-L-lysin and BSA could not do so. These findings clearly indicate that mutant fibroblasts cannot produce the tenascin inducing factor without tenascin-C support.

We further noted that when GLMT1 were injected underneath the skin of mutant and control females, the tumors in the wild-type females did not develop as well as those in mutants (Fig. 2). The tumors in ovariectomized control mice, however, developed as well as those in control males. Therefore, the following test was performed to know whether or not GLMT1 can respond to the estrogen. GLMT1 were seeded on a precoated culture plate with several ECM molecules. It was noted that these cells could respond to estrogen only when they were seeded onto a tenascin-C-coated plate. It was of interest that these cancer cells expressed a smaller variant form of tenascin-C in response to estrogen (74). These data clearly demonstrated that cell attachment to tenascin-C makes it possible for both the cancer cell and the surrounding mesenchyme to behave properly.

As described above, we demonstrated the tenascin-C function using tenascin-C knockout mutant. Thus, we would propose that a tumor environment can be created by the following process: 1. a cancerous cell secretes a growth factor such as transforming growth factor (TGF)-β, and this factor stimulates the surrounding matrix fibroblasts; 2. migrating fibroblasts then begin to produce tenascin-C; 3. this fibroblast derived tenascin-C fibroblast stimulates itself to produce the tenascin-C-inducing factor for the cancer cell; 4. the cancer cell is stimulated by this factor and expresses tenascin-C; 5. finally, the cancer cell derived tenascin-C is utilized as a ligand and improves the cancer cell condition; 6. in a normal female, the cells can respond to estrogen after they are exposed to tenascin-C; 7. these cells thus express smaller isoform; 8. this isoform may inhibit cell proliferation (Fig. 4).

Taken together, these findings suggest that tenascin-C may

play a fundamental role in the conduct of cell response to other factors such as cytokines and hormones, and that tenascin-C is involved in the signal transduction in the interface between cancer cell and matrix.

SUMMARY

    Cells are known to normally require attachment and spreading *in vitro* and *in vivo* to proliferate and express their differentiated properties. Experimental verification has focused attention on the nature of such adhesive macromolecules as ECM and on the characteristics and consequences of their interactions with their receptors on the cell surface. Cells that ordinarily remain attached *in vivo* must also be capable of diminishing adhesive forces for rounding and dividing; therefore, it is apparent that the interaction between cells and their environment must be a dynamic one. It also seems that, during tissue remodeling such as histogenesis, carcinogenesis, and wound healing, cells require a number of cell divisions and movement with formation and destruction of the ECM.

    Recent advances in the research on the boundary between cancer and surrounding connective tissue research areas have made it increasingly clear that the relationship between malignant tissue and its stroma is a very complex one. Although no ECM molecule has yet been discovered as a cancer-related gene product, the discovery of the oncofetal ECM glycoprotein, tenascin-C, has introduced us to a new field in cancer biology. The possible role of the cancer matrix in oncogenesis has thus become a phenomenon of great interest.

REFERENCES

    1. Grobstein, C. (1953) *Nature* **172**, 869.
    2. Grobstein, C. (1967) *Natl. Cancer Inst. Monogr.* **26**, 279–299.
    3. Orr, J.W. (1961) *Acta Union Int. Cancrum.* **17**, 64–71.
    4. Hay, E.D. (ed.) (1991) *Cell Biology of Extracellular Matrix*, 2nd ed.,

Plenum Press, New York.

5. Kreis, T. and Vale, R. (eds.) (1993) *Guidebook to the Extracellular Matrix and Adhesion Proteins*, Oxford Univ. Press, Oxford.

6. Bourdon, M.A., Wikstrand, C.J., Furthmayr, H., Matthews, T.J., and Binger, D.D. (1983) *Cancer Res.* **43**, 2796–2807.

7. Chiquet-Ehrismann, R., Mackie, E.J., Pearson, C.A., and Sakakura, T. (1986) *Cell* **47**, 131–139.

8. Inaguma, Y., Kusakabe, M., Mackie, E.J., Pearson, C.A., Chiquet-Ehrismann, R., and Sakakura, T. (1988) *Dev. Biol.* **128**, 245–255.

9. Mackie, E.J., Chiquet-Ehrismann, R., Pearson, C.A. *et al.* (1989) *Proc. Natl. Acad. Sci. USA* **84**, 4621–4625.

10. Natali, P.G., Nicotra, M.R., Bartolazzi, A. *et al.* (1990) *Int. J. Cancer* **46**, 586–590.

11. Mackie, E.J., Halfter, W., and Liverani, D. (1988) *J. Cell Biol.* **107**, 2757–2767.

12. Kusakabe, M., Sakakura, T., Sano, M., and Nishizuka, Y. (1985) *Dev. Biol.* **110**, 382–391.

13. Kusakabe, M., Sakakura, T., Sano, M., and Nishizuka, Y. (1984) *Dev. Growth Diff.* **26**, 263–271.

14. Hosick, H.L., Inaguma, Y., Kusakabe, M., and Sakakura, T. (1988) *Dev. Growth Diff.* **30**, 229–240.

15. Sakakura, T., Nishizuka, Y., and Dawe, C.J. (1976) *Science* **194**, 1439–1441.

16. Pfeffer, L., Lipkin, M., Stutman, O., and Kopelovich, L. (1976) *J. Cell. Physiol.* **89**, 29–37.

17. Schor, S.L., Schor, A.M., Rushton, G., and Smith, L. (1985) *J. Cell. Sci.* **73**, 221–234.

18. Schor, S.L., Schor, A.M., Durning, P., and Rushton, G. (1985) *J. Cell. Sci.* **73**, 235–244.

19. Schor, S.L., Haggie, J.A., Durning, P. *et al.* (1986) *Int. J. Cancer* **37**, 831–836.

20. Sakakura, T., Sakagami, Y., and Nishizuka, Y. (1981) *J. Natl. Cancer Inst.* **66**, 953–957.

21. Erickson, H.P. and Inglesias, J.L. (1984) *Nature* **311**, 267–269.

22. Chiquet, M., Vrucinic-Filipi, N., Schenk, S., Beck, K., and Chiquet-Ehrismann, R. (1991) *Eur. J. Biochem.* **199**, 379–388.

23. Erickson, H.P. and Bourdon, M.A. (1989) *Annu. Rev. Cell Biol.* **5**, 71–92.

24. Taylor, H.C., Lightner, V.A., Beyer, W.F. Jr., McCaslin, D., Briscoe, G., and Erickson, H.P. (1983) *J. Cell Biochem.* **41**, 71–90.

25. Jones, F.S., Hoffman, S., Cunningham, B.A., and Edelman, G.M. (1989) *Proc. Natl. Acad. Sci. USA* **86**, 1905–1909.

26. Pearson, C.A., Pearson, D., Shibahara, S., Hofsteenge, J., and Chiquet-Ehrismann, R. (1989) *EMBO J.* **7**, 2977–2982.

27. Nies, D.E., Hemesath, T.J., Kim, J.-H., Gulcher, J.R., and Stefansson, K.

(1991) *J. Biol. Chem.* **266**, 2818–2823.

28. Siri, A., Carnemolla, B., Saginati, M. *et al.* (1991) *Nucl. Acids Res.* **19**, 525–531.

29. Weller, A., Beck, S., and Ekblom, P. (1991) *J. Cell Biol.* **112**, 355–362.

30. Saga, Y., Tsukamoto, T., Jing, N., Kusakabe, M., and Sakakura, T. (1991) *Gene* **104**, 177–185.

31. Nishi, T., Weinstein, J., Gillepsie, W.M., and Paulson, J.C. (1991) *Eur. J. Biochem.* **202**, 643–648.

32. Onda, H., Poulin, M.L., Tassava, R.A., and Chiu, I.-M. (1991) *Dev. Biol.* **148**, 219–232.

33. Taylor, H.C., Lightner, V.A., Beyer, W.F. Jr., McCaslin, D., Briscoe, G., and Erickson, H.P. (1989) *J. Cell Biochem.* **41**, 71–90.

34. Gulcher, J.R., Allexakos, M.J., LeBeau, M.M., Lemons, R.S., and Stefansson, K. (1990) *Genomics* **6**, 616–622.

35. Rocchi, M., Archidiacono, N., Romeo, G., Saginati, M., and Zardi, L. (1991) *Hum. Genet.* **86**, 621–623.

36. Pilz, A., Moseley, H., Peters, J., and Abbott, C. (1992) *Mamm. Genome* **3**, 247–249.

37. Awata, T., Yamaguchi, H., Kumagai, M., and Yasue, H. (1995) *Cytogenet. Cell Genet.* **69**, 33–34.

38. Garrido, J.J., Lahbib-Mansais, Y., Geffrotin, C., Yerle, M., and Vaiman, M. (1995) *Mamm. Genome* **6**, 221.

39. Matsumoto, K., Ishihara, N., Ando, A., Inoko, H., and Ikemura, T. (1992) *Immunogenetics* **36**, 400–403.

40. Matsumoto, K., Arai, M., Ishihara, N., Ando, A., Inoko, H., and Ikemura, T. (1992) *Genomics* **12**, 485–491.

41. Min, J., Shukla, H., Kozono, H., Bronson, S.K., Weissman, S.M., and Chaplin, D.D. (1995) *Genomics* **30**, 149–156.

42. Morel, Y., Bristow, J., Gitelman, S.E., and Miller, W.L. (1989) *Proc. Natl. Acad. Sci. USA* **86**, 6582–6586.

43. Gitelman, S.E., Bristow, J., and Miller, W.L. (1992) *Mol. Cell. Biol.* **12**, 2124–2134.

44. Norenberg, U., Willie, H., Wolff, J.M., Frank, R., and Rathjen, F.G. (1992) *Neuron* **8**, 849–863.

45. Norenberg, U., Hubert, M., Brummendorf, T., Tarnok, A., and Rathjen, F.G. (1995) *J. Cell Biol.* **130**, 473–484.

46. Zisch, A.H., D'Alessandri, L., Ranscht, B., Falchetto, R., Winterhalter, K.H., and Vaughan, L. (1992) *J. Cell Biol.* **119**, 203–213.

47. Hay, E.D. (1981) In *Cell Biology of Extracellular Matrix* (Hay, E.D., ed.), pp. 379–409, Plenum Press, New York.

48. Kleinman, H.K., Cannon, F.B., Laurie, G.W. *et al.* (1985) *J. Cell Biochem.* **27**, 317–325.

49. Yamada, K.M., Humphries, M.J., Hasegawa, T. *et al.* (1985) In *The Cell*

in Contact (Edelman, G.M. and Thiery, J.-P., eds.), pp. 303–332, Wiley, New York.

50. Chiquet-Ehrismann, R. (1990) *FASEB J.* **4**, 2598–2604.
51. Erickson, H.P. and Lightner, V.A. (1988) *Adv. Cell Biol.* **2**, 55–90.
52. Herlyn, M., Graeven, U., Speicher, D. *et al.* (1991) *Cancer Res.* **51**, 4853–4858.
53. Yoshida, J., Wakabayashi, T., Kimura, S., Washizu, K., Kiyosawa, K., and Mokuno, K. (1994) *J. Neurol. Neurosurg. Psychiatry* **57**, 1212–1215.
54. Schienk, S., Lienard, D., Gerain, J. *et al.* (1995) *Int. J. Cancer* **63**, 665–672.
55. Sugawara, I., Hirakoshi, J., Masunaga, A., Itoyama, S., and Sakakura, T. (1991) *Invasion Metast.* **11**, 325–331.
56. Ishihara, A., Yatani, R., and Sakakura, T. (1993) *Clin. Pathol.* **41**, 1099–1107.
57. Chiquet-Ehrismann, R. (1991) *Curr. Opin. Cell Biol.* **3**, 800–804.
58. Sage, H.E. and Bornstein, P. (1991) *J. Biol. Chem.* **266**, 14831–14834.
59. Tautz, D. (1992) *BioEssays* **14**, 263–266.
60. Erickson, H.P. (1993) *J. Cell Biol.* **120**, 1079–1081.
61. Joyner, A.L., Herrup, K., Auerbach, B.A., Davis, C.A., and Rossant, J. (1991) *Science* **251**, 1239–1243.
62. Donehower, L.A., Harvey, M., Slagle, B.L. *et al.* (1992) *Nature* **356**, 215–221.
63. Saga, Y., Yagi, T., Ikawa, Y., Sakakura, T., and Aizawa, S. (1992) *Gen. Dev.* **6**, 1821–1831.
64. Campos-Ortega, J.A. and Kunst, E. (1990) *Annu. Rev. Genet.* **24**, 387–407.
65. Steindler, D.A., Settles, D., Erickson, H.P. *et al.* (1995) *J. Neurosci.* **15**, 1971–1983.
66. Fukamauchi, F., Mataga, N., Wang, Y.-J., Sato, S., Yoshiki, A., and Kusakabe, M. (1996) *Biochem. Biophys. Res. Commun.* **221**, 151–156.
67. Mitrovic, N. and Schachner, M. (1995) *J. Neurosci. Res.* **42**, 710–717.
68. Settles, D.L., Kusakabe, M., Steindler, D.A., Fillmore, H., and Erickson, H.P. (1997) *J. Neurosci. Res.* **47**, 109–117.
69. Sakai, T., Kawakatsu, H., Furukawa, Y., and Saito, M. (1995) *Int. J. Cancer* **63**, 720–725.
70. Sakai, T., Ohta, M., Furukawa, Y. *et al.* (1995) *J. Cell Physiol.* **165**, 18–29.
71. Forsberg, E., Hirsch, E., Frohlich, L. *et al.* (1996) *Proc. Natl. Acad. Sci. USA* **93**, 6594–6599.
72. Kusakabe, M., Hiraiwa, N., and Sakakura, T. (1996) *Basic Invest. Breast Carcinoma* **5**, 41–45 (in Japanese).
73. Hiraiwa, N., Kida, H., Sakakura, T., and Kusakabe, M. (1993) *J. Cell. Sci.* **104**, 289–296.
74. Kusakabe, M., Yoshiki, A., Ike, F., Takaku, K., Sakakura, T., and Hiraiwa, N. (1996) Proceedings of 21st Meeting of the International Association for Breast Cancer Research, France.

Extracellular Matrix-Cellular Interaction: Molecules to Diseases (Y. Ninomiya et al., eds.), pp. 109–121,
Japan Sci. Soc. Press, Tokyo / S. Karger, Basel (1998)

Correlation of Elastin Expression and Vascular Smooth Muscle Cell Proliferation *In Vitro*

SHINGO TAJIMA

Department of Dermatology, National Defense Medical College, Tokorozawa, Saitama 359-0042, Japan

The proliferation of vascular smooth muscle cells (VSMCs) is a frequent consequence following endothelial cell injury, and is thought to be an important early pathogenic event in the evolution of atherosclerotic plaque (*1*). VSMC proliferation under normal circumstances has been reported to be regulated by many mitogens, including epidermal growth factor (EGF) (*2, 3*), angiotension II (*4*), platelet-derived growth factor (PDGF) (*5*), and transforming growth factor β (TGFβ) (*6*).

Elastin is a major constituent of extracellular matrix in the aortic wall and plays a crucial role in tissue elasticity. Overproduction of elastin is thought to be responsible for the development of atherosclerosis (*7*). The synthesis of elastin has been demonstrated to be regulated by many growth factors or cytokines: TGFβ (*8*), insulin-like growth factor (*9*), EGF (*4*), and interleukin 1 (*10*). These potent regulators of elastin synthesis play a key role in elastogenesis during atherosclerosis.

We attempted to clarify the relationship between cell prolif-

eration and elastin expression, and examined whether cell proliferation can influence elastin expression or *vice versa*.

I. ELASTIN EXPRESSION IS RECIPROCALLY MODULATED BY THE MODULATORS OF VSMC PROLIFERATION

VSMCs were isolated from aortas of 20-day-old chick
embryos by serial enzyme digestion (*11*). The cells were plated
at a density of 2.2×10^5 cell/cm² in 35 mm diameter tissue culture
dishes and cultured in Dulbecco's modified Eagle's medium
(DMEM) supplemented with 10% fetal bovine serum (FBS). The
cells of primary culture were used in this experiment. To bring
the cells to quiescence, cultures of confluent cells were placed in
a serum starved condition (0.5% FBS) for 48 hr (quiescent cells).
To bring the cells to a proliferative state quiescent cultures were
subsequently placed in 10% FBS medium for 24 hr (proliferating
cells). Quiescent cultures were treated for 48 hr with potent
stimulators of VSMC proliferation including EGF, high potas-
sium salt, angiotensin II, or 12-O-tetradecanoylphorbol 13-
acetate (TPA). Proliferating cultures were treated for 24 hr with
potent inhibitors of VSMC proliferation like minoxidil, heparin, or retinoic acid.

Potent growth-stimulating factors for VSMCs such as EGF
(*3*), angiotensin II (*12*), high K^+ concentration (*13*), or TPA
(*14*) were found to inhibit elastin synthesis in VSMCs, and
minoxidil (*15*), heparin (*16*), and retinoic acid (*17*) which are
potent inhibitors for VSMC proliferation were shown to stimulate elastin synthesis (Table I). These data strongly suggest that
elastin expression and VSMC proliferation are coupled tightly
and inversely: potent stimulators of cell proliferation may potentially inhibit elastin expression and potent inhibitors for cell
proliferation can stimulate elastin expression. The effect of
heparin or TPA on elastin expression and VSMC proliferation
was found to be mediated by the modulation of protein kinase

TABLE I

Various Correlating Factors between Elastin Expression and Smooth Muscle Cell Proliferation

	Cell proliferation	Elastin expression	PKC activity
Minoxidil	↓	↑	
Heparin	↓	↑	↓
Retinoic Acid	↓	↑	
EGF	↑	↓	
K⁺	↑	↓	
Angiotensin II	↑	↓	
TPA	↑	↓	↑

C (PKC) activity using various antagonists of kinases including H-7, W-7, and HA1004 (*14, 16*).

II. ELASTIN EXPRESSION IS MODULATED BY CELL PROLIFERATIVE STATE (*18*)

To confirm the possibility that the modulation of elastin expression is mediated by the modulation of VSMC proliferation, we studied the effects of the cell proliferative state on elastin expression using three different culture systems: deprivation or readdition of serum in confluent cells, suspension culture independent of serum-derived growth factors or adhesion, and cell synchronization with double thymidine treatment (*19*).

1. Expression of Elastin in Serum-deprived Quiescent and Serum-induced Proliferating Cells

Cells were considered quiescent (G0) after 48 hr of serum deprivation, based on the cessation of DNA synthesis and a reduction in c-myc mRNA expression (*20*) (Fig. 1). Upon readdition of serum, cells left the G0 phase, and DNA synthesis and c-myc mRNA level were stimulated within 24 hr up to the level of the proliferating state (Fig. 1). Under these conditions,

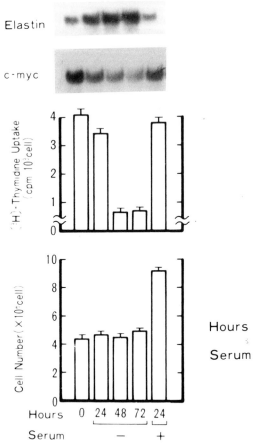

Fig. 1. Expression of elastin, c-myc, and GAPDH mRNAs in smooth muscle cells entering into a quiescent and proliferative state with deprivation and readdition of serum. Confluent cells in 10% FBS medium were shifted to 0.5% FBS medium for 24, 48, and 72 hr and again placed in medium containing 10% FBS for 24 hr. RNA was isolated from the cells at the indicated times and subjected to Northern blot analysis. Cell number and thymidine incorporation were determined at the termination of incubation.

the elastin mRNA level was increased 3.5-fold by 48 and 72 hr of serum deprivation and reduced to the level of the proliferative state by 24 hr of serum addition (Fig. 1).

2. Expression of Elastin in Suspension-arrested Cells or during Cell Cycle

Studies with the growth control systems using a serum deprivation/readdition method in confluent culture do not properly address the question of the relationship between cell growth and elastin expression, since growth factors derived from serum including EGF are known to control this expression (3). To address this question adequately, smooth muscle cells, anchorage-dependent cells, were made quiescent using a suspension culture system. Suspension of smooth muscle cells in methylcellulose culture for 72 hr caused the majority of cells to enter the G0 phase because of the cessation of DNA synthesis and a reduction in c-myc mRNA expression. Under this condition, the elastin mRNA level was markedly enhanced after 72 hr, suggesting that elastin was maximally expressed at a G0 state (Fig. 2). This was confirmed by the results obtained from cell synchronization experiments. The cells started to divide at the 10 hr time point after hydroxyurea treatment and their number increased 2-fold over the basal level at the 11 hr time point (Fig. 3b). Incorporation of thymidine was sharply elevated at the 3 hr time point and rapidly declined to a basal level at 6 hr, indicating that the cells were synchronized at the G1/S-interphase at zero-time and were in the S and M phases at the 3 and 10–11 hr time points, respectively (Fig. 3b). The elastin mRNA level as measured by Northern blot hybridization was found to be unaltered during the G1/S phases and to have declined at the G2/M phase (Fig. 3a). The elastin mRNA level of the cells which were brought to the G0 phase by the treatment of density-arrested cells with serum deprivation for 72 hr (lane D) or by suspension culture in 10% methylcellulose for 72 hr (lane S) was simultaneously determined in the same filters. The elastin mRNA level in the G0 phase appeared to be maximally ex-

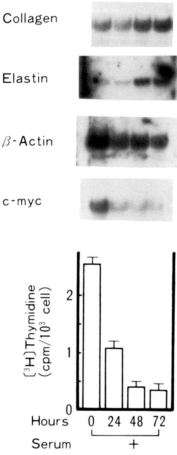

Fig. 2. Expression of collagen, elastin, β-actin, and c-myc mRNAs in smooth muscle cells entering into the quiescent state by suspension culture. Exponentially growing cells were trypsinized and transferred to the suspension condition containing 10% methylcellulose in the presence of 10% FBS for 72 hr. RNA was extracted from the cells before (lane 1) and after 24 (lane 2), 48 (lane 3), and 72 hr (lane 4) of suspension cultures and hybridized with ^{32}P-labelled collagen, elastin, β-actin, and c-myc cDNAs. DNA synthesis at the same time points was measured.

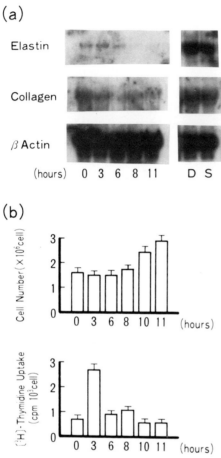

Fig. 3. Expression of elastin, collagen, and β-actin mRNA in synchronized cells. Cells were treated with 2.5 mM thymidine for 24 hr and later
with 1 mM hydroxyurea for 16 hr. After the treatment, cells were incubated
for 0, 3, 6, 8, 10, and 11 hr in 10% FBS DMEM. At the termination of
incubation, RNA was extracted from cells and elastin, collagen and
β-actin mRNA levels were assayed by Northern blot analysis. RNAs at
the G0 phase by the treatment of density-arrested cells with serum deprivation for 72 hr (lane D) or by the suspension culture in 10% methylcellulose
for 72 hr (lane S) were simultaneously assayed (a). Cells were labeled with
[^{3}H]thymidine for the final 30 min and the number of cells and thymidine
incorporation were determined after trypsin treatment (b).

pressed compared with that in the G1-S-G2-M phases. The collagen mRNA level in the G0 phase was slightly higher than that in any phase of the cell cycle. These results suggest that the elastin expression and cell cycle are closely regulated in VSMCs and that the elastin gene could be a cell cycle-related gene. The reason why the expression of elastin, an extracellular matrix protein, is regulated by the cell cycle is unclear at present. Under normal conditions, the vast majority of VSMCs in the medial layer of the artery, elastin-rich tissue, are believed to be in a stationary phase ("contractile" state) under constraints of the extracellular matrices (*1*). The proliferation of VSMCs has been reported to be selectively inhibited on α-elastin-coated dishes but not on type I- or fibronectin-coated dishes (*21*). This suggests that VSMCs can express a maximum level of elastin in the medial layer and that elastic fibers accumulated in the layer, in turn, can hinder the rapid proliferation of VSMCs. Thus, preferential expression of elastin at the G0 phase appears to be regulated, in part, by the elastin molecule present in the extracellular space.

III. CELL PROLIFERATION IS REGULATED BY ELASTIN FRAGMENTS (*22*)

Elastin has a unique repeating sequence in the hydrophobic region: tetrapeptide VPGX (X-G or A), pentapeptide VPGVG, hexapeptide XPGVGV (X = A or V), and nonapeptide VPGX-GVGAG (X = L or F). Pentapeptide VPGVG is the only repeating sequence present in the elastin molecules of all animal species analyzed including human, bovine, porcine, and chicken (*23-27*). VGVAPG is a hexapeptide repeated multiple times in the human, bovine, and porcine elastin molecules but is not present in chicken elastin molecule (*23-27*). This sequence is active as a chemoattractant for monocytes (*28*), elastin-producing fibroblasts (*29*) and tumor cells (*30*), and modulates PKC activity in lung carcinoma cells (*31*). We synthesized a pentapeptide VPGVG which is present in chicken elastin molecule and repeats multiple times.

The elastin peptides, VPGVG monomer and polymer (Fig. 4c and e) enhanced VSMC proliferation (1.5-fold) to the same extent as TPA (1.6-fold) (Fig. 4a). The peptides, monomer and polymer of VAPGVG, showed no significant effect on cell

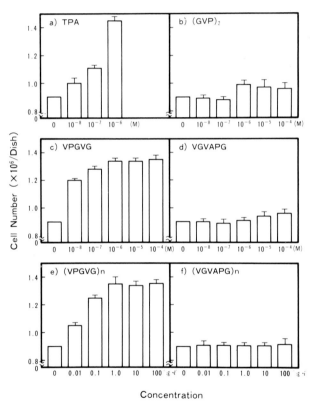

Fig. 4.   Effect of TPA and various elastin peptides on cell proliferation. Quiescent cultures were treated with TPA (a) or various elastin peptides (b–f) at the concentrations indicated for 48 hr. At the termination of incubation, cells were trypsinized and the number of cells was counted. Values are mean±deviations of duplicate experiments.

proliferation (Fig. 4d and f). Treatment of the cells with $(VGV)_2$ resulted in a slight increase in cell proliferation but to a lesser extent than VPGVG monomer or polymer (Fig. 4b). Since both VPGVG monomer and polymer showed the same extent of enhancement on cell proliferation, further experiments were performed using VPGVG monomer alone and VAPGVG monomer as a control. Elastin synthesis was inhibited by the treatment with VPGVG dose-dependently. Relative elastin synthesis measured by autoradiograms demonstrated that maximum inhibition of one third of control was achieved at the concentration of 10^{-5} M during 48 hr treatment (Fig. 5). In contrast, VAPGVG

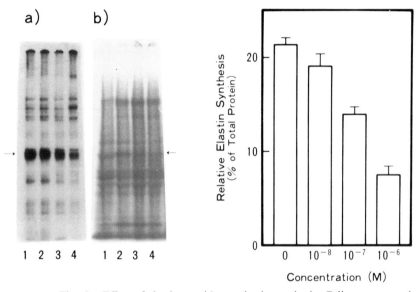

Fig. 5. Effect of elastin peptide on elastin synthesis. Cells were treated with VPGVG for 24 hr at the concentration of 0, 10^{-8}, 10^{-7}, and 10^{-6} M. Proteins from medium (a) and cell layer (b) were extracted and resolved on 4–15% sodium dodecyl sulfate-polyacrylamide gel electrophoresis (SDS-PAGE) followed by autoradiography. Elastin synthesis relative to total protein synthesis was estimated from autoradiograms of medium and cell layer fractions (right panel).

exhibited no significant effect on elastin synthesis (not shown).

The results demonstrated that exogenously added elastin fragment, VPGVG, stimulated cell proliferation while simultaneously inhibiting elastin expression. Because the modulation of elastin expression was found to be controlled by the cell growth state as described above, the affected elastin expression by elastin fragment may therefore be related to its stimulatory effect on cell proliferation. The inhibition of elastin expression by elastin fragments may reflect the negative feedback regulatory mechanism by which elastin synthesis is controlled under normal and diseased states. The modulation seems reasonable because excess elastin fragments probably generated from accumulated elastic fibers in the elastogenic tissues inhibit elastin synthesis to balance the normal metabolism of elastin.

The modulations were specifically found in the sequence VPGVG but not in VAPGVG; the VPGVG sequence is a ubiquitous pentapeptide found in the elastin molecules of all animal species analyzed. Unlike other repeating elastin fragments, it has a unique elastic property *in vitro* (*32, 33*), suggesting that VPGVG plays an essential role in elastin metabolism in normal and diseased states in chick aortic tissues.

SUMMARY

Elastin expression in cultured VSMC was enhanced by potent inhibitors of VSMC proliferation including minoxidil, heparin, and retinoic acid. In contrast, elastin expression was reduced by potent stimulators of VSMC proliferation like EGF, high K^+, angiotension II, and phorbol ester. To elucidate the relationship between elastin expression and cell proliferation, the elastin expression in the different cell growth states brought by cell-synchronizing culture or suspension culture, a culture system independent of potent modulators of VSMC proliferation was investigated. Elastin was found to be expressed maximally at G0 and minimally at G2/M phases during cell cycling, suggesting that its expression is regulated by cell growth state.

Synthetic elastin peptide VPGVG or its polymeric form in the reduction $(VPGVG)_n$ enhanced VSMC proliferation, resulting in the reduction of elastin expression. The results suggest that the elastin fragment regulates VSMC proliferation. These correlations between elastin expression and cell growth state may play an essential role in elastin metabolism under normal and diseased conditions.

REFERENCES

1. Ross, R. (1986) *N. Engl. J. Med.* **314**, 488–500.
2. Berk, B.C., Brock, T.A., Webb, R.C. *et al.* (1985) *J. Clin. Invest.* **75**, 1083–1086.
3. Tokimitsu, I., Tajima, S., and Nishikawa, T. (1990) *Biochem. Biophys. Res. Commun.* **168**, 850–856.
4. Taubman, M.B., Derk, B.C., Izumo, S., Tsuda, T., Alexander, R.W., and Nadal-Ginard, B. (1989) *J. Biol. Chem.* **264**, 526–530.
5. Roe, M.W., Hepler, J.R., Harden, T.K., and Herman, B. (1989) *J. Cell. Physiol.* **130**, 100–108.
6. Majack, R.A. (1987) *J. Cell Biol.* **105**, 465–471.
7. Ross, R. (1981) *Arteriosclerosis* **1**, 293–311.
8. Liu, J.M. and Davidson, J.M. (1988) *Biochem. Biophys. Res. Commun.* **154**, 895–901.
9. Wolfe, B.L., Rich, C.B., Goud, H.D. *et al.* (1993) *J. Biol. Chem.* **268**, 12418–12426.
10. Mauviel, A., Chen, Y.Q., Katlari, V.M. *et al.* (1993) *J. Biol. Chem.* **268**, 6520–6524.
11. Oakes, B.W., Batty, A.C., Handley, C.J., and Sandberg, L.B. (1982) *Eur. J. Cell Biol.* **27**, 34–46.
12. Tokimitsu, I., Kato, H., Wachi, H., and Tajima, S. (1994) *Biochim. Biophys. Acta* **1207**, 68–73.
13. Tokimitsu, I. and Tajima, S. (1994) *J. Biochem.* **115**, 536–539.
14. Tajima, S., Wachi, H., and Seyama, Y. Submitted for publication.
15. Hayashi, A., Suzuki, T., Wachi, H. *et al.* (1994) *Arch. Biochem. Biophys.* **315**, 137–141.
16. Wachi, H., Seyama, Y., and Tajima, S. (1995) *J. Biochem.* **118**, 582–586.
17. Hayashi, A., Suzuki, T., and Tajima, S. (1995) *J. Biochem.* **117**, 132–136.
18. Wachi, H., Seyama, Y., Yamashita, S., and Tajima, S. (1995) *Biochem. J.* **309**, 575–579.
19. Adams, R.L.P. (ed.) (1990) *Laboratory Techniques in Biochemistry and Molecular Biology: Cell Culture for Biochemists*, 2nd ed., Elsevier, Amster-

"dimer-trimer switch model" for the selective and replaceable laminin heterotrimer assembly.

Acknowledgments

We thank Dr. Y. Yamada (National Institute of Dental Research, National Institutes of Health, Bethesda, MD) for p1284 and p1238; Dr. A. Yoshimura (Institute of Cancer Research, Faculty of Medicine, Kagoshima University, Kagoshima, Japan) for pEFneoER; Dr. K. Miki (Nagoya University BioScience Center) for pLAM and suggestions about F9 cell culture; and Dr. T. Kadowaki (Nagoya University BioScience Center, Nagoya, Japan) for comments on the manuscript. This work was supported in part by a Grant-in-Aid for Scientific Research from the Ministry of Education, Science, Sports and Culture of Japan (07308073).

REFERENCES

1. Martin, G.R., Timpl, R., and Kühn, K. (1988) *Adv. Protein Chem.* **39**, 1–50.
2. Timpl, R. (1989) *Eur. J. Biochem.* **180**, 487–502.
3. Engel, J. (1992) *Biochemistry* **31**, 10633–10651.
4. Beck, K., Hunter, I., and Engel, J. (1990) *FASEB J.* **4**, 148–160.
5. Burgeson, R.E., Chiquet, M., Deutzmann, R. *et al.* (1994) *Matrix Biol.* **14**, 209–211.
6. Engel, J., Odermatt, E., Engel, A. *et al.* (1981) *J. Mol. Biol.* **150**, 97–120.
7. Engel, J. (1993) In *Molecular and Cellular Aspects of Basement Membranes* (Rohrbach, D.H. and Timpl, R., eds.), pp. 147–176, Academic Press, San Diego.
8. Paulsson, M., Deutzmann, R., Timpl, R., Dalzoppo, D., Odermatt, E., and Engel, J. (1985) *EMBO J.* **4**, 309–316.
9. Graf, J., Iwamoto, Y., Sasaki, M. *et al.* (1987) *Cell* **48**, 989–996.
10. Charonis, A.S., Skubitz, A.P., Kolialos, G.G. *et al.* (1988) *J. Cell Biol.* **107**, 1253–1260.
11. Iwamoto, Y., Robey, F.A., Graf, J. *et al.* (1987) *Science* **238**, 1132–1134.
12. Tashiro, K., Sephel, G.C., Weeks, B. *et al.* (1989) *J. Biol. Chem.* **264**, 16174–16182.
13. Schittny, J.C. and Yurchenco, P.D. (1990) *J. Cell Biol.* **110**, 825–832.
14. Sung, U., O'Rear, J.J., and Yurchenco, P.D. (1993) *J. Cell Biol.* **123**, 1255–1268.

15. Nomizu, M., Kim, W.H., Yamamura, K. *et al*. (1995) *J. Biol. Chem.* **270**, 20583-20590.
16. Colognato-Pyke, H., O'Rear, J.J., Yamada, Y., Carbonetto, S., Chen, Y.-S., and Yurchenco, P.D. (1995) *J. Biol. Chem.* **270**, 9378-9406.
17. Sasaki, M. and Yamada, Y. (1987) *J. Biol. Chem.* **262**, 17111-17117.
18. Sasaki, M., Kato, S., Kohno, K., Martin, G.R., and Yamada, Y. (1987) *Proc. Natl. Acad. Sci. USA* **84**, 935-939.
19. Sasaki, M., Kleinman, H.K., Huber, H., Deutzmann, R., and Yamada, Y. (1988) *J. Biol. Chem.* **263**, 16536-16544.
20. Beck, K., Dixon, T.W., Engel, J., and Parry, D.A.D. (1993) *J. Mol. Biol.* **31**, 311-323.
21. Kammerer, R.A., Antonsson, P., Schulthess, T., Fauser, C., and Engel, J. (1995) *J. Mol. Biol.* **250**, 64-73.
22. Antonsson, P., Kammerer, R.A., Schulthess, T., Hänisch, G., and Engel, J. (1995) *J. Mol. Biol.* **250**, 74-79.
23. Bernier, S.M., Utani, A., Sugiyama, S., Doi, T., Polistina, C., and Yamada, Y. (1994) *Matrix Biol.* **14**, 447-455.
24. Galliano, M.-F., Aberdam, D., Aguzzi, A., Ortonne, J.-P., and Meneguzzi, G. (1995) *J. Biol. Chem.* **270**, 21820-21826.
25. Liu, J. and Mayne, R. (1996) *Matrix Biol.* **15**, 433-437.
26. Miner, J.H., Lewis, R.M., and Sanes, J.R. (1995) *J. Biol. Chem.* **270**, 28523-28526.
27. Durkin, M.E., Gautam, M., Loechel, F. *et al*. (1996) *J. Biol. Chem.* **271**, 13407-13416.
28. Utani, A., Kopp, J.B., Kozak, C.A. *et al*. (1995) *Lab. Invest.* **72**, 300-310.
29. Sugiyama, S., Utani, A., Yamada, S., Kozak, C.A., and Yamada, Y. (1995) *Eur. J. Biochem.* **228**, 120-128.
30. Nissinen, M., Vuolteenaho, R., Boot-Handford, R., Kallunki, P., and Tryggvason, K. (1991) *Biochem. J.* **276**, 369-379.
31. Vuolteenaho, R., Nissinen, M., Sainio, K. *et al*. (1994) *J. Cell Biol.* **124**, 381-394.
32. Zhang, X., Vuolteenaho, R., and Tryggvason, K. (1996) *J. Biol. Chem.* **271**, 27664-27669.
33. Ryan, M.C., Tizard, R., VanDevanter, D.R., and Carter, W.G. (1994) *J. Biol. Chem.* **269**, 22779-22787.
34. Vidal, F., Baudoin, C., Miquel, C. *et al*. (1995) *Genomics* **30**, 273-280.
35. Iivaninen, A., Sainio, K., Sariola, H., and Tryggvason, K. (1995) *FEBS Lett.* **365**, 183-188.
36. Richards, A., Al-Imara, L., and Pope, F.M. (1996) *Eur. J. Biochem.* **238**, 813-821.
37. Pikkarainen, T., Eddy, R., Fukushima, Y. *et al*. (1987) *J. Biol. Chem.* **262**, 10454-10462.
38. Wewer, U.M., Gerecke, D.R., Durkin, M.E. *et al*. (1994) *Genomics* **24**, 243-

252.
39. Iivaninen, A., Vuolteenaho, R., Sainio, K. *et al.* (1994) *Matrix Biol.* **14**, 489–497.
40. Gerecke, D.R., Wagman, D.W., Champliaud, M.-F., and Burgeson, R.E. (1994) *J. Biol. Chem.* **269**, 11073–11080.
41. Pikkarainen, T., Kallunki, T., and Tryggvason, K. (1988) *J. Biol. Chem.* **263**, 6751–6758.
42. Kallunki, P., Sainio, K., Eddy, R. *et al.* (1992) *J. Cell Biol.* **119**, 679–693.
43. Vailly, J., Verrando, P., Champliaud, M.-F. *et al.* (1994) *Eur. J. Biochem.* **219**, 209–218.
44. Miner, J.H., Patton, B.L., Lentz, S.I. *et al.* (1997) *J. Cell Biol.* **137**, 685–701.
45. Morita, A., Sugimoto, E., and Kitagawa, Y. (1985) *Biochem. J.* **229**, 259–264.
46. Aratani, Y. and Kitagawa, Y. (1988) *J. Biol. Chem.* **263**, 16163–16169.
47. Tokida, Y., Aratani, Y., Morita, A., and Kitagawa, Y. (1990) *J. Biol. Chem.* **265**, 18123–18129.
48. Hunter, I., Schulthess, T., Bruch, M., Beck, K., and Engel, J. (1990) *Eur. J. Biochem.* **188**, 205–211.
49. Hunter, I., Schulthess, T., and Engel, J. (1992) *J. Biol. Chem.* **267**, 6006–6011.
50. Utani, A., Nomizu, M., Timpl, R., Roller, P.P., and Yamada, Y. (1994) *J. Biol. Chem.* **269**, 19167–19175.
51. Utani, A., Nomizu, M., Sugiyama, S., Miyamoto, S., Roller, P.P., and Yamada, Y. (1995) *J. Biol. Chem.* **270**, 3292–3298.
52. Nomizu, M., Otaka, A., Utani, A., Roller, P.P., and Yamada, Y. (1994) *J. Biol. Chem.* **269**, 30386–30392.
53. Nomizu, M., Utani, A., Beck, K. *et al.* (1996) *Biochemistry* **35**, 2885–2893.
54. Martin, G.R., Wiley, L.M., and Damjanov, I. (1977) *Dev. Biol.* **61**, 230–244.
55. Fessler, L.I., Campbell, A.G., Duncan, K.G., and Fessler, J.H. (1987) *J. Cell Biol.* **105**, 2383–2391.
56. Kusche-Gullberg, M., Garrison, K., MacKrell, A.J., Fessler, L.I., and Fessler, J.H. (1992) *EMBO J.* **11**, 4519–4527.
57. Montell, D.J. and Goodman, C.S. (1988) *Cell* **53**, 463–473.
58. Chi, H.-C. and Hui, C.-F. (1989) *J. Biol. Chem.* **264**, 1543–1550.
59. Montell, D.J. and Goodman, C.S. (1989) *J. Cell Biol.* **109**, 2441–2453.
60. Echalier, G. (1976) In *Invertebrate Cell Culture, Applications in Medicine, Biology, and Agriculture* (Kurstak, E. and Maramorosch, K., eds.), pp. 131–150, Academic Press, New York.
61. Fessler, J.H., Nelson, R.E., and Fessler, L.I. (1994) In *Drosophila melanogaster: Practical Uses in Cell and Molecular Biology* (Goldstein, L.S.B. and Fyrberg, E.A., eds.), pp. 303–328, Academic Press, New York.

Extracellular Matrix-Cellular Interaction: Molecules to Diseases (Y. Ninomiya et al., eds.), pp. 71–85, Japan Sci. Soc. Press, Tokyo/S. Karger, Basel (1998)

Aggrecan: Structure and Role in Genetic Disorders

HIDETO WATANABE AND YOSHIHIKO YAMADA

Craniofacial Developmental Biology and Regeneration Branch, National Institute of Dental Research, National Institutes of Health, Bethesda, Maryland 20892-4370, U.S.A.

Cartilage is a highly specialized tissue important to bearing compressive loads in joints. It also serves as the precursor for most bone tissues during development. Since cartilage has limited repair capacity, its deterioration results in major problems in joint diseases such as rheumatoid arthritis and osteoarthritis. Therefore, the study of cartilage formation is important to elucidate the mechanisms of skeletal development and diseases.

Aggrecan, a large chondroitin sulfate proteoglycan, is one of the major structural macromolecules of cartilage (*1*), forming huge aggregates of approximately 0.2 μm by binding to both hyaluronan and link protein. The protein core of aggrecan consists of three globular domains (G1, G2, and G3) and two glycosaminoglycan attaching domains (KS and CS) (Fig. 1) (*2, 3*). Extensive hydration of the chondroitin sulfate chains attached to the protein core results in the unique gel-like property and resistance to deformation characteristic of cartilage. Thus, aggrecan plays a major role in the maintenance of cartilage structure and functions.

I. FUNCTIONAL DOMAINS AND GENE STRUCTURE OF AG-GRECAN

The genomic structure of aggrecan core protein has been reported for rat (*4*), mouse (*5*), and chicken (*6*). The gene for mouse aggrecan spans more than 61 kb from the transcriptional start site to the polyadenylation site and contains 18 exons (Fig. 1). Exon 1 encodes the 5′-untranslated sequence and the translation starts in exon 2. The coding sequence contains 6,545 bases

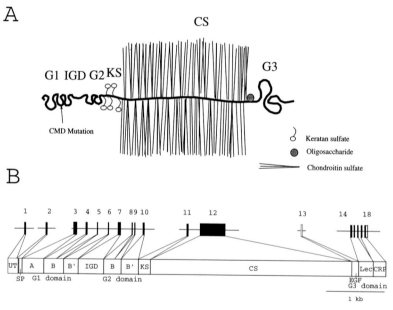

Fig. 1.   Aggrecan and its genomic structure. A, a schematic presentation of an aggrecan molecule with location of cartilage matrix deficiency (CMD) mutation. B, a schematic presentation of the exon-intron organization and relationship of exons and structural domains. Exons are numbered. The dashed box shows the alternatively spliced exon (UT, untranslated region; SP, signal peptide; IGD, interglobular domain; KS, keratan sulfate domain; CS, chondroitin sulfate domain; Lec, lectin-like domain; CRP, complement regulatory protein like domain).

for a core protein of 2,132 amino acids with a calculated $M_r =$ 259,131 including a signal peptide.

The sequences at the intron-exon boundaries are in agreement with general consensus splice sequences. Most of the 18 exons begin with a split codon. These exons include: 3–12, 15, 17, and 18. The introns vary in size, the longest being intron 1 which spans more than 21 kb; intron 8 is the shortest with 190 base pairs (bp). Sequence comparisons of aggrecan between species reveal that the overall nucleotide identity of the coding sequence between mouse and rat is 92.7%, between mouse and human is 75.9%, and between mouse and chicken is 65.3%. The sequence differences occur most frequently in the third position of the codons. Overall amino acid identities of aggrecan core protein are 95.0% between mouse and rat, 85.9% between mouse and human, and 67.9% between mouse and chicken. The conservation of nucleotide and amino acid sequences varies, however, in different domains. Significant correspondence between exons and structural domains is observed in the aggrecan gene and these features are described in detail below.

1. N-terminal Globular Domains, G1 and G2

The two N-terminal globular domains, G1 and G2, show similar disulfide-bonded structural motifs. G1 consists of three loop-like subdomains, loops A, B, and B′ whose structure is similar to link protein (7, 8). The B and B′ loops form a so-called proteoglycan tandem repeat (PTR), a structure critical for hyaluronan-binding activity. The A loop shares structural homology to an immunoglobulin fold and interacts with the A loop of link protein, which stabilizes the interaction of aggrecan and hyaluronan. G2 consists of B and B′ subdomains without an A subdomain and lacks hyaluronan binding. G1 and G2 are highly conserved among species in both nucleotide and protein sequences. Each subdomain of mouse G1 domain shows 96–100% homology to that of rat and human. Similar structural motifs occur in other hyaluronan-binding molecules such as human versican (9, 10), rat neurocan (11), and CD44 (12, 13).

The G1 region of mouse aggrecan shows 54.8% and 54.3% amino acid sequence identity to that of human versican and rat neurocan, respectively. Human CD44, which has only one B subdomain, shows 20.6% amino acid sequence identity to the B subdomain of the G1 region of mouse aggrecan. A pairwise alignment of the homologous B and B' sequences of mouse aggrecan G1 and G2 domains, mouse link protein, and human versican reveals that, within the two subgroups, the sequence of the B subdomain is more highly conserved than the B' subdomain. The sequence of the B and B' subdomains of G1 of mouse aggrecan shows more similarity to human versican and to rat neurocan than to mouse link protein, suggesting that aggrecan is evolutionarily closer to the two proteoglycans than to link protein. Analysis of the gene structure reveals that the B loop of the G1 and G2 domains of aggrecan is encoded by two exons: B of the G1 domain by exons 4 and 5, and B of the G2 domain by exons 8 and 9. In contrast, the B' loop is encoded by a single exon: B' of G1 domain by exon 6 and that of G2 domain by exon 10. It is interesting to note that each of the B and B' loops of link protein is coded for by a single exon. By analysis of intron-exon boundaries, the patterns of the split of codons are the same among the B and B' subdomains, supporting the finding of the high conservation of these subdomains in aggrecan, link protein, and versican.

2. Interglobular Domain

The interglobular domain (IGD) is located between the two globular domains and is encoded by exon 7. IGD shows less homology among species. For example, chick has an additional amino acid residue in the carboxyl portion of the domain. IGD is susceptible to several proteinases including collagenases, gelatinases, stromelysin, putative metalloproteinase (PUMP), cathepsin B, and leukocyte elastase (*14*). Studies of aggrecan fragments in cartilage and in synovial fluid indicate that a proteinase, aggrecanase, that cleaves aggrecan at a specific site in the IGD may play an important role in aggrecan turnover and in progression of destructive joint diseases (*15, 16*). A consensus sequence for the aggrecanase cleavage site is conserved among species, whereas the metalloproteinase cleavage site is conserved in mammalian species but is not present in chicken. Porcine aggrecan contains a putative KS binding region with the sequence of TIQTVT located within the IGD (*17*). This hexameric sequence is conserved among mouse, rat, human, and chick.

3. Glycosaminoglycan-attachment Domains, KS and CS

Aggrecan core protein has two glycosaminoglycan-attachment domains, KS and CS. The KS domain is located at the carboxyl terminus of the G2 domain, and is coded for by exon 11. The KS domain shows low levels of similarity among species. The size of the domain also varies in different species. The protein sequence identity of the KS domain is 93.8% between mouse and rat, 76.6% between mouse and human, and 45.2% between mouse and chicken. Putative hexameric KS-attachment sites have been reported as E-(E, K)-P-F-P-S and E-E-P-(S, F)-P-S by analysis of the human (*3*) and bovine sequences (*18*). Absence of these sequences in rodents may account for the observation that rodent aggrecan contains no keratan sulfate.

The CS-domain is the largest domain encoded by a single exon of 3,482 bp, and is located in the middle of the aggrecan core protein. This domain can be divided into two subdomains, CS1 and CS2, based on a difference in specific repeated sequences. The CS domain shows 88.3% amino acid sequence identity between the mouse and rat, although the mouse sequence has a 10 amino acid deletion and two additions of 1 and 3 amino acids at different sites. The mouse CS domain contains 120 Ser-Gly repeat sequences. Studies on CS-attachment sites using xylosyl activity of peptides suggested that the putative CS-recognition sequence is S-G-X-G (*19*). However, there are only four S-G-X-G sequences in the mouse CS-domain. Partial proteolytic digestion followed by HF deglycosylation of chick aggrecan demonstrated the CS-attachment sequence, (D, E)-X-S-G. The mouse and rat CS-domain contains 45 repeats of this latter tetrapeptide sequence. Together with the data that rat chondrosarcoma aggrecan has 80–100 CS chains, (D, E)-X-S-G may be a candidate for the CS recognition site rather than S-G-X-G, although some other sequences may also be involved in the CS attachment. Like the rat aggrecan CS domain, the mouse CS domain also contains 11 complete or partial repeats of a 40-residue unit and four complete and two partial repeats of a 100-residue sequence ((4X) S-G (2X)-S-G)$_7$(30X), indicative of a common root sequence, (4X)SG(2X)SG.

4. C-terminal Globular Domain, G3

The most carboxyl terminal domain, G3, is a composite of three structural motifs: epidermal growth factor (EGF)-like, lectin-like, and complement regulatory protein-like domain (CRP)-like motifs. The coding region for EGF-like modules is alternatively spliced. Two EGF-like modules, EGF1 (*20*) and EGF2 (*21*), have been reported by cDNA sequencing. The human aggrecan gene apparently encodes both EGF1 and EGF2. EGF2 is found in several other species such as bovine, dog, and rat with highly conserved nucleotide sequences. The other two motifs (lectin-like and CRP-like motifs) are also well conserved among species. The identity of the nucleotide sequence for mouse shows 94.6%, 88.6%, and 78.1%, to rat, human, and chicken, respectively. The amino acid sequence identities of these motifs are 98.1% , 88.3% , and 78.9% to rat, human, and chicken, respectively. The lectin-like subdomain is encoded by exons 14, 15, and 16, and the CRP-like subdomain by exons 17 and 18. Introns of this region are relatively short and five exons are contained within a 3.6 kb region. A similar gene structure for this domain is observed in chicken. The intron-exon boundaries are also conserved between the two species. The lectin-like domain is most similar to C-type carbohydrate-recognition domains (CRDs). The lectin-like subdomain of aggrecan is encoded by three separate exons, similar to the asialoglycoprotein receptor, CD23, and Kupffer cell receptor. Versican and neurocan also contain similar structural motifs at their carboxyl termini.

5. Transcription Start Site and Promoter Sequence

Primer extension revealed four transcription start sites which are clustered within 70 bp. The positions of the transcription start are different from those of the rat aggrecan gene, although the gene structure is quite similar (4). Numbering the most upstream site as $+1$, the $G+C$ content of the 523 bp promoter sequence exceeds 65%, similar to the promoter regions of the rat link protein and type II collagen genes. Within this 523 bp upstream sequence, there was no TATAA sequence. Two glucocorticoid receptor-binding sequences (TGTTCT/C) and one GGGCGG sequence (Sp-1 site) are located at -517 to -512, -380 to -375, and -426 to -421, respectively. There are several homologous direct repeat sequences including sequences between -317 to -308 and -274 to -285, and between -226 to -212 and -178 to -202. In addition, a region between -54 to -111 shows sequence homology to a sequence of the rat type II collagen promoter (-103 to -132). This sequence is highly conserved in both the rat and mouse type II collagen genes (22) and is important for type II collagen gene promoter activity. Another stretch of sequence from -287 to -259 shows homology to a sequence of the rat link protein promoter (-82 to -60). These sequences may play a role in cartilage-specific gene expression.

II. CARTILAGE MATRIX DEFICIENCY (CMD) AS A MOUSE MODEL OF AGGRECAN GENE DEFECTS

Several mutations have been described which result in defects in cartilage tissues. The cartilage matrix deficiency (cmd/cmd), an autosomal recessive lethal mutation in mice, results in a syndrome including dwarfism, short snout, and cleft palate (23). Heterozygous mice are born normal, while homozygous mice die just after birth due to respiratory failure. The cartilage of homozygous mice appears as tightly packed chondrocytes with little matrix, unlike the extensive matrix seen in normal mice. Biochemical and immunological studies have demonstrated an absence of aggrecan in the cartilage matrix of these mice, although normal levels of link protein and type II collagen were detected (*24, 25*).

1. Mapping of Aggrecan Gene, Agc and Identification of a Mutation Responsible for Cmd (26)

Mouse aggrecan cDNA was used to type the progeny of two multilocus crosses for restriction enzyme polymorphisms: (NFS/ N or C58/J × *Mus musculus musculus*) × *M. m. musculus* and (NFS/N × *M. spretus*) × *M. spretus* or C58/J. Inheritance of the aggrecan gene, *Agc*, was compared with over 575 other genetic markers typed in these crosses and distributed over 19 autosomes and the X chromosome. The aggrecan gene, *Agc*, mapped to the *cmd* locus, which strongly suggests that this is responsible for the *cmd* mutation.

DNA sequencing of genomic clones for *cmd* aggrecan revealed a single 7 bp deletion in exon 5 of the *cmd* gene. The mutation was confirmed by polymerase chain reaction (PCR) analysis with primers from exon 5 using genomic DNA from heterozygous *cmd* mice. Two different sizes of PCR products with equal molar amounts were detected. Sequencing of these two products revealed that one represented the wild type sequence and the other the 7 bp deletion. These results demonstrated that the 7 bp deletion was present not only in a single genomic clone but also in the genomic DNA of *cmd* mice.

The deleted sequence of exon 5 in the *cmd* mouse is ACCTATG or CCTATGA in which A can be placed either at the 5′ or the 3′ position because the deletion start and end point in either case is A in the normal gene. The deletion occurred in the B subdomain of the amino terminal globular G1 domain which binds to hyaluronan. The potentially truncated polypeptide created by this mutation has a predicted molecular mass of dam.

20. Zetterberg, A. and Larson, O. (1985) *Proc. Natl. Acad. Sci. USA* **82**, 5365–5369.
21. Yamamoto, M., Yamamoto, K., and Nomura, T. (1993) *Exp. Cell Res.* **204**, 121–129.
22. Wachi, H., Seyama, Y., Yamashita, S. *et al*. (1995) *FEBS Lett.* **368**, 215–219.
23. Bressan, G.M., Argos, P., and Stanley, K.K. (1987) *Biochemistry* **26**, 1497–1503.
24. Raju, K. and Anwar, R.A. (1987) *J. Biol. Chem.* **262**, 5755–5762.
25. Foster, J.A., Bruenger, E., Gray, W.R., and Sandberg, L.B. (1973) *J. Biol. Chem.* **248**, 2876–2879.
26. Sandberg, L.B., Leslie, J.G., Leach, C.T., Alvarez, V.L., Torres, A.R., and Smith, D.W. (1985) *Pathol. Biol.* **33**, 266–274.
27. Indik, Z., Yeh, H., Omstein-Goldstein, N. *et al*. (1987) *Proc. Natl. Acad. Sci. USA* **84**, 5680–5684.
28. Fulop, T. Jr., Jacob, M.P., Varga, Z., Foris, G., Leovey, A., and Robert, L. (1986) *Biochem. Biophys. Res. Commun.* **141**, 92–98.
29. Senior, R.M., Griffin, G.L., Mecham, R.P., Wrenn, D.S., Prasad, K.U., and Urry, D.W. (1984) *J. Cell Biol.* **99**, 870–874.
30. Blood, C.H., Sasse, J., Brodt, P., and Zetter, B.R. (1988) *J. Cell Biol.* **107**, 1987–1993.
31. Blood, C.H. and Zetter, B.R. (1989) *J. Biol. Chem.* **264**, 10614–10620.
32. Urry, D.W., Okamoto, K., Harris, R.D., Hendrix, C.F., and Long, M.M. (1976) *Biochemistry* **15**, 4083–4089.
33. Okamoto, K., Rapaka, R.S., Long, M.M., and Urry, D.W. (1983) In *Peptide Chemistry: Polypeptide Models of Elastin: Properties of Synthetic Cross-linked Polypeptides Reviewed* (Sakakibara, S., ed.), pp. 289–294, Protein Research Foundation, Osaka.

Extracellular Matrix-Cellular Interaction: Molecules to Diseases (Y. Ninomiya et al., eds.), pp. 123–140,
Japan Sci. Soc. Press, Tokyo/S. Karger, Basel (1998)

Topological Distribution of Collagen Binding Sites on Fibroblasts Cultured within Collagen Gels

MASAYUKI YAMATO* AND TOSHIHIKO HAYASHI

Department of Life Sciences, Graduate School of Arts and Sciences, The University of Tokyo, Meguro-ku, Tokyo 153-0041, Japan

Purified type I collagen molecules are reassembled into fibrils under a physiological condition, concomittantly forming gels (*1*). Cells mixed with the collagen solution can be trapped in the network of collagen fibrils. The cells can be cultured in a three-dimensional collagen gel (*2*). The collagen gels populated with mesenchyme-derived cells such as fibroblasts and smooth muscle cells contract without significant loss or degradation of the collagen, resulting in increased density of collagen fibrils up to one thousand times the initial density (*3–5*). The cellular activity to contract collagen gel is in turn one of the most sensitive indices for the cell aging of fibroblasts, in that the contractility decreases gradually by repeating passages of fibroblast culture (*6, 7*).

---

*Present address: Institute of Biomedical Engineering, Tokyo Women's Medical University, Shinjuku-ku, Tokyo 162-0054, Japan.

I. THE EFFECTS OF COLLAGEN FIBRILS ON FIBROBLAST FUNCTIONS

The integrin $\alpha2\beta1$, which directly binds to collagen, is shown to be at least somewhat involved in collagen gel contraction from the finding that preincubation of cells with anti-$\alpha2$ or anti-$\beta1$ integrin antibody inhibited this contraction (8–10). Integrin $\alpha2\beta1$ is selectively upregulated at the beginning of the culture of fibroblasts within collagen gels (9). Transfection of $\alpha2$ integrin cDNA into cells that express little $\alpha2$ integrin enhances collagen gel contraction (10).

Cellular activities are greatly influenced by the adhesion of the fibroblasts through the entire surface to surrounding collagen fibrils (11). Cellular activities including growth rates, mitogenic responses to growth factors, and collagen synthesis of normal diploid fibroblasts cultured within reconstituted collagen fibrils are quantitatively distinct from those of the cells cultured on two-dimensional surfaces of collagen gels, on collagen-coated dishes, or on plastic dishes (Table I). Synthesis of type I collagen and cell proliferation are repressed by the surrounding collagen fibrils (11–17). Interestingly, the growth of transformed cells that have less ability to contract collagen gels is not inhibited in the gel (16). Reduction in growth rates of fibroblasts in collagen gels apparently depends on the density of the surrounding collagen fibrils (18, 19). Cells are sparsely

TABLE I
DNA Synthetic Activity of Human Fibroblasts

Culture condition	^{3}H-deoxythymidine incorporation (10^3 dpm/μg DNA/hr)
On plastic	30 ± 2
On collagen gel	29 ± 3
In collagen gel	14 ± 1
In contracted collagen gel	1 ± 0

distributed three-dimensionally in collagen gels and hence the inhibition of cell growth in three-dimensional collagen matrix is not due to direct cell-to-cell contact. Response to soluble growth factors is often greatly depressed in contracted collagen gels compared with that of the cells on plastic dishes (20–29), although binding of a growth factor results in phosphorylation of the receptor (30). In the environment surrounded by collagen fibrils, the cells show a characteristically elongated shape similar to the fibroblasts *in vivo* (31–33).

II. GRINNELL'S "TENSION HYPOTHESIS"

Frederick Grinnell, who is one of the most intensive investigators engaged in elucidating the mechanism of the effects of collagen fibrils, has suggested that mechanical tension spanning throughout collagen gels is a key (34, 35). Collagen gels cast on plastic dishes initially attach to the dish surface. When fibroblasts begin to contract the collagen fibrils, the gels attaching to the dishes resist contraction and stay as they are, or spontaneously detach and float in the culture medium. He discriminated two types of gels: one called "attached gel" and the other "floating gel". Other investigators including ourselves call these two gel states: "gel" and "contracted gel", respectively (19). Gels attaching to the dishes are under a tension that decreases gradually with the increasing distance from the bottom layer. In floating gels, the tension, if any, is essentially isotropical. Hence, Grinnell and his colleagues later called the attached gel and floating gel "stressed" and "relaxed", respectively, to emphasize the difference in the mechanical tension loaded on collagen gels (21). Fibroblasts showed profoundly different phenotypes of growth potential in the two collagen gel states: fibroblasts in attached or stressed collagen gels are under tension and respond to serum growth factors as strongly as the cells on the collagen gel, while the cells in floating or relaxed gels without tension respond little to any growth factors (21).

Grinnell and his colleagues concluded that the difference in regulatory effects on cell functions between the two gel states could not be attributed to the collagen concentration around cells based on the following observation (21). The collagen concentration in attached gels increased from 1.5 to about 7 mg/ml after 1 day of contraction and to about 22 mg/ml after 4 days of contraction. In floating gels the density was about 28 mg/ml after 1 day and 55 mg/ml after 4 days. However, DNA synthesis by fibroblasts occurred at similar rates regardless of whether the gels were contracted for 1 or 4 days, even though DNA synthesis in floating gels was only 15% of that in attached gels. Moreover, DNA synthesis was much higher in attached gels that were contracted 4 days than in floating gels that were contracted for 1 day, even though the extent of gel contraction was similar in these samples.

We previously reported a more detailed comparison of the collagen density and growth inhibition for a longer period (Fig. 1) (11). The reduction in growth of fibroblasts in contracting collagen gels became more prominent as the density of collagen fibrils increased owing to collagen gel contraction. Thus, the

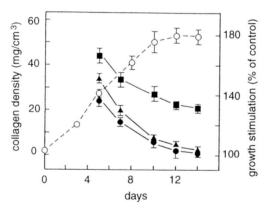

Fig. 1. Comparison of the effects of growth factors on the growth stimulation of human fibroblasts in contracting collagen gels with the density of collagen fibrils in the gel. (■) PDGF; (▲) bFGF; (●) EGF.

collagen density and growth inhibition were well-correlated, at least after the fourth day of culture.

III. DISTRIBUTION OF COLLAGEN FIBRILS IN THE CONTRACTED COLLAGEN GELS

Our interpretation of the apparent discrepancy between Grinnell's findings and ours is that not only the density obtained from gel volume, but also the microscopic locations of cell-collagen binding sites are crucial for these regulatory effects of extracellular matrix (ECM) on cellular functions. A body of knowledge has been accumulated on the signal transduction triggered by cell-ECM binding. Many cells cultured on a planar surface adhere tightly to the underlying substrate through limited regions of the ventral plasmalemma, referred to as focal adhesions or adhesion plaques (36–38). Focal adhesion plaques are formed by a number of specific proteins including the integrins (39) as well as vinculin (40, 41), talin (42, 43), α-actinin (44), paxillin (45, 46), zyxin (47), and tensin that has a src homology 2 domain (48, 49). Some other proteins are involved in the transduction of extracellular signals (50) such as p125[FAK] (51, 52), phospholipase C-γ (53), and type 3 phosphokinase C (54, 55).

Scanning electron microscopy revealed three regions of different collagen fibril densities in the contracted gels: one region contains thickly condensed fibrils, another has moderately packed collagen fibrils, and the third region has a sparse density of collagen fibrils (Fig. 2a) (11, 56). Highly condensed fibrils with an approximate average thickness of 1 μm were found in the vicinity of holes. We assumed that the holes would have corresponded to the spaces occupied by cell bodies or pseudopodia. The cell surface exhibited rough bristles which might correspond to putative interacting sites with collagen fibrils and a site where a fibril with a diameter of about 80 nm merged with the cell surface, suggesting the direct connection of collagen fibrils with the cell surface (Fig. 2b). The surface

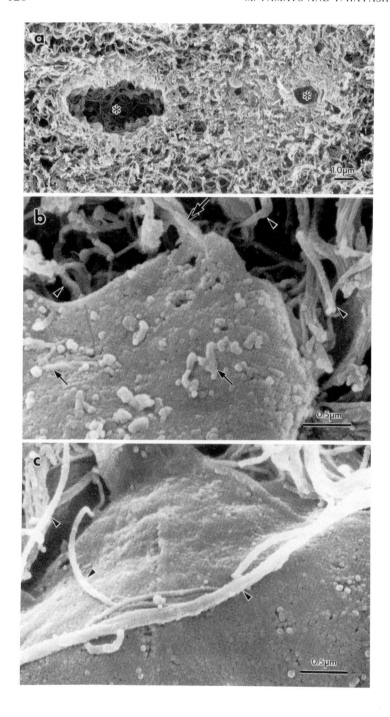

roughness is quite a contrast to the smoothness of the surface cultured on collagen gels (Fig. 2c). Upper surfaces of fibroblasts cultured on reconstituted collagen gel were smooth and no bristles were observed.

Condensation of collagen fibrils into the direct vicinity of fibroblasts under three-dimensional collagen gel culture was also observed by fluorescence microscopy (56). When fluorescein isothianate (FITC)-covalently prelabeled type I collagen was used as reconstituted collagen gels, condensation of the fluorescence was observed. Fibroblasts cultured within collagen gels were bipolar and highly elongated in shape (31-33). In the fibroblasts cultured on collagen gels, actin filaments ran along the inner surface of the basal plasma membrane (33). By double fluorescence microscopy of stress fibers with rhodamine-phalloidin staining and of FITC-prelabeled collagen, the relative position of collagen fibrils to the cell membrane was examined along with culture time. Collagen fibrils apparently began to be "tugged" by a cell at 12 hr of culture (Fig. 3a). The picture shows that fibroblast wound collagen fibrils around the cell body. A consequence of the condensation of collagen fibrils over the entire surface may thus have resulted in the packed collagen fibrils shown in Fig. 2a. At 24 hr of culture, FITC-prelabeled collagen fibrils were condensed, encompassing the

---

←Fig. 2. Scanning electron micrographs of collagen gels populated with human dermal fibroblasts. (a) Holes found on cracked surface of the gels. Densely packed collagen fibrils, approximately 80 nm in diameter, run spiral to form the wall of holes in the dense network (∗). They might have been occupied by cell bodies or filopodia which were retracted through freeze-cracking in liquid nitrogen, leaving collagen fibrils. (b) Scanning electron micrograph showing a fibroblast cultured in collagen gel on 7 days culture. Bristles can be seen on the surface of the cell. Note a collagen fibril anchored to the cell surface (arrowhead). Double-headed arrow: collagen fibril merging sites. Arrowhead: collagen fibril (approximately 80 nm in diameter). Arrow: bristle. (c) Scanning electron micrograph of a fibroblast cultured on top of a collagen gel. No bristles can be seen. Arrowhead: collagen fibrils.

pseudopodia and cell body (Fig. 3b, c, d). A layer consisting of condensed collagen fibrils spanning about 1 μm was formed very close to the cell membrane. Rigid actin filaments ran along the inner surface, implying strong interaction between fibroblasts and collagen fibrils. This time course of the condensation of collagen fibrils onto the direct vicinity of fibroblasts and resultant heterogeneous distribution of collagen may explain the apparent disagreement between the collagen density and inhibitory effects of collagen gels on cell growth at an earlier phase of culture.

IV. DISTRIBUTION OF $\alpha2\beta1$ INTEGRIN AND VINCULIN ON THE ELONGATED FIBROBLASTS IN COLLAGEN GELS

Double fluorescent staining of the exogenous type I collagen and F-actin revealed that collagen fibrils had been reorganized onto the entire surface of the cells, and were superimposed on F-actin organized underneath the plasma membrane of the elongated bipolar cells. The observation raised the question whether the focal contacts can also be found in the cells within the three-dimensional matrix. Ubiquitous distribution of condensed collagen fibrils surrounding the entire surface of the cell does not appear to cause particular focal contacts with collagen fibrils in a large area, like the focal contacts seen on the cells on plastic dishes. The distributions of F-actin and $\beta1$ integrin of elongated fibroblasts cultured in three-dimensional collagen gels were examined simultaneously by double staining (57). Fibroblasts not spread on a planar surface but surrounded by collagen matrix three-dimensionally over the entire surface of cells showed distinct localizations of actin filaments (Fig. 4a) and $\beta1$ integrin (Fig. 4b). The distribution was quite different from that of cells cultured on a planar surface. In elongated bipolar cells, actin filaments ran parallel along the long axis of the elongated cells from end to end (Fig. 4a). Focal contacts were not observed (Fig. 4b). The spots containing $\beta1$ integrins were dispersed over the entire surface of the cells in a bird's-eye view. Instead of focal

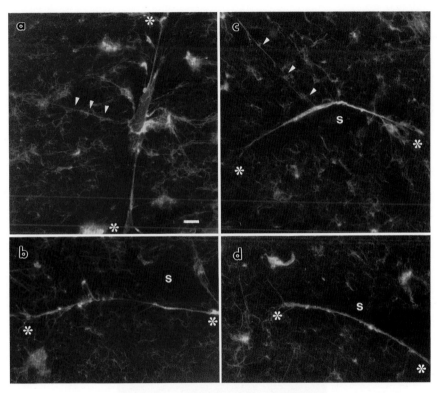

Fig. 3. Condensation of collagen fibrils by bipolar fibroblasts. Fluorescent micrographs show the distribution of FITC-labeled collagen fibrils, in relation to rhodamine-phalloidin-labeled actin filaments at 12 hr (a) and 24 hr (b–d). Condensation of FITC-labeled collagen fibrils (yellow green) is seen on the cell surface that is marked by the rhodamine-phalloidin-labeled actin filaments (red). High fluorescence intensity from both FITC and rhodamine with unchanged intermediate intensity in the outer region. A region left with sparse collagen-fluorescence is designated as S. Note that many collagen fibrils appeared to be under tension beneath the cells (arrowhead). Asterisk: ends of an elongated bipolar cell. The scale bar corresponds to 10 μm.

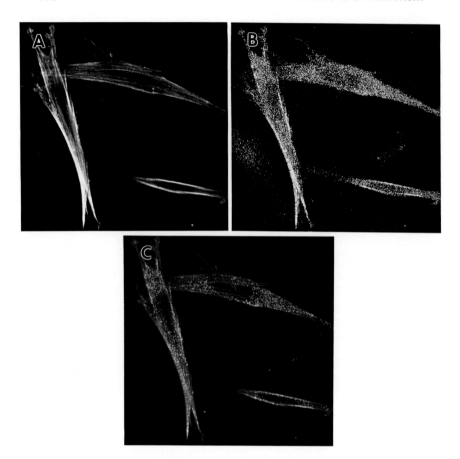

Fig. 4. Double fluorescence microscopy of actin filament and β1 integrin of the elongated fibroblasts cultured in three-dimensional collagen gel. Human diploid fibroblasts were cultured for 24 hr in reconstituted collagen gel and the distributions of actin filaments (A) and β1 integrin (B) were examined with a confocal laser scanning microscope after staining with rhodamine-labeled phalloidin and anti-β1 integrin monoclonal antibody. Actin filaments and β1 integrins were displayed simultaneously in red and blue, respectively (C). No correlated distribution of actin filaments and β1 integrins was observed. Actin filaments ran parallel to the long axis of the elongated cells, while the spots containing β1 integrins were dispersed as spots over the entire surface of the cells.

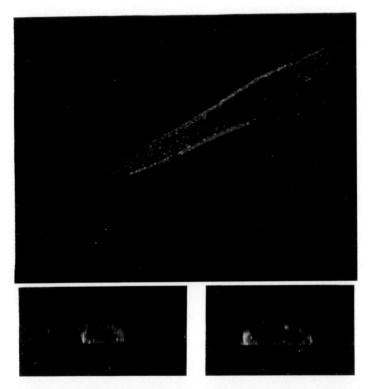

Fig. 5. Computerized tomography of elongated fibroblasts cultured in three-dimensional collagen gel. Human diploid fibroblasts were cultured for 24 hr in reconstituted collagen gel and the distributions of actin microfilaments and $\beta 1$ integrin were examined with a confocal laser scanning microscope after staining with rhodamine-labeled phalloidin and anti-$\beta 1$ integrin monoclonal antibody. Actin filaments and $\beta 1$ integrins are displayed simultaneously in red and green, respectively, in a bird's eye view (upper) and a cross-sectional view (lower). The cross-sections of distal (left) and proximal (right) portions of the cell body were obtained by computerized tomography from this view. Both actin filaments and $\beta 1$ integrins were localized along the plasma membrane.

contacts with large areas, $\beta1$ integrin was immunolocalized as tiny spots. The spots were distributed over the entire surface of the bipolar elongated cells from end to end. No particular co-localization of $\beta1$ integrin with actin filaments was observed (Fig. 4c). By computerized tomography with a confocal laser scanning microscope, a cross-view of elongated cells from a viewing angle perpendicular to the original optical axis was obtained (Fig. 5). Actin filaments (Fig. 5a) and $\beta1$ integrin (Fig. 5b) were stained at the cell periphery in the cross-sectional view of both the cell body and pseudopodia. The staining patterns were superimposed on each other (Fig. 5c), implying that both actin filament and $\beta1$ integrin were localized only along the plasma membrane over the entire surface of the cell, not in the cytoplasmic area at either the cell body or pseudopodia. $\alpha2$ integrin (Fig. 6a) and vinculin (Fig. 6b) were also dispersed over the entire surface of the cells similar to $\beta1$ integrin, suggesting that $\alpha2\beta1$ integrin and vinculin are co-distributed as binding sites to collagen fibrils over the entire surface of fibroblasts cultured in three-dimensional collagen gels.

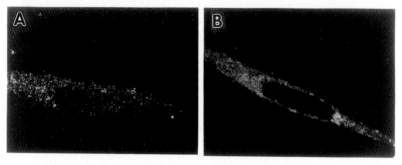

Fig. 6.   Fluorescence microscopy of $\alpha2$ integrin (A) and vinculin (B) at the central region of elongated fibroblasts cultured in three-dimensional collagen gel. Human diploid fibroblasts were cultured for 24 hr in reconstituted collagen gel and their distribution was examined with a confocal laser scanning microscope.

V. SIGNIFICANCE OF TOPOLOGICAL DISTRIBUTION AND DENSITY OF CELL-COLLAGEN BINDING SITES

From the morphological observations above mentioned, we point out that not only the density but also the localization of cell-collagen binding sites must be crucial for these regulatory effects of collagen fibrils on cellular functions. In attached gels, collagen fibrils anchor to the dish surfaces directly or indirectly. Therefore, condensation of collagen fibrils in the vicinity of fibroblasts is disturbed and inhibited. With FITC-prelabeled type I collagen we demonstrated that the distribution of collagen fibrils was rearranged by the cells to result in three regions with different densities: condensed, sparse, and unchanged, when collagen gels were prevented from macroscopic contraction by their strong adhesion to external surfaces (56). The fluorescence-sparse region often spread on only one side of the elongated cells in a cross-sectional view, for example, on the upper side of the cell in Fig. 3b and d, or on the lower side of the cell in Fig. 3c, as if the cells had rolled up a carpet composed of a collagen fibril network. In contrast, when the gel was free to contract, collection of the collagen fibrils by direct adhesion of cell processes was concomitantly accompanied by pulling of the fibrils beyond through interfibrillar adhesion or entanglement as revealed by scanning electron microscopy (11, 56). Microscopically, contraction of collagen gels resulted in a heterogeneous distribution of the density of collagen fibrils. Therefore, densities of these fibrils surrounding the cells would be even more different between attached and floating gels than those calculated, assuming that collagen fibril distribution is homogeneous. The picture of the consequence of interactions between fibroblasts and collagen fibrils is somewhat like a bag worm with the entire surface of the body covered with fibrous materials. In cocoons of collagen fibrils, fibroblasts establish a number of binding sites to the fibrils that might be called "buttoned adhesions", over the entire cell surface (57). Integrins might bind to these fibrils like buttons and line the cell surface.

The ubiquitous distribution of cell-collagen adhesion sites may contribute to the influence of the collagen fibrils on cellular functions as well as the specific adhesion structure or signalling. On two-dimensional surfaces of collagen gels, fibroblasts show the same growth rate as they do on plastic dishes (Table I). Densities of collagen fibrils are not greatly different between "on gel" and "in gel" cultures at an early phase of culture, even if little condensation occurs. Therefore, the inhibitory effect of "in gel" culture on cell growth is attributable to collagen-adhesion sites over the entire cell surface. In "in gel" culture, the binding sites are distributed all over the surface of the cells as small patches less than 1 μm in diameter (Figs. 4–6) (57), while they are restricted to the focal adhesion plaques on the basal surface of the cells under "on gel" culture. Ingber's group recently utilized microfabricated surfaces to demonstrate that the cells spreading in the multiple small dots had less tendency to undergo apoptosis and thrived compared with the more rounded cells on the single dots, even if the total cell-matrix contact area was constant (58). If chemical signals from the integrins are the predominant factor, the cells should survive and grow equally whether attached to a large dot or several small dots with the same combined area. They concluded that the cell shape determined by the distribution of cell-ECM binding sites governs whether individual cells grow or die. That is, there is local geometric control of cell growth. In this sense, the finding that the global distribution of collagen-binding sites with tiny patches at a light microscopic level was established only under "in gel" culture suggests that the topologically ubiquitous distribution of collagen binding sites plays a crucial role in regulating cell functions when fibroblasts are cultured within three-dimensional collagen gels.

SUMMARY

Collagen molecules are reassembled into fibrils under a physiological condition, composing a network in which cells can be trapped and cultured. Collagen gels populated with fibroblasts contract resulting in an increased density of collagen fibrils equal to that in dermis. In the three-dimensional collagen gels, normal fibroblasts display distinct phenotypes rather different from those cultured on flat surfaces, including growth inhibition and diminished responses to growth factors. Scanning electron microscopy of collagen fibrils in the contracted gels showed a disproportional distribution as well as an increased density: a highly condensed region in the vicinity of the cells spanning about 1 μm and a moderately condensed region beyond. It was concluded that a major portion of the densely packed collagen fibrils in the direct vicinity of the cells came from the reconstituted collagen fibrils, since the initially homogeneous distribution of fluorescein-prelabeled collagen fibrils was condensed into elongated cores comprised of cells at the center. Fibroblasts are bipolarly elongated in shape when they are surrounded by the three-dimensional collagen fibrils. The elongated cells showed spotted distributions of $\alpha2\beta1$ integrin and vinculin over their entire surface. Stress fibers ran parallel to the long axis of the elongated cells, as if the actin microfilaments contouring the skeletal structure of the elongated shape were connected with the extracellular collagen fibrils which ran in parallel with the longitudinal cell surface. Cytoskeletal filaments and collagen fibrils appeared to be connected through aligned button-like adhesion spots comprised of $\alpha2\beta1$ integrin and vinculin. We propose that the increased density of collagen fibrils and the ubiquitous distribution of cell-collagen binding sites might be crucial for the potent regulation of cell functions by collagen fibrils.

Acknowledgments

This study was supported in part by the Ministry of Education, Culture, Sports and Science Grants-in-Aid for Scientific Research on Priority Areas (09229219, Functionally Graded Materials; 09217210, Supramolecular Structure), and a Grant-in-Aid for Developmental Scientific Research (07558249).

REFERENCES

1. Gross, J. and Kirk, D. (1958) *J. Biol. Chem.* **233**, 355–360.
2. Elasdale, T. and Bard, J. (1972) *J. Cell Biol.* **54**, 626–637.
3. Bell, E., Invarsson, B., and Merrill, C. (1979) *Proc. Natl. Acad. Sci. USA* **76**, 1274–1278.
4. Nishiyama, T., Tominaga, N., and Hayashi, T. (1988) *Collagen Rel. Res.* **8**, 259–273.
5. Yamamoto, M., Nakamura, H., Yamato, M., Aoyagi, M., and Yamamoto, K. (1996) *Exp. Cell Res.* **225**, 12–21.
6. Yamamoto, M., Yamamoto, K., and Hayashi, T. (1992) *Connect. Tiss.* **24**, 157–162.
7. Yamato, M., Yamamoto, K., and Hayashi, T. (1993) *Mech. Ageing Dev.* **67**, 149–158.
8. Gullberg, D., Borg, T.K., Terracio, L., and Rubin, K. (1990) *Exp. Cell Res.* **190**, 254–264.
9. Klein, C.E., Dressel, D., Steinmayer, T. *et al.* (1991) *J. Cell Biol.* **115**, 1427–1436.
10. Schiro, J.A., Chan, B.M., Roswit, W.T. *et al.* (1991) *Cell* **67**, 403–410.
11. Hayashi, T., Nishiyama, T., and Adachi, E. (1991) In *Fundamental Investigations on the Creation of Biofunctional Materials* (Okamura, S. *et al.*, eds.), pp. 55–64, Kagaku-Dojin, Kyoto.
12. Nusgens, B., Merrill, C., Lapière, C.M., and Bell, E. (1984) *Collagen Rel. Res.* **4**, 351–364.
13. Mauch, C., Hatamochi, A., Scharffetter, K., and Krieg, T. (1988) *Exp. Cell Res.* **178**, 493–503.
14. Paye, M., Nusgens, B.V., and Lapière, C.M. (1987) *Eur. J. Cell Biol.* **45**, 44–50.
15. Schor, S.L. (1980) *J. Cell. Sci.* **41**, 159–175.
16. Buttle, D.J. and Ehrlich, H.P. (1983) *J. Cell. Physiol.* **116**, 159–166.
17. Yoshizato, K., Taira, T., and Yamamoto, N. (1985) *Biomed. Res.* **6**, 61–71.
18. Sarber, R., Hull, B., Merrill, C., Soranno, T., and Bell, E. (1981) *Mech. Ageing Dev.* **17**, 107–117.
19. Nishiyama, T., Tsunenaga, M., Nakayama, Y., Adachi, E., and Hayashi, T.

(1989) *Matrix* **9**, 193–199.

20. Mochitate, K., Pawelek, P., and Grinnell, F. (1991) *Exp. Cell Res.* **193**, 198–207.

21. Nakagawa, S., Pawelek, P., and Grinnell, F. (1989) *Exp. Cell Res.* **182**, 572–582.

22. Kono, T., Tanii, T., Furukawa, M. *et al.* (1990) *Arch. Dermatol. Res.* **282**, 258–262.

23. Kono, T., Tanii, T., Furukawa, M. *et al.* (1990) *J. Dermatol.* **17**, 2–10.

24. Mauch, C., Hatamochi, A., Scharffetter, K., and Krieg, T. (1988) *Exp. Cell Res.* **178**, 493–503.

25. Hey, K.B., Jutley, J.K., Cunliffe, W.J., and Wood, E.J. (1990) *Biochem. Soc. Trans.* **18**, 899–900.

26. Barr, R.M., Symonds, P.H., Akpan, A.S., and Greaves, M.W. (1992) *Exp. Cell Res.* **198**, 321–327.

27. Colige, A., Nusgens, B., and Lapière, C.M. (1988) *Arch. Dermatol. Res.* **280**, s42–s46.

28. Nishiyama, T., Horii, I., Nakayama, Y., Ozawa, T., and Hayashi, T. (1990) *Matrix* **10**, 412–419.

29. Nishiyama, T., Akutsu, N., Horii, I., Nakayama, Y., Ozawa, T., and Hayashi, T. (1991) *Matrix* **11**, 71–75.

30. Lee, T.-L., Lin, Y.-C., Mochitate, K., and Grinnell, F. (1993) *J. Cell. Sci.* **105**, 167–177.

31. Tomasek, J.J., Hay, E.D., and Fujiwara, K. (1982) *Dev. Biol.* **92**, 107–122.

32. Tomasek, J.J. and Hay, E.D. (1984) *J. Cell Biol.* **99**, 536–549.

33. Nishiyama, T., Tsunenaga, M., Akutsu, N. *et al.* (1993) *Matrix* **13**, 447–455.

34. Grinnell, F. (1991) In *Fundamental Investigations on the Creation of Biofunctional Materials* (Okamura, S. *et al.,* eds.), pp. 33–43, Kagaku-Dojin, Kyoto.

35. Grinnell, F. (1994) *J. Cell Biol.* **124**, 401–404.

36. Burridge, K., Fath, K., Kelly, T., Nuckolls, G., and Turner, C. (1988) *Annu. Rev. Cell Biol.* **4**, 487–525.

37. Turner, C. and Burridge, K. (1991) *Curr. Opin. Cell Biol.* **3**, 849–853.

38. Luna, E.J. and Hitt, A.L. (1992) *Science* **258**, 955–964.

39. Hynes, R.O. (1992) *Cell* **69**, 11–25.

40. Tapley, P., Horwitz, A., Buck, C., Duggan, K., and Rohrschneider, L. (1989) *Oncogene* **4**, 325–333.

41. Feltkamp, C.A., Pijnenburg, M.A.P., and Roos, E. (1991) *J. Cell. Sci.* **100**, 579–587.

42. Horwitz, A., Duggan, K., Buck, C., Beckerle, M.C., and Burridge, K. (1986) *Nature* **320**, 531–533.

43. Nuckolls, G.H., Romer, L.H., and Burridge, K. (1992) *J. Cell. Sci.* **102**, 753–762.

44. Otey, C.A., Pavalko, F.M., and Burridge, K. (1990) *J. Cell Biol.* **111**, 721–

730.

45. Turner, C.E., Glenney, J.R. Jr., and Burridge, K. (1990) *J. Cell Biol.* **111**, 1059–1068.
46. Burridge, K., Turner, C.E., and Romer, L.H. (1992) *J. Cell Biol.* **119**, 893–903.
47. Crawford, A.W., Michelsen, J.W., and Beckerle, M.C. (1992) *J. Cell Biol.* **116**, 1381–1393.
48. Davis, S., Lu, M.L., Lo, S.H. *et al.* (1991) *Science* **252**, 712–715.
49. Bockholt, S.M. and Burridge, K. (1993) *J. Biol. Chem.* **268**, 14565–14567.
50. Damsky, C.H. and Werb, Z. (1992) *Curr. Opin. Cell Biol.* **4**, 772–781.
51. Lipfert, L., Haimovich, B., Schaller, M.D., Cobb, B.S., Parsons, J.T., and Brugge, J.S. (1992) *J. Cell Biol.* **119**, 905–912.
52. Schaller, M.D., Borgman, C.A., Cobb, B.S., Vines, R.R., Reynolds, A.B., and Parsons, J.T. (1992) *Proc. Natl. Acad. Sci. USA* **89**, 5192–5196.
53. McBride, K., Rhee, S.G., and Jaken, S. (1991) *Proc. Natl. Acad. Sci. USA* **88**, 7111–7115.
54. Woods, A. and Couchman, J.R. (1992) *J. Cell. Sci.* **101**, 277–290.
55. Jaken, S., Leach, K., and Klauck, T. (1989) *J. Cell Biol.* **109**, 697–704.
56. Yamato, M., Adachi, E., Yamamoto, K., and Hayashi, T. (1995) *J. Biochem.* **117**, 940–946.
57. Yamato, M., Yamamoto, K., Adachi, E., and Hayashi, T. Submitted.
58. Chen, C.S., Mrksich, M., Huang, S., Whitesides, G.M., and Ingber, D.E. (1997) *Science* **276**, 1425–1428.

Extracellular Matrix-Cellular Interaction: Molecules to Diseases (Y. Ninomiya et al., eds.), pp. 141–168,
Japan Sci. Soc. Press, Tokyo/S. Karger, Basel (1998)

Signaling Pathways Downstream of PDGF Receptors Involved in Regulation of Cell Growth and Motility

KOUTARO YOKOTE, SEIJIRO MORI, AND
YASUSHI SAITO

Second Department of Internal Medicine, Chiba University
School of Medicine, Chiba 260-0856, Japan

I. PDGF IS INVOLVED IN VARIOUS PHYSIOLOGICAL AND PATHOLOGICAL FUNCTIONS

The platelet-derived growth factor (PDGF) was first purified from platelets in 1979 as a potent mitogen for cultured cells
of mesenchymal origin (*1, 2*). Later it was shown that PDGF is
secreted by various types of cells such as endothelial cells,
vascular smooth muscle cells, and macrophages. PDGF shows
not only mitogenic but also chemotactic effects on cultured cells.
It also stimulates production of extracellular matrix components
and their modulators, *e.g.*, fibronectin, collagen, thrombospondin, hyaluronic acid, and collagenase (*3–5*).

PDGF is a family of homo- or hetero-dimeric proteins
composed of disulfide-linked A and B polypeptide chains. It
elicits its effect by binding to two structurally related cell surface
receptors designated α- and β-receptors (*6–10*).

A study on "knockout" mice lacking a functional PDGF

B-chain gene suggested a role for the B-chain in development of certain types of smooth muscle cells in embryo (*11*). Animals displayed subcutaneous hemorrhage and dilatation of the large vessels, most likely due to a defect in smooth muscle cell development. In addition, the smooth muscle cell-related mesangial cells in the kidney glomeruli were lacking in the knockout mice. A majority of the knockout-mice died perinatally. PDGF β-receptor-knockout mice showed a slightly more severe, but strikingly similar, phenotype to that of the B-chain-knockout mice (*12*). The knockout studies as well as other *in vivo* studies prove the biological significance of PDGF in embryonic development.

PDGF also plays important roles in wound healing, and in the development of several pathological conditions such as tumorigenesis and atherosclerosis (reviewed in refs. *13, 14*).

PDGF increases the rate of wound healing without altering the sequence of events seen in normal repair *in vivo* (*15, 16*). This is in agreement with the notion that PDGF stimulates many cellular responses involved in the wound healing process, such as cell migration, proliferation, and extracellular matrix production, as described above. PDGF was found to be capable of improving wound healing in a diabetic model (*17*), which suggests its potential application as a therapeutic agent in conditions with an impaired repair process, *e.g.*, skin ulcers in patients suffering from diabetes or Werner's syndrome.

Aberrant activation of the PDGF system is considered to be involved in the development of various pathological conditions. Abnormal expression of either or both PDGF and PDGF receptors has been reported in fibrotic diseases of lung, liver, skin, and bone marrow (*18, 19*). PDGF has also been suggested to be involved in the development of inflammatory diseases, such as certain types of glomerulonephritis and rheumatoid arthritis (*20–22*).

The importance of PDGF and PDGF receptors in the establishment and development of malignancies was first realized when the PDGF B-chain was found to be a cellular counterpart of the v-*sis* oncogene product, and that transformation by *sis* occurs in an autocrine manner involving stimulation of PDGF receptors (*14*). Since then, coexpression of both PDGF and PDGF receptors has been detected in different types of malignant tumor cells, *e.g.*, glioma, fibrosarcoma, gastric carcinoma, and thyroid carcinoma (reviewed in ref. *13*). Recently a chromosomal rearrangement which causes a fusion between the genes corresponding to the tyrosine kinase domain of PDGF β-receptor and Tel, an Ets-like transcription factor, was identified in acute myelomonocytic leukemia (*23*). Functional analysis of the gene product may provide new insight into the mechanism of tumorigenesis.

Finally, PDGF has been suggested to play a significant role in the etiology of atherosclerosis (*24*). In the earliest stage of this disease, mechanically or non-mechanically injured endothelial cells allow increased adhesion and activation of monocytes and their subsequent migration into the subendothelial (intimal) space. The monocytes phenotypically change into macrophages, accumulate lipids and become foam cells. The endothelial cells, macrophages, and in certain instances, platelet thrombi become sources of PDGF and other substances. Secretion of PDGF in the intimal lesion leads to migration of vascular smooth muscle cells from the medial layer into the intimal layer through lamina elastica. Intimal smooth muscle cells acquire the activated (synthetic) phenotype which can respond to PDGF in an autocrine manner. Thus, smooth muscle cells in the intimal layer become major players in the ongoing fibroproliferative process. Migration and proliferation of smooth muscle cells are also crucial for post-PTCA (percutaneous transluminal coronary angioplasty) restenosis. Administration of recombinant PDGF-BB into carotid-injured rats facilitated the smooth muscle cell migration from the medial layer to the intimal layer, thus resulting in an accelerated intimal thickening (*25*). Infusion of an anti-PDGF antibody showed suppressed intimal accumulation of smooth muscle cells in a balloon injury model of rat carotid artery (*26*).

To discover the mechanism of signaling initiated by PDGF receptors will help our understanding in the development of certain diseases and may give us clues for novel therapeutic approaches.

In this paper we will describe our data on the signaling mechanism involved in regulation of cell motility induced by the PDGF receptors, as well as an overview concerning recent progress in the research field of PDGF signaling.

II. PDGF RECEPTOR SIGNALING IS MEDIATED BY INTRACEL-LULAR MOLECULES CONTAINING SH2 DOMAINS

PDGF α- and β-receptors consist of an extracellular domain with five sets of immunoglobulin-like folds, a single transmembrane domain, and an intracellular domain. The intracellular domain contains a tyrosine kinase domain which is split into two parts by a non-catalytic insert of about 100 amino acids. The region in between the transmembrane domain and the tyrosine kinase domain is denoted as the juxtamembrane domain and the region carboxyterminal of the kinase domain, the carboxyterminal tail. The similarity between the amino acid sequences of the PDGF α- and β-receptors is high in the kinase domain and in the juxtamembrane domain, whereas it is relatively low in the extracellular domain, the kinase insert and in the carboxyterminal tail (Fig. 1).

The PDGF α-receptor binds to either of PDGF A- or B-chains, whereas the β-receptor can only bind B-chain (27, 28). Ligand-binding leads to formation of receptor dimers (reviewed in ref. 29). Due to the specificity of ligand-receptor combinations, PDGF-AA recruits homodimers of PDGF α-receptors ($\alpha\alpha$-receptor dimer) only. PDGF-AB induces formation of $\alpha\alpha$- or $\alpha\beta$-receptor dimers, whereas PDGF-BB can induce formation of all possible forms of receptor dimers ($\alpha\alpha$-, $\alpha\beta$-, or $\beta\beta$).

Dimerization of PDGF receptors leads to induction of their intrinsic tyrosine kinase activities, probably through specific conformational changes (5, 30, 31). The activated PDGF recep-

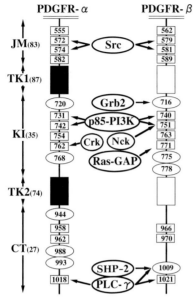

Fig. 1. Schematic illustration of the location of tyrosine residues in the intracellular domains of the PDGF α- and β-receptors, and their interacting SH2 domain-containing proteins. The numbers stand for positions of the tyrosine residues; numbers in ovals represent uniquely positioned tyrosine residues, whereas numbers in squares indicate tyrosine residues positioned in regions conserved between the two types of receptors. The shaded squares indicate that the tyrosine residues are conserved in their position, but have unique surrounding motifs. JM, juxtamembrane domain; TK1, 1st part of the tyrosine kinase domain; KI, kinase insert; TK2, 2nd part of the tyrosine kinase domain; CT, carboxyterminal tail. The percent amino acid identity between the corresponding domains of the receptors is indicated in the parentheses. Intracellular signaling molecules and their binding sites in the receptors are also shown.

tors phosphorylate tyrosine residues in the intracellular domains of the receptors themselves (autophosphorylation) as well as in various intracellular proteins.

Currently, two major roles are suggested for the receptor autophosphorylation. Firstly, autophosphorylation of a certain tyrosine residue in the kinase domain is implicated in upregulation of the intrinsic kinase activity. In the case of the PDGF β-receptor, Tyr-857 belongs to this category. Thus, replacement of Tyr-857 with a phenylalanine residue results in impaired kinase activation compared to the wild-type receptor (*32*). This tyrosine residue is well conserved in all tyrosine kinase receptors. Three-dimensional structural analysis of the insulin receptor kinase domain revealed that phosphorylation of this residue (Tyr-1162 in the human insulin receptor) plays a key role in kinase activation by stabilizing the activated conformation (*33*). Tyr-857 in the PDGF β-receptor as well as its homologue in the PDGF α-receptor, Tyr-849, probably has the same role in receptor activation. Secondly, certain autophosphorylated tyrosine residues are known to serve as binding sites for intracellular signaling molecules containing so-called Src homology 2 (SH2) domains (reviewed in ref. *34*). SH2 domains are conserved non-catalytic stretches of about 100 amino acids which are found in certain signal transduction molecules, and which bind to phosphorylated tyrosine residues in proteins (reviewed in ref. *35*). A number of studies, including analysis of the three-dimensional structure of SH2 domains, have shown that SH2 domains specifically recognize amino acid sequences carboxyterminal of the phosphorylated tyrosine residue in a phosphopeptide (*36–38*).

Nine and five autophosphorylation sites have so far been identified in the PDGF β- and α-receptors, respectively (reviewed in ref. *39*). For the PDGF β-receptor, two autophosphorylation sites in the juxtamembrane domain (Tyr-579 and Tyr-581) mediate the binding of Src family kinase. The four autophosphorylation sites in the kinase insert serve for binding of Grb2 (Tyr-716), the regulatory subunit (p85) of phosphatidylinositol 3′ kinase (PI3-kinase; Tyr-740 and Tyr-751), Nck (Tyr-751), and the GTPase activating protein (GAP) of Ras (Tyr-771). Two autophosphorylation sites have been identified in the carboxyterminal tail (Tyr-1009 and Tyr-1021) which mediate binding of the SH2 domain-containing phosphatase Syp/PTP1D and phospholipase C-γ (PLC-γ), respectively. In the case of the PDGF α-receptor, tyrosine residues 754, 762, 768, 988, and 1018 have been identified as autophosphorylation sites (*40, 41*). Among them, Tyr-1018 has been shown to mediate association of PLC-γ (*40*). Tyr-731 and Tyr-742 are important for the binding of PI3-kinase (*42*), although data showing that they are autophosphorylation sites has not been presented.

SH2 domain-containing proteins can be classified into subgroups according to their functions, *i.e.,* molecules with enzymatic activities, adaptors which bridge two or more molecules, and transcription factors (Table I). The SH2 domain-containing enzymes are proposed to become activated by binding to the autophosphorylated receptors via their SH2 domains through different mechanisms as outlined below.

1) Binding to the phosphopeptide could give rise to a conformational change in the SH2 domain, leading to an increase in the enzymatic activity of the SH2 domain-containing protein (suggested for SHP-2 by an *in vitro* study, see ref. *43*).

2) Translocation of an SH2 domain-containing enzyme to the vicinity of the plasma membrane could allow increased access to downstream signaling molecules (*e.g.,* Grb2-Sos complex, see below).

3) Phosphorylation of tyrosine residues in an SH2 domain-containing enzyme could lead to upregulation of the enzymatic activity (*e.g.,* phospholipase C-γ).

Other structures than SH2 domain are also implicated in protein-protein interactions, independent of tyrosine phosphorylation (reviewed in refs. *34, 35*). The SH3 (Src homology 3) domain is a structure of $\sim$60 amino acids, which is known to associate with so-called proline-rich regions (*44*). An example of this type of interaction is seen between the SH3 domain of the adaptor molecule Grb2 (growth factor receptor binding 2)/Ash (abundant Src homology) and the proline-rich region of the guanine nucleotide exchange factor (GEF), Sos (see ref. *45*; also described further below). Another structure is denoted as PH (pleckstrin homology) domain. The role of PH domains in signaling pathways is still not clear. Some studies implicate their

TABLE I

SH2 Domain-containing Proteins and Their Biochemical Characters

Function	Name	Tyrosine phos-phorylation	SH3 domain	PH domain
Enzyme	PI3-kinase (p85-p110 complex)	Yes	Yes	No
	PLC-γ1&2	Yes	Yes	Yes
	RasGAP	Yes	Yes	Yes
	SHP-1	Yes	Yes	No
	SHP-2	Yes	Yes	No
	Src family kinases	Yes	Yes	No
Adaptor	Crk	Yes	Yes	No
	Grap	No	Yes	No
	Grb2	No	Yes	No
	Nck	Yes	Yes	No
	Shb	No	No	No
	Shc	Yes	No	No
	SLAP	Unknown	Yes	No
Transcription factor	Stat1-6	Yes	Yes	No
Structural protein	Tensin	Yes	No	No
Unknown (adaptor?)	Grb7	Yes	No	Yes
	Grb10	No	No	Yes
	Grb14	No	No	Yes

involvement in interaction with phospholipids, such as phosphatidylinositol 4,5-bisphosphate (PIP$_2$, or PI(4, 5)P$_2$ and inositol 1,4,5-trisphosphate (IP$_3$), as well as in interaction with protein structures such as $\beta\gamma$ subunits of heterotrimeric G proteins (*46–48*).

III. Ras-DEPENDENT AND -INDEPENDENT PATHWAYS ARE IMPLICATED IN REGULATION OF CELL GROWTH

When cells are treated with pM to nM concentrations of PDGF, transcriptional activation of early response genes occurs within minutes to hours (*49*). This is followed by DNA synthesis and cell division. Both PDGF α- and β-receptors are able to induce mitogenicity. Whether activation of the two PDGF receptor types leads to the same magnitude of mitogenic response or not is still controversial. When similar numbers of α- or β-receptors were expressed in porcine aortic endothelial (PAE) cells, both types of receptors mediated dose-dependent increases in [^{3}H]thymidine-incorporation in response to PDGF-BB with higher maximal response in the α-receptor-expressing cells (50). Another study suggested a higher transforming potential of the PDGF β-receptor compared to that of the α-receptor by cotransfection of ligand and receptors (51). One group reported that activation of the α-receptor induced protein synthesis, but not DNA synthesis (52). However, in Swiss 3T3 cells, PDGF-AA, which only activates PDGF α-receptor, was shown to be as effective as PDGF-BB, which stimulates both α- and β-receptors, in stimulating DNA synthesis (53).

The small GTP-binding protein (G protein), Ras, was first identified as a product of a retroviral oncogene (54, 55). Similar to the "classical" trimeric G proteins, Ras cycles between an "active" GTP-bound form and an "inactive" GDP-bound form, thus functioning as a molecular "switch". Mutated constitutively active Ras has been found in various types of malignant tumors, indicating its role in cell proliferation. Consequently, Ras was pointed out as a crucial downstream mediator of mitogenic signaling by tyrosine kinase receptors (56, 57).

The mechanism by which tyrosine kinase receptors activate Ras became dramatically comprehensive when the adaptor protein, Grb2/Ash was identified (58, 59). Grb2 is a 24-kDa protein with a single SH2 domain positioned in between two SH3 domains. Grb2 binds to a proline-rich region in Sos (son of sevenless), a GEF, via its SH3 domains (45, 60). Sos activates Ras by mediating release of GDP thereby allowing binding of GTP (54). Upon activation of the PDGF β-receptor, Grb2 binds via its SH2 domain to autophosphorylated Tyr-716 in the kinase insert of the receptor (62). In parallel, Grb2 associates with phosphorylated tyrosine residues in the SH2 domain-con-

taining protein-tyrosine phosphatase, SHP-2 (*63*), and in Shc
(*61*). SHP-2 can associate with autophosphorylated Tyr-1009 in
the β-receptor (*64*), and Shc interacts with several different
tyrosine residues in the β-receptor (*61*). Therefore, Grb2 forms
a complex with the autophosphorylated PDGF β-receptor
directly or indirectly. Recruitment of the Grb2-Sos complex to
the vicinity of the plasma membrane leads to juxtaposition of
Sos to membrane-bound Ras, which is believed to result in Ras
activation. One report described that another adaptor molecule,
Nck, which binds to Tyr-751 in the activated PDGF β-receptor,
can also bind Sos via its SH3 domains (*65*). This exemplifies an
alternative mechanism for the translocation of Sos. Activation of
the PDGF α-receptor also leads to formation of a Grb2-Shc
complex. However, a direct interaction between Grb2 and the
α-receptor has not been reported. Another possible modulator
of Ras activity downstream of tyrosine kinase receptors is
RasGAP (Ras GTPase activating protein) (*54*). RasGAP upregulates the GTPase activity of Ras, thus negatively regulating Ras
by increasing the amount of the GDP-bound form. Upon activation of the PDGF β-receptor, RasGAP binds to autophosphorylated Tyr-771 in the receptor and becomes tyrosine phosphorylated (*66*). Replacement of Tyr-771 in the β-receptor abolished
binding and phosphorylation of RasGAP, but Ras activation as
well as mitogenic capacity of the mutant receptor remained
unaffected. The wild-type PDGF α-receptor, which is capable of
inducing Ras activation and mitogenicity, does not bind or
tyrosine phosphorylate RasGAP upon ligand stimulation (*41*).
Thus, despite its obvious enzymatic character, it is unclear how
RasGAP is involved in PDGF receptor signaling. Studies on the
cells derived from recently established RasGAP knockout mice
may give a clue to the function of RasGAP (*67*).

Stimulation of DNA synthesis in the nucleus is an essential
event for cell growth. Then, how does the tyrosine kinase receptor signaling initiated at the plasma membrane eventually reach
the nucleus? Signaling pathways downstream of Ras have also
been identified in the past few years. Activated Ras binds

directly to a MAP (mitogen-activated protein) kinase kinase kinase, Raf1, via its effector domain (*68, 69*). Although the precise mechanism of Raf1 activation is not known, activated Raf1 phosphorylates MAP kinase kinase/MEK on two serine residues. Activated MEK in turn phosphorylates MAP kinase/ ERK on a tyrosine and a threonine residue. The phosphorylated ERK translocates into the nucleus, where it phosphorylates and thereby activates transcription factors such as Elk-1 (*70*). A pathway related to MAP kinase pathway has been described lately. It is initiated by various stresses, such as exposure to heat-shock and to ultraviolet light, and involves the stress-activated protein (SAP) kinase (*71*). SAP kinase is also called c-*jun* N-terminal kinase (JNK), which activation eventually results in transcriptional activation of c-Jun. This pathway is partly Ras-dependent but its major activators are the Rho family small G proteins, Rac and Cdc42 (*72, 73*). The significance of the SAP kinase in tyrosine kinase receptor signaling is yet to be investigated.

Ras-independent pathways also seem to be involved in regulation of cell growth. Activation of PDGF α- and β-receptors leads to binding of Src family tyrosine kinases, *e.g.*, Src, Fyn, and Yes (*74*). The binding sites are identified in the juxtamembrane region of the receptors (*75*). Upon association with the receptor, the Src family kinases become tyrosine-phosphorylated and their kinase activities are increased. Microinjection studies showed that PDGF-induced S-phase entry was inhibited by a neutralizing antibody for the Src kinases (*76*), and that this inhibition could be rescued by overexpression of c-Myc (*77*), implicating c-Myc as a downstream mediator of the Src pathway. However, the mutant PDGF receptors lacking Src-binding sites could still induce intact mitogenicity (*75*).

PI3-kinase (see the next section) seems to be involved in PDGF receptor-mediated mitogenicity, at least in certain cell types. Thus, a mutant PDGF β-receptor, lacking the PI3-kinase binding sites, was shown to be defective in inducing DNA synthesis when expressed in CHO cells (*36, 78*). However, in other studies, PI3-kinase binding-deficient PDGF α- and β-receptor mutants were both able to mediate a mitogenic response (*42, 79*). Thus, there seems to be functional differences in PI3-kinase-mediated pathways depending on the cell type. Recently, several intracellular molecules which are implicated in mitogenic signaling, such as Akt kinase, p70 S6 kinase, and protein kinase C (PKC)-ζ, have been localized downstream of PI3-kinase (*80–83*). These pathways are of interest to pursue. However, many of the studies were performed utilizing wortmannin, an agent which has been recognized as a specific inhibitor for PI3-kinase, but which recently has also been found to affect other enzymatic activities (*84*). Therefore, one should be alert to the possibility that these pathways might also involve other wortmannin-sensitive molecules. It is also possible that PI3-kinase is involved in mitogenicity via activation of Ras (*85*).

The receptors for cytokines and interferons do not possess intrinsic kinase activities. Ligand-activation of these receptors instead leads to activation of members of a cytoplasmic tyrosine kinase family, Jak, which associate with the cytoplasmic domains of the receptors (reviewed in ref. *86*). Jak in turn tyrosine-phosphorylates a family of SH2 domain-containing transcription factors, Stat. The phosphorylated Stats form homo- or hetero-dimeric complexes which translocate into the nucleus to regulate the expression of specific sets of genes. Recently, Stat1 was shown to be tyrosine-phosphorylated by PDGF receptors as well as the epidermal growth factor (EGF) receptor (*87, 88*). It remains to be determined how important the Jak-Stat pathway is in PDGF receptor-mediated mitogenicity compared to, *e.g.*, the Ras-dependent pathway.

IV. PHOSPHATIDYLINOSITOL 3'-KINASE IS A MAJOR ROLE PLAYER IN CELL MOTILITY RESPONSES INDUCED BY THE PDGF β-RECEPTOR

It is well established that PDGF influences the modulation of cell motility. For example, cells expressing PDGF β-receptor migrate towards PDGF-BB depending on the concentration gradient. This type of direction-oriented cell migration is called "chemotaxis", and can be examined by utilizing a method such as modified Boyden chamber assay. On the cytoskeletal level, PDGF-BB induces reorganization of actin filaments, such as breakdown of stress fibers and formation of membrane ruffles. Two types of membrane ruffles, namely circular ruffle and edge ruffle (or lamellipodia) are formed by activation of PDGF β-receptor (Fig. 2, ref. *89*). However, the actual role of membrane ruffling in cell movement has not been established.

In contrast to the PDGF β-receptor, activation of the wild type fibroblast growth factor (FGF) receptor-1 does not induce formation of ruffles when expressed in PAE cells. However, the chimeric FGF receptor constructed by replacing the endogenous kinase insert of FGF receptor-1 with the PDGF β-receptor kinase insert could induce the ruffle formation (Fig. 3). Thus it is likely that the kinase insert of the PDGF β-receptor is critical for mediating a signal for membrane ruffle formation (*90*). One of the proteins which bind to the kinase insert of the autophosphorylated PDGF β-receptor but not to the insert of FGF receptor-1 is PI3-kinase. As a matter of fact, the chimeric FGF receptor was able to bind to and activate PI3-kinase upon ligand stimulation.

PI3-kinase consists of an 85-kDa regulatory subunit (p85) containing two SH2 domains and a single SH3 domain, and a 110-kDa subunit (p110) which is equipped with catalytic activity. The region in between the two SH2 domains of p85 contains the p110 binding site (*91*). Upon activation of PDGF α- or β-receptors, the p85 subunit associates via its SH2 domains directly with the autophosphorylated receptors. In the case of the PDGF β-receptor, autophosphorylated Tyr-740 and Tyr-751 in the kinase insert are responsible for the binding of p85 (*36*). The homologous tyrosine residues in the PDGF α-receptor, Tyr-731 and Tyr-742, have been shown to provide a p85 binding site (*42*). Tyrosine phosphorylation of p85 follows its binding to the receptor. Recruitment of the p85-p110 PI3-kinase complex to the

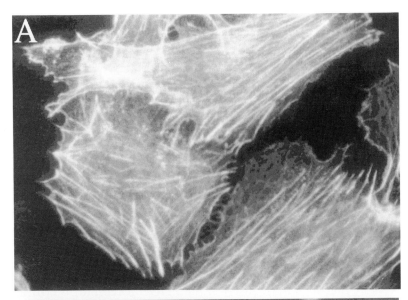

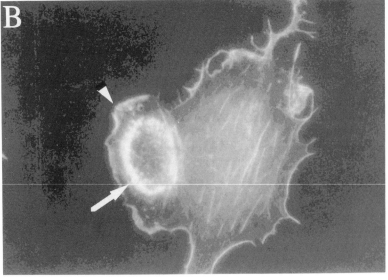

inner surface of plasma membrane, and possibly a conformational change brought about by binding of the SH2 domains to
the phosphorylated tyrosine residues in the receptor, lead to
PI3-kinase activation. PI3-kinase catalyzes the phosphorylation
of the D-3 position of the inositol ring. A major substrate for

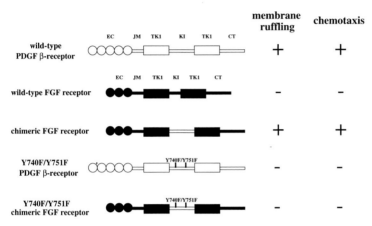

Fig. 3.   Schematic illustration of the wild-type and mutated PDGF β-
receptor (open boxes) and FGF receptor-1 (closed boxes), and their ability
to induce cell motility responses (79). The different regions in the receptors are indicated. EC, extracellular domain; JM, juxtamembrane domain;
TK1, 1st part of the tyrosine kinase domain; KI, kinase insert; TK2, 2nd
part of the tyrosine kinase domain; CT, carboxyterminal tail. Y740F/
Y751F mutants are unable to bind to or activate PI3-kinase. Whether each
receptor is able to induce the cell motility response upon ligand stimulation (+) or not (−) is indicated.

←Fig. 2.   Effect of PDGF-BB on actin reorganization in porcine aortic
endothelial (PAE) cells expressing the wild-type PDGF β-receptor. PAE
cells expressing the wild-type PDGF β-receptor were incubated in the
absence (A) or in the presence of 50 ng/ml PDGF-BB (B) for 7 min at
37°C. Actin filaments were visualized using rhodamine-conjugated phalloidin. Stress fibers are observed in (A), whereas loss of stress fiber as well
as formation of circular ruffle (arrow) and edge ruffle (arrowhead) were
induced upon PDGF stimulation (B).

this enzyme is PI(4, 5)P$_2$, thus giving rise to phosphatidylinositol 3,4,5-trisphosphate (PI(3, 4, 5)P$_3$, or PIP$_3$). Recent data
show that PI3-kinase is a family of related proteins, containing
variants of p85 (α, β) as well as p110 (α, β) ($92\text{-}96$). However,
p110γ, which is another member of the p110 family, does not
bind to p85 and is activated by heterotrimeric G proteins (97).
It has also been reported that the catalytic activity of p110α can
be regulated by Ras (98). Features such as complex formation
between different p85s and p110s, and the potential functional
differences in the PDGF receptor signaling have not yet been
elucidated. A purified p85 has been shown to bind to the SH3
domains of Src family tyrosine kinases, which resulted in their
activation (99). However, whether this type of interaction is also
of importance during signaling via the PDGF receptors is
unknown.

To understand the importance of PI3-kinase in mediating a
signal for actin reorganization, mutated chimeric FGF receptor
or PDGF β-receptor, in which tyrosine residues responsible for
the binding of PI3-kinase were replaced with phenylalanine
residues, were constructed (Y740F/Y751F mutants, see Fig. 3).
When these receptors were expressed in PAE cells, they failed to
associate with or activate PI3-kinase upon ligand stimulation
(Fig. 3). Examined by rhodamine-conjugated phalloidin staining, the mutant PDGF β-receptor as well as the mutant chimeric
FGF receptor were no longer able to induce ruffling upon
activation. The result implies that the activation of PI3-kinase is
necessary for PDGF-induced ruffle formation.

Ability of the mutated receptors to mediate chemotaxis was
also examined. Chemotaxis assays were performed using a
modified Boyden chamber. Briefly, the cells were seeded on 150
μm thick nitrocellulose filters coated with type I collagen.
Complete medium with fetal calf serum (FCS) was placed above
(upper chamber) as well as below the filter (lower chamber).
Either PDGF-BB or FGF-2 was added to the lower chamber at
various concentrations as possible chemoattractant. By diffusion, a concentration gradient was created in the filter and the

cells started to migrate towards the source of the chemo-attractant. The filter was removed after 6 hr, fixed, stained, and examined by microscopy. Migration was measured as the distance of the two furthest migrating PAE cell nuclei from the point of origin of migration. The migration distance due to random migration, with Ham's F-12, 10% FCS on both sides of the filters, served as a control and was set to 100% migration. PAE cells expressing the PDGF β-receptor showed a clear increase in chemotaxis towards PDGF-BB with a maximal response of up to 170% at 10 ng/ml of PDGF-BB. Cells expressing the wild-type FGF receptor did not show migration towards FGF-2 unless the concentration of FGF-2 was higher than 50 ng/ml. In contrast, cells expressing the chimeric FGF receptor with the wild-type PDGF β-receptor kinase insert showed potent chemotactic response at low concentration of the ligand, which was strikingly similar to that of the wild-type PDGF β-receptor-expressing cells. Interestingly, cells expressing the mutant PDGF β-receptor or the mutant chimeric FGF receptor which lack PI3-kinase binding site failed to induce chemotactic response (Fig. 3).

These findings strongly suggest that activation of PI3-kinase is involved in PDGF-dependent induction of both membrane ruffles and chemotaxis.

Another report also showed that PDGF-mediated migration was dependent on the PI3-kinase binding site when the receptors were expressed in dog kidney epithelial cells (*100*). However, this group claimed that the binding sites for phospholipase C-γ (PLC-γ) and RasGAP were also involved in modulation of chemotaxis. It is thus possible that different types of cells require different sets of signaling molecules to mediate certain biological responses.

The mechanism by which PI3-kinase regulates cell motility is not yet understood. It has been shown that a Rho family small G protein, Rac, is a mediator of membrane ruffle formation (*101*). As a matter of fact, activation of PI3-kinase leads to an increase in the GTP-bound, activated, form of Rac (*102*).

Therefore, PI3-kinase activation and the consequent Rac activation appear to be a pathway for PDGF-induced actin rearrangement. However, the precise mechanism by which Rac activation leads to a change in actin filament organization is not known. Whether Rac is involved in chemotaxis, and how ruffle formation and chemotaxis are related to each other are not known either. The SAP kinase pathway has also been positioned downstream of Rac (*72, 73*). In addition, Rac as well as one of its relatives, Rho, the mediator of stress fiber formation (*103*), were shown to be involved in induction of cell cycle progression (*104*). Thus, PDGF-induced activation of the PI3-kinase-Rac pathway can, in principle, elicit a large spectrum of biological effects.

We recently discovered another signaling component, focal adhesion kinase (FAK), downstream of PI3-kinase (*105, 106*). FAK is a cytoplasmic tyrosine kinase localized to focal adhesions (*107*). When cells are plated on fibronectin, FAK becomes tyrosine-phosphorylated in conjunction with cell-spreading, a process which is believed to involve clustering of integrins. Src family tyrosine kinases, PI3-kinase and Grb2 have been shown to bind to tyrosine-phosphorylated FAK (*108–110*). Fibroblasts derived from FAK-knockout mice showed decreased cell motility (*111*). These findings imply the importance of FAK in maintaining normal cell motility. It has been shown that treatment of Swiss 3T3 cells with a low dose of PDGF-BB induces tyrosine phosphorylation of both FAK and its associated structural protein, paxillin (*112*). Tyrosine phosphorylation of FAK induced by activation of the PDGF β-receptor was abolished by treatment of the cells with wortmannin, an inhibitor of PI3-kinase. Moreover, the mutant PDGF β-receptor which lacks the PI3-kinase binding site failed to mediate increase in tyrosine phosphorylation of FAK. In contrast, the mutant receptor lacking the binding site for Src intactly induced FAK phosphorylation. Therefore, PI3-kinase activity is required in the activation process of FAK downstream of the PDGF receptor. However, the biological consequence of PDGF-induced FAK phosphorylation remains to be elucidated.

V. PDGF α-RECEPTOR MEDIATES NEGATIVE SIGNALING FOR CHEMOTAXIS

The PDGF α-receptor activates PI3-kinase as efficiently as the β-receptor does; however, the former seems to mediate chemotaxis only in certain cell types. For example, it has been reported that in Swiss 3T3 cells or in human granulocytes, PDGF-AA (which specifically activates PDGF α-receptor) induces chemotaxis efficiently (53, 113). A hematopoietic cell line, 32D, when transfected with the wild-type PDGF α-receptor cDNA, was also found to migrate towards PDGF-BB (9). However, human foreskin fibroblasts, human monocytes, rat vascular smooth muscle cells, and baboon vascular smooth muscle cells failed to migrate towards PDGF-AA (113–115). PAE cells stably expressing the wild-type PDGF α-receptor also failed to migrate towards PDGF-BB, whereas PAE cells transfected with the wild-type PDGF β-receptor migrated efficiently (50). As described in the previous section, PDGF β-receptor-mediated chemotaxis is critically dependent on activation of PI3-kinase (79, 100). It is interesting that activation of the PDGF α-receptor fails to induce cell migration in certain cell types in spite of activating PI3-kinase. This has been hypothesized as being due to the presence of negative signaling which inhibits signal transduction for cell movement downstream of PI3-kinase. In fact, PDGF-AA has been shown to inhibit PDGF-BB-induced chemotactic activity in such cell types (113, 114). PDGF-AA showed an inhibitory effect on fibronectin-induced chemotaxis as well, but not on chemotaxis induced by fMLP (113, 114). It is to be noted that the specific inhibitory effect of PDGF-AA on chemotaxis is observed especially in primary cultured cells rather than in cell lines. The PDGF α-receptor-mediated inhibition of chemotaxis may have an important role in *in vivo* modulation of cell migration, *e.g.*, in smooth muscle cell migration in the aortic wall during atherogenesis.

As described in the earlier sections, interactions between autophosphorylated tyrosine residues and SH2 domain-containing signaling molecules are thought to play important roles in receptor tyrosine kinase signaling. In the closely related PDGF α- and β-receptors, certain tyrosine residues are positioned in regions conserved between the two receptors, whereas other tyrosine residues are unique for each receptor. We reasoned that the uniquely positioned tyrosine residues could be involved in α-receptor-specific signaling, *e.g.*, in the transduction of a negative effect on chemotaxis. Thus, the codons for the α-receptor-unique tyrosine residues 720, 768, 944, 988, 993, and 1018 (see Fig. 1) of the wild-type human PDGF α-receptor cDNA, were changed individually to phenylalanine codons by site-directed mutagenesis, generating Y720F, Y768F, Y944F, Y988F, Y993F, and Y1018F receptor mutants. PAE cells expressing the wild-type or mutant PDGF α-receptors were subjected to a modified Boyden chamber assay for chemotaxis. PAE cells expressing Y768F, Y993F, or Y1018F PDGF α-receptors (referred to as chemotactic mutants) showed a dose-dependent chemotactic response towards PDGF-BB, whereas cells expressing either wild-type PDGF α-receptor, mutant receptors Y720F, Y944F, or Y988F, as well as non-transfected PAE cells showed no chemotactic response (*41*).

PI3-kinase and PLC-γ have been suggested to be critical for chemotactic signaling elicited by the PDGF β-receptor (*79, 100*). However, the wild-type α- and β-receptors as well as the chemotactic mutant PDGF α-receptors showed similar PI3-kinase activity. Furthermore, no difference in PLC-γ activation was observed between the wild-type and the chemotactic mutant PDGF α-receptors. RasGAP, which has been implicated as a negative modulator of chemotaxis (*100*), did not associate with or become phosphorylated by either the wild-type or the mutant α-receptors (*41*).

To investigate the molecular mechanisms for the gain of

chemotactic capacity of the mutant PDGF α-receptors, we first examined whether Tyr-768 is an autophosphorylation site in the PDGF α-receptor. Thus, cells expressing wild-type or Y768F mutant PDGF α-receptors were [^{32}P]orthophosphate-labeled and stimulated by PDGF-BB; receptors were collected by anti-α-receptor immunoprecipitation and subjected to tryptic digestion. The digests were analyzed by two-dimensional phosphopeptide mapping as well as immunoprecipitation using an antiserum raised against the Tyr-768 region, followed by Edman degradation analysis. Both results suggested that Tyr-768 as well as Tyr-762 is autophosphorylated *in vivo*; Ser-767 was also found to be phosphorylated *in vivo*, and the phosphorylation of this residue increased upon replacement of Tyr-768 with a phenyl-alanine residue. Using a similar experimental procedure, Tyr-993 was, in contrast, shown not to be an autophosphorylation site. However, further analyses including a detailed comparison between tryptic digests derived from the carboxyterminal tail of [^{32}P]-orthophosphate-labeled, PDGF-BB-stimulated wild-type and Y993F mutant PDGF α-receptors indicated that Tyr-988 is phosphorylated with increased stoichiometry following muta-tion of Tyr-993 to a phenylalanine residue. Tyr-1018 has already been reported to be an autophosphorylation site (*41*).

Our data indicate that several mechanisms are operating in negative modulation of PDGF α-receptor signaling. We suggest that autophosphorylated tyrosine residues 768 and 1018 serve to mediate negative regulation of chemotaxis downstream of PI3-kinase. In contrast, it is possible that Tyr-988 mediates a signal for positive regulation (Fig. 4).

VI. FUTURE PERSPECTIVES

The discovery of the SH2 domain and characterization of its function have promoted our understanding of receptor tyrosine kinase signaling mechanisms significantly. Mitogenicity and motogenicity are major cell functions regulated by tyrosine kinase receptors. The small G-protein Ras and its regulators

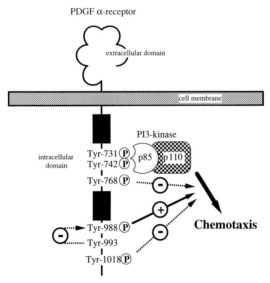

Fig. 4. Schematic representation of possible modes of regulation of chemotactic signaling by the PDGF α-receptor (*41*). Numbers indicate the positions of tyrosine residues. (P) indicates that the tyrosines are phosphorylatable. (+), positive modulation; (−), negative modulation; PI3-K, phosphatidylinositol 3′-kinase. Mutation of the phosphorylatable tyrosine residues 768 or 1018 to phenylalanine residues relieves a putative negative regulatory mechanism, whereas mutation of Tyr-993 causes an increased autophosphorylation of Tyr-988 possibly resulting in an enhanced chemotactic signaling.

seem to play a central role in mitogenic signaling downstream of tyrosine kinases. We further described that PI3-kinase is involved in cell motility response mediated by the PDGF β-receptor. In addition, certain tyrosine residues in the PDGF α-receptor are implicated for positive and negative signaling for chemotaxis. It is of interest to find out what kind of molecules are involved in the PDGF α-receptor-specific modulatory signals. We have been trying to identify the molecules which interact specifically with the PDGF α-receptor using phosphopeptide affinity purification. Through this approach, we recently identified a group of molecules which binds specifically to the kinase insert of the PDGF α-receptor as Crk family proteins (*116*). We are currently investigating the biological significance of Crk proteins as a downstream mediator of PDGF receptor signaling.

It is of major importance to apply the knowledge on signal transduction to future clinical practices. By further analyzing the molecular mechanisms of biological functions mediated by the PDGF receptors as well as other receptor tyrosine kinases, we might be able to specifically manipulate cellular functions at the molecular level. In conjunction with the currently applied gene therapeutic techniques, for example, introducing dominant negative mutant forms of signaling molecules into pathological tissues such as tumors and atherosclerotic lesions *in vivo*, this will establish a new frontier in the biomedical field.

SUMMARY

PDGF, a potent mitogen for mesenchymal cells, is involved in formation of atherosclerotic vascular diseases and neoplasms as well as in normal embryonic development. PDGF elicits its effects by binding to cell surface tyrosine kinase receptors designated as α- and β-receptors. Ligand-induced dimerization of the receptors triggers initiation of intracellular signaling through induction of their intrinsic kinase activity and recruitment of various SH2 domain-containing proteins. Ras and PI3 kinase play key roles in regulating mitogenicity and motogenicity, respectively, which are the two major cell functions modulated by PDGF. In addition, PDGF α-receptor mediates negative signals for chemotaxis due to its unique structure.

REFERENCES

1. Antoniades, H.N., Scher, C.D., and Stiles, C.D. (1979) *Proc. Natl. Acad. Sci. USA* **76**, 1809–1812.
2. Heldin, C.-H., Westermark, B., and Wasteson, Å. (1979) *Proc. Natl. Acad.*

Sci. USA **76**, 3722–3726.

3. Bauer, E.A., Cooper, T.W., Huang, J.S., Altman, J., and Deuel, T.F. (1985) *Proc. Natl. Acad. Sci. USA* **82**, 4132–4136.
4. Blatti, S.P., Foster, D.N., Ranganathan, G., Moses, H.L., and Getz, M.J. (1988) *Proc. Natl. Acad. Sci. USA* **85**, 1119–1123.
5. Heldin, P., Laurent, T.C., and Heldin, C.-H. (1989) *Biochem. J.* **258**, 919–922.
6. Claesson-Welsh, L., Eriksson, A., Morén, A. *et al*. (1988) *Mol. Cell. Biol.* **8**, 3476–3486.
7. Claesson-Welsh, L., Eriksson, A., Westermark, B., and Heldin, C.-H. (1989) *Proc. Natl. Acad. Sci. USA* **86**, 4917–4921.
8. Gronwald, R.G.K., Grant, F.J., Haldeman, B.A. *et al*. (1988) *Proc. Natl. Acad. Sci. USA* **85**, 3435–3439.
9. Matsui, T., Heidaran, M., Miki, T. *et al*. (1989) *Science* **243**, 800–803.
10. Yarden, Y., Escobedo, J.A., Kuang, W.-J. *et al*. (1986) *Nature* **323**, 226–232.
11. Levéen, P., Pekny, M., Gebre-Medhin, S., Swolin, B., Larsson, E., and Betsholtz, C. (1994) *Genes Dev.* **8**, 1875–1887.
12. Soriano, P. (1994) *Genes Dev.* **8**, 1888–1896.
13. Heldin, C.-H. and Westermark, B. (1995) In *Guidebook to Cytokines and Their Receptors* (Nicola, N., ed.), Oxford University Press, Oxford.
14. Westermark, B., Betsholtz, C., Johnsson, A., and Heldin, C.-H. (1987) In *Viral Carcinogenesis* (Kjeldgaard, N.O. and Forchhammer, J., eds.), pp. 445–457, Munksgaard, Copenhagen.
15. Pierce, G.F., Mustoe, T.A., Senior, R.M. *et al*. (1988) *J. Exp. Med.* **167**, 974–987.
16. Robson, M.C., Phillips, L.G., Thomason, A., Robson, L.E., and Pierce, G.F. (1992) *Lancet* **339**, 23–25.
17. Greenhalg, D.G., Sprugel, K.H., Murray, M.J., and Ross, R. (1990) *Am. J. Pathol.* **136**, 1235–1246.
18. Martinet, Y., Bitterman, P.B., Mornex, J.-F., Grotendorst, G.R., Martin, G.R., and Crystal, R.G. (1986) *Nature* **319**, 158–160.
19. Peterson, T.C. and Isbrucker, R.A. (1992) *Hepatology* **15**, 191–197.
20. Floege, J., Eng, E., Young, B.A. *et al*. (1993) *J. Clin. Invest.* **92**, 2952–2962.
21. Johnson, R.J., Raines, E.W., Floege, J. *et al*. (1992) *J. Exp. Med.* **175**, 1413–1416.
22. Sano, H., Engleka, K., Mathern, P. *et al*. (1993) *J. Clin. Invest.* **91**, 553–565.
23. Golub, T.R., Barker, G.F., Lovett, M., and Gilliland, D.G. (1994) *Cell* **77**, 307–316.
24. Ross, R. (1993) *Nature* **362**, 801–809.
25. Jawen, A., Bowen-Pope, D.F., Lindner, V., Schwarts, S.M., and Clowes, A.W. (1992) *J. Clin. Invest.* **89**, 507–511.
26. Ferns, G.A.A., Raines, E.W., Sprugel, K.H., Motani, A.S., Reidy, M.A., and

Ross, R. (1991) *Science* **253**, 1129-1132.

27. Hart, C.E., Forstrom, J.W., Kelly, J.D. *et al*. (1988) *Science* **240**, 1529-1531.
28. Heldin, C.-H., Bäckström, G., Östman, A. *et al*. (1988) *EMBO J.* **7**, 1387-1393.
29. Heldin, C.-H. (1995) *Cell* **80**, 213-223.
30. Bishayee, S., Majumdar, S., Scher, C.D., and Khan, S. (1988) *Mol. Cell. Biol.* **8**, 3696-3702.
31. Keating, M.T., Escobedo, J.A., and Williams, L.T. (1988) *J. Biol. Chem.* **263**, 12805-12808.
32. Kazlauskas, A., Durden, D.L., and Cooper, J.A. (1991) *Cell Regul.* **2**, 413-425.
33. Hubbard, S.R., Wei, L., Ellis, L., and Hendrickson, W.A. (1994) *Nature* **372**, 746-754.
34. Pawson, T. (1995) *Nature* **373**, 573-580.
35. Pawson, T. and Schlessinger, J. (1993) *Curr. Biol.* **3**, 434-442.
36. Fantl, W.J., Escobedo, J.A., Martin, G.A. *et al*. (1992) *Cell* **69**, 413-423.
37. Songyang, Z., Shoelson, S.E., Chaudhuri, M. *et al*. (1993) *Cell* **72**, 767-778.
38. Waksman, G., Kominos, D., Robertson, S.C. *et al*. (1992) *Nature* **358**, 646-653.
39. Claesson-Welsh, L. (1994) *J. Biol. Chem.* **269**, 32023-32026.
40. Eriksson, A., Nånberg, E., Rönnstrand, L. *et al*. (1995) *J. Biol. Chem.* **270**, 7773-7781.
41. Yokote, K., Mori, S., Siegbahn, A. *et al*. (1996) *J. Biol. Chem.* **271**, 5101-5111.
42. Yu, J.-C., Heidaran, M.A., Pierce, J.H. *et al*. (1991) *Mol. Cell. Biol.* **11**, 3780-3785.
43. Lechleider, R.J., Sugimoto, S., Bennett, A.M. *et al*. (1993) *J. Biol. Chem.* **268**, 21478-21481.
44. Cicchetti, P., Mayer, B.J., Thiel, G., and Baltimore, D. (1992) *Science* **257**, 803-806.
45. Buday, L. and Downward, J. (1993) *Cell* **73**, 611-620.
46. Abrams, C.S., Wu, H., Zhao, W., Belmonte, E., White, D., and Brass, L.F. (1995) *J. Biol. Chem.* **270**, 14485-14492.
47. Inglese, J., Koch, W.J., Touhara, K., and Lefkowitz, R.J. (1995) *Trends Biochem. Sci.* **20**, 151-156.
48. Lemmon, M.A., Ferguson, K.M., O'Brien, R., Sigler, P.B., and Schlessinger, J. (1995) *Proc. Natl. Acad. Sci. USA* **92**, 10472-10476.
49. Cochran, B.H., Reffel, A.C., and Stiles, C.D. (1983) *Cell* **33**, 939-947.
50. Eriksson, A., Siegbahn, A., Westermark, B., Heldin, C.-H., and Claesson-Welsh, L. (1992) *EMBO J.* **11**, 543-550.
51. Heidaran, M.A., Beeler, J.F., Yu, J.-C. *et al.* (1993) *J. Biol. Chem.* **268**, 9287-9295.
52. Inui, H., Kitami, Y., Tani, M., Kondo, T., and Inagami, T. (1994) *J. Biol.*

Chem. **269**, 30546–30552.

53. Hosang, M., Rouge, M., Wipf, B., Eggimann, B., Kaufmann, F., and Hunziker, W. (1989) *J. Cell. Physiol.* **140**, 558–564.
54. Boguski, M.S. and McCormick, F. (1993) *Nature* **366**, 643–654.
55. Bourne, H.R., Sanders, D.A., and McCormick, F. (1990) *Nature* **348**, 125–132.
56. Mulcahy, L.S., Smith, M.R., and Stacey, D.W. (1985) *Nature* **313**, 241–243.
57. Smith, M.R., DeGudicubus, S.J., and Stacey, D.W. (1986) *Nature* **320**, 540–543.
58. Matuoka, K., Shibata, M., Yamakawa, A., and Takenawa, T. (1992) *Proc. Natl. Acad. Sci. USA* **89**, 9015–9019.
59. Lowenstein, E.J., Daly, R.J., Batzer, A.G. *et al.* (1992) *Cell* **70**, 431–442.
60. Egan, S.E., Giddings, B.W., Brooks, M.W., Buday, L., Sizeland, A.M., and Weinberg, R.A. (1993) *Nature* **363**, 45–51.
61. Yokote, K., Mori, S., Hansen, K. *et al.* (1994) *J. Biol. Chem.* **269**, 15337–15343.
62. Arvidsson, A.-K., Rupp, E., Nånberg, E. *et al.* (1994) *Mol. Cell. Biol.* **14**, 6715–6726.
63. Li, W., Nishimura, R., Kashishian, A. *et al.* (1994) *Mol. Cell. Biol.* **14**, 509–517.
64. Kazlauskas, A., Feng, G.-S., Pawson, T., and Valius, M. (1993) *Proc. Natl. Acad. Sci. USA* **90**, 6939–6943.
65. Hu, Q., Milfay, D., and Williams, T. (1995) *Mol. Cell. Biol.* **15**, 1169–1174.
66. Kazlauskas, A., Ellis, C., Pawson, T., and Cooper, J.A. (1990) *Science* **247**, 1578–1581.
67. Henkemeyer, M., Rossi, D.J., Holmyard, D.P. *et al.* (1995) *Nature* **377**, 695–701.
68. Moodie, S.A., Willumsen, B.M., Weber, M.J., and Wolfman, A. (1993) *Science* **260**, 1658–1661.
69. Vojtek, A.B., Hollenberg, S.M., and Cooper, J.A. (1993) *Cell* **74**, 205–214.
70. Treisman, R. (1994) *Curr. Opin. Genet. Dev.* **4**, 96–101.
71. Kyriakis, J., Banerjee, P., Nikolakaki, E. *et al.* (1994) *Nature* **369**, 156–160.
72. Coso, O.A., Chiariello, M., Yu, J.-C. *et al.* (1995) *Cell* **81**, 1137–1146.
73. Minden, A., Lin, A., Claret, F.-X., Abo, A., and Karin, M. (1995) *Cell* **81**, 1147–1157.
74. Kypta, R.M., Goldberg, Y., Ulug, E.T., and Courtneidge, S.A. (1990) *Cell* **62**, 481–492.
75. Mori, S., Ronnstrand, L., Yokote, K. *et al.* (1993) *EMBO J.* **12**, 2257–2264.
76. Twamley-Stein, G.M., Pepperkok, R., Ansorge, W., and Courtneidge, S.A. (1993) *Proc. Natl. Acad. Sci. USA* **90**, 7696–7700.
77. Barone, M.V. and Courtneidge, S.A. (1995) *Nature* **378**, 509–512.
78. Satoh, T., Fantl, W.J., Escobedo, J.A., Williams, L.T., and Kaziro, Y. (1993) *Mol. Cell. Biol.* **13**, 3706–3713.

79. Wennström, S., Siegbahn, A., Yokote, K. *et al.* (1994) *Oncogene* **9**, 651–660.
80. Burgering, B.M. and Coffer, P.J. (1995) *Nature* **376**, 553–554.
81. Chung, J., Grammer, T.C., Lemon, K.P., Kazlauskas, A., and Blenis, J. (1994) *Nature* **370**, 71–75.
82. Franke, T.F., Yang, S.I., Chan, T.O. *et al.* (1995) *Cell* **81**, 727–736.
83. Nakanishi, H., Brewer, K.A., and Exton, J.H. (1993) *J. Biol. Chem.* **268**, 13–16.
84. Cross, M.J., Hodgkin, M.N., Stewart, A., Kerr, D.J., and Wakelam, M.J.O. (1995) *J. Biol. Chem.* **270**, 25352–25355.
85. Valius, M. and Kazlauskas, A. (1993) *Cell* **73**, 321–334.
86. Ihle, J.N. (1995) *Nature* **377**, 591–594.
87. Silvennoinen, O., Schindler, C., Schlessinger, J., and Levy, D.E. (1993) *Science* **261**, 1736–1739.
88. Vignais, M.L., Sadowski, H.B., Watling, D., Rogers, N.C., and Gilman, M. (1996) *Mol. Cell. Biol.* **16**, 1759–1769.
89. Mellström, K., Heldin, C.-H., and Westermark, B. (1988) *Exp. Cell Res.* **177**, 347–359.
90. Wennström, S., Hawkins, P., Cooke, F. *et al.* (1994) *Curr. Biol.* **4**, 385–393.
91. Klippel, A., Escobedo, J.A., Hu, Q., and Williams, L.T. (1993) *Mol. Cell. Biol.* **13**, 5560–5566.
92. Escobedo, J.A., Navankasattusas, S., Kavanaugh, W.M., Milfay, D., Fried, V.A., and Williams, L.T. (1991) *Cell* **65**, 75–82.
93. Hiles, I.D., Otsu, M., Volinia, S. *et al.* (1992) *Cell* **70**, 419–429.
94. Hu, P., Mondino, A., Skolnik, E.Y., and Schlessinger, J. (1993) *Mol. Cell. Biol.* **13**, 7677–7688.
95. Otsu, M., Hiles, I., Gout, I. *et al.* (1991) *Cell* **65**, 91–104.
96. Skolnik, E.Y., Margolis, B., Mohammadi, M. *et al.* (1991) *Cell* **65**, 83–90.
97. Stoyanov, B., Volinia, S., Hanck, T. *et al.* (1995) *Science* **269**, 690–693.
98. Rodriguez-Viciana, P., Warne, P.H., Dhand, R. *et al.* (1994) *Nature* **370**, 527–532.
99. Pleiman, C.M., Hertz, W.M., and Cambier, J.C. (1994) *Science* **263**, 1609–1612.
100. Kundra, V., Escobedo, J.A., Kazlauskas, A. *et al.* (1994) *Nature* **367**, 474–476.
101. Ridley, A.J., Paterson, H.F., Johnston, C.L., Diekmann, D., and Hall, A. (1992) *Cell* **70**, 401–410.
102. Hawkins, P.T., Eguinoa, A., Qiu, R.-G. *et al.* (1995) *Curr. Biol.* **5**, 393–403.
103. Ridley, A.J. and Hall, A. (1992) *Cell* **70**, 389–399.
104. Olson, M.F., Ashworth, A., and Hall, A. (1995) *Science* **269**, 1270–1272.
105. Saito, Y., Mori, S., Yokote, K., Kanzaki, T., Saito, Y., and Morisaki, N.

(1996) *Biochem. Biophys. Res. Commun.* **224**, 23–26.

106. Rankin, S., Hooshmand-Rad, R., Claesson-Welsh, L., and Rozengurt, E. (1995) *J. Biol. Chem.* **271**, 7829–7843.

107. Schaller, M.D., Borgman, C.A., Cobb, B.S., Vines, R.R., Reynolds, A.B., and Parsons, J.T. (1992) *Proc. Natl. Acad. Sci. USA* **89**, 5192–5196.

108. Chen, H.C. and Guan, J.L. (1994) *Proc. Natl. Acad. Sci. USA* **91**, 10148–10152.

109. Schlaepfer, D.D., Hanks, S.K., Hunter, T., and van der Geer, P. (1994) *Nature* **372**, 786–791.

110. Xing, Z., Chen, H.C., Nowlen, J.K., Taylor, S.J., Shalloway, D., and Guan, J.L. (1994) *Mol. Biol. Cell* **5**, 413–421.

111. Ilic, D., Furuta, Y., Kanazawa, S. *et al.* (1995) *Nature* **377**, 539–544.

112. Rankin, S. and Rozengurt, E. (1994) *J. Biol. Chem.* **269**, 704–710.

113. Siegbahn, A., Hammacher, A., Westermark, B., and Heldin, C.-H. (1990) *J. Clin. Invest.* **85**, 916–920.

114. Koyama, N., Hart, C.E., and Clowes, A.W. (1994) *Circ. Res.* **75**, 682–691.

115. Koyama, N., Morisaki, N., Saito, Y., and Yoshida, S. (1992) *J. Biol. Chem.* **267**, 22806–22812.

116. Yokote, K., Hellman, U., Ekman, S. *et al.* (1998) *Oncogene* **16**, 1229–1239.

Extracellular Matrix-Cellular Interaction: Molecules to Diseases (Y. Ninomiya et al., eds.), pp. 169–183,
Japan Sci. Soc. Press, Tokyo / S. Karger, Basel (1998)

Membranous Structure Immediately Above, and Tubular Stomas in Grooves between Endothelial Cell Clusters

MOTOHARU HASEGAWA AND YU-ICHI TAKARADA

Department of Clinical Physiology, Toho University School of Medicine, Ota-ku, Tokyo 143-0015, Japan

Under pulsatile blood circulation, the arterial aperture, inner pressure, and blood flow rate and volume all follow repeated time-sequence changes, which are maximal at the systolic acme and minimal at the diastolic acme. When an attempt is made to understand such circulatory dynamics from a hydrodynamic viewpoint, problems may arise from rheological characteristics between clusters of endothelial cells and blood flow and fatigue of constituents resulting from mechanical stress and distortion. In consideration of the characteristics of systemic blood pressure distribution, it can be said that the main arteries in the traverse direction in the presence of pulsatile blood flow are involved in the unique characteristics of circulatory dynamics, in which much higher systolic blood pressure and much lower diastolic blood pressure are seen in the elbow, knee, and in the carotid artery system, than in the origin of the aorta (*1*). Because of this, the aorta should have a rough luminal surface, with grooves and folds, along the direction of blood flow. This structure can absorb turbulences and prevent their occurrence, resulting in a relative decrease in the viscosity coefficient up to 10% (2-4). It can also be inferred that a membranous structure, which uniformly covers the rough surface of clusters of endothelial cells, exists to enhance the effects of the unique structures mentioned above.

There is an essential and more important issue concerning the potentiality of nutritional supply to the arterial wall, where fatigue of constituents is accelerated by mechanical stress and distortion. It is known that the outer one-third of the arterial wall is supplied with nutrition from the vasa vasorum (5, 6), whereas several possibilities have been considered about the nutritional supply to the inner two-thirds of the aortic wall (7, 8) but as yet there is no well-established theory. At any rate, no palliative mechanism, such as diffusion, sufficiently explains parietal nutrition. It is expected that some tubular structure, which is far superior to the vasa vasorum and serves as a system of transporting humoral elements from blood, must be present as the best mechanism of nutritional supply to the inner area. Based on these ideas, we conducted the present study using human and rabbit aortae, to determine the presence or absence of membranous structure, fold/groove structure, and tubular structure on the aortic luminal surface. Several useful findings were obtained as a result of improved preparation techniques for scanning electron microscopic examination and photographic techniques, in addition to precise observation, analysis, and evaluation: the adventitial surface of each aortic segment, which was immersed in phosphate buffer to avoid production of artifacts by drying, was glued onto a plastic plate. The flattened aortic segment was then postfixed with 1% OsO_4 buffered with 0.1 M phosphate buffer for 1 hr, dehydrated with analytical-grade ethanol, dried with t-butanol using a Hitachi freeze-dryer ES-2030, and then coated with Pt and Pb using a Hitachi ion sputter E-1010. The resulting segment was observed on a Hitachi S 450 at an accelerating voltage of 15 kv. Thin sections of 4 $f\hat{E}m$ prepared from conventional paraffin-embedded aortic tissue specimens using the method we developed (9) were also observed under a scanning electron microscope.

I. MEMBRANOUS STRUCTURE OVERLYING THE SURFACE OF ENDOTHELIAL CELLS

The aortic luminal surface in the control rabbits, in which

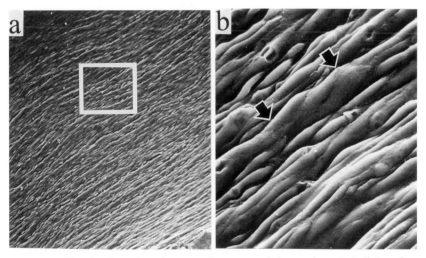

Fig. 1. Scanning electron microscopy of the aortic endothelial surface (rabbit, prefixed with 0.5% PFA). a: ×30. b: ×1,000. Higher magnification of a. The boundary of the endothelial cells appears to be fused in some places (arrow), suggesting the presence of a membranous structure that covers the cell layer. Japanese white rabbits, about 25 months old and weighing about 4 kg, were used. Animals were fed rabbit pellets, at 100 g/ day, and were freely given tap water. After their sacrifice, full circumference segments of the thoracic portion of the descending aorta were dissected, with care taken to minimize damage to the tissue. Segments 0.5 cm² in size were removed at the level of the 5th intercostal space from the body and quickly washed in saline. Each segment was placed in a fixative containing 2.5% glutaraldehyde (GA) and 4% paraformaldehyde (PFA) for 24 hr, and was then washed in 0.1% phosphate buffer for 1 hr, followed by processing for conventional electron microscopy.

the resected aortic tissue was fixed with 2.5% GA and 4% PFA,
showed a regular arrangement of clusters of endothelial cells in
conformance with the path of blood flow, and clusters of endothelial cells were properly arranged to suit distributors at
branches. Individual endothelial cells generally were spindle-
shaped and measured about 20 fÊm in length and about 2–4
fÊm in width, and these cells were densely and closely arranged

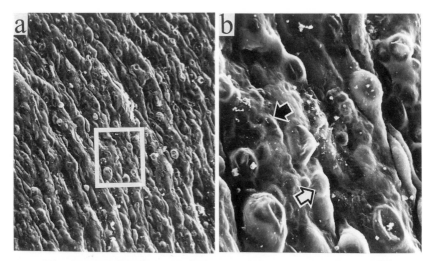

Fig. 2. Scanning electron microscopy of the aortic endothelial surface
(atherosclerotic rabbit, prefixed with 0.5% PFA). a: ×380. b: ×1,800.
Disrupted endothelial cells with destroyed membranous structure are
shown by arrows. Atherosclerosis was induced by the following method:
100 g/day (329 kcal) of atherogenic rabbit pellets, containing alpha starch
(200 kcal), proteins (45.1 kcal), lipids (74.1 kcal), and salt (1.3 g), was
given to animals for 5 months; the animals were also intramuscularly
injected daily with 0.5% norepinephrine (0.5 mg/day). After their sacrifice,
full circumference segments of the thoracic portion of the descending aorta
were dissected, with care taken to minimize damage to the tissue. Segments
0.5 cm² in size were removed at the level of the 5th intercostal space from
the body and quickly washed in saline. Each segment was placed in a
fixative containing 2.5% GA and 4% PFA for 24 hr, and was then washed
in 0.1% phosphate buffer for 1 hr, followed by processing for conventional
electron microscopy.

(Fig. 1a). When the surface of clusters of endothelial cells was observed in more detail, a membranous structure covering the surface of the endothelial cell layer and closely adhering to the cells was seen (Fig. 1b).

In the atherosclerotic rabbits, individual endothelial cells were markedly deformed compared with the control rabbits; they had lost their spindle shape and showed an irregular arrangement and an uneven surface (Fig. 2a, b). Some endothelial cells were destroyed, and the membranous structure was partially destroyed or defective.

In human specimens, the surface of the endothelial layer was also covered with a thick membranous structure, which was densely overlaid with fibrils of about 200 to 300 nm diameter (Fig. 3a). Stomas with a diameter of several $f\hat{E}m$ were also

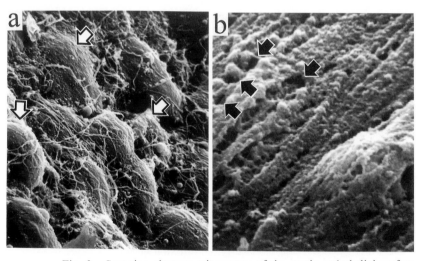

Fig. 3. Scanning electron microscopy of the aortic endothelial surface (human, prefixed with 0.5% PFA). a: ×2,500. A 63-year-old woman. Protruding endothelial cells (8–12 $f\hat{E}m$ × 5–6 $f\hat{E}m$) with a membranous structure were covered with numerous fibrils (200- to 300-nm diameter). b: ×15,000. A 39-year-old woman. Holes 1–2 $f\hat{E}m$ in the long axis are marked.

observed (Fig. 3b). In the control rabbits fixed by supravital perfusion (see the legend of Fig. 2), clusters of endothelial cells formed folds about 10 to 20 fÊm in width, between which grooves served as borders (Fig. 4a, b). Two or three endothelial cells were present in each fold, and the tunica elastica interna was thought to be present in the lower layer. These folds and grooves were regularly arranged along the path of blood flow; the surface was smooth and glossy; a membranous structure appeared to cover the entire luminal surface, rather than individual endothelial cells. Both Figs. 1 and 4 show this aortic luminal surface. In the former figure, clusters of endothelial cells were

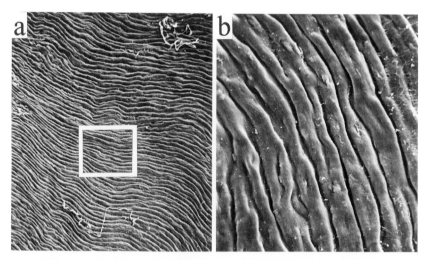

Fig. 4. Scanning electron microscopy of the aortic endothelial surface (rabbit). a: ×100. Undulation-forming folds and grooves (10–20 fÊm wide) are visible and oriented in the direction of blood flow. b: ×400. Higher magnification of a. Under deep anesthesia, about 50 ml of saline was injected through a cannula placed in the abdominal aorta, over a period of 2 min, and then a fixative composed of 2.5% GA and 4% PFA was slowly injected over a period of 5 min, while the displaced blood was suctioned from the left ventricle through an injector needle. The animal's heart continued to beat for 2–3 min after the start of this procedure.

clearly visible in silhouette, but the membranous structure was not very obvious. In the latter figure, clusters of endothelial cells were not clearly silhouetted, but the membranous structure covering folds and grooves was more clearly seen. In 4 fÊm longitudinal sections of an aortic tissue specimen, which was obtained from a control rabbit and treated with formalin fixation and paraffin embedding, the longitudinal cut surface dis-

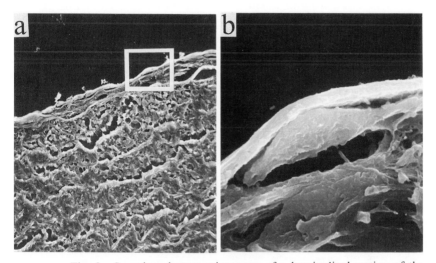

Fig. 5. Scanning electron microscopy of a longitudinal section of the endothelium (rabbit, thin section (4 fÊm) of paraffin-embedded aortic tissue). a: ×1,500. An endothelial cell about 2–3 fÊm in thickness is situated in the bead-like arrangement. Note the continuous layer of membranous structure with strong electron reflection (1 fÊ, thick) over the endothelial cells. b: ×15,000. Higher magnification of a. Japanese white rabbits, about 25 months old and weighing about 4 kg, were used. Animals were fed rabbit pellets, at 100 g/day, and were freely given tap water. After their sacrifice, full circumference segments of the thoracic portion of the descending aorta were dissected, with care taken to minimize damage to the tissue. Segments 0.5 cm² in size were removed at the level of the 5th intercostal space from the body and quickly washed in saline. Each segment was placed in a fixative containing 2.5% GA and 4% PFA for 24 hr, and was then washed in 0.1% phosphate buffer for 1 hr, followed by processing for conventional electron microscopy.

closed a layer of endothelial cell clusters which were arranged like chains of beads near the luminal surface (Fig. 5a). Immediately above these endothelial cells, a membranous structure about 1 ƒÊm thick was clearly seen to uniformly cover the entire luminal surface (Fig. 5b).

II. TUBULAR STOMAS ASSOCIATED WITH GROOVES

A resected human aorta was fixed with 0.5% PFA and its luminal surface is shown in Fig. 6a and b. A membranous structure and fibrils were again present over the entire luminal surface (Fig. 6a). Higher magnification of the region between the membranous structure and fibrils showed another structure thought to be composed of several tubular holes of 2–6 ƒÊm. A

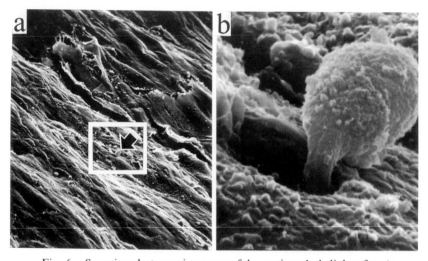

Fig. 6.   Scanning electron microscopy of the aortic endothelial surface (a 32-year-old man, prefixed with 0.5% PFA). a: ~500. Membranous structure over the endothelium is seen in the center. Beneath the fibrils the membranous structure is visible. b: ~3,000. Higher magnification of a. Several holes of 1–7 ƒÊ, in long axis are visible. Note a cell (5–6 ƒÊ, in long axis) migrating through this hole.

monocyte entered an inlet of one hole. In the atherosclerotic rabbits (Fig. 7a), tissue damage was generally remarkable, and folds and grooves like those observed in Fig. 4 were markedly disarranged. The membranous structure overlying deformed and irregularly protruded folds was apparently swollen. Grooves

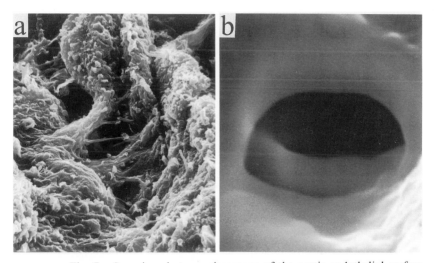

Fig. 7. Scanning electron microscopy of the aortic endothelial surface (atherosclerotic rabbit, prefixed with 0.5% PFA). a: atherosclerotic rabbit ∼4,000. b: Atherosclerotic rabbit ∼30,000. Folds (2-4 $f\hat{E}$, in width) and hollows (2-5 $f\hat{E}$, in width) are visible, and their disarrangement is marked. Several holes of 3-5 $f\hat{E}$, in long axis are situated in the hollow structure. Atherosclerosis was induced by the following method: 100 g/day (329 kcal) of atherogenic rabbit pellets, containing alpha starch (200 kcal), proteins (45.1 kcal), lipids (74.1 kcal), and salt (1.3 g), was given to animals for 5 months; the animals were also intramuscularly injected daily with 0.5% norepinephrine (0.5 mg/day). After their sacrifice, full circumference segments of the thoracic portion of the descending aorta were dissected, with care taken to minimize damage to the tissue. Segments 0.5 cm² in size were removed at the level of the 5th intercostal space from the body and quickly washed in saline. Each segment was placed in a fixative containing 2.5% GA and 4% PFA for 24 hr, and was then washed in 0.1% phosphate buffer for 1 hr, followed by processing for conventional electron microscopy.

were observed in detail showed sequential tubular holes which appeared to be separated by many partitions (Fig. 7a). In our observation, it was difficult to demonstrate such tubular stomas in grooves among aortic tissue specimens without atherosclerosis because of the presence of a membranous structure on the surface. The inner aspect of tubular stoma was more clearly visualized by adjusting the angle of photography when looking at aortic tissue specimens (Fig. 7b). Tubular stomas with partitions could be demonstrated in both rabbits and humans if severe arteriosclerosis or damage on the tunica intima was present.

III. THE NUTRITIONAL SUPPLY TO THE INNER AREA OF THE AORTIC WALL

We first proved the presence of a membranous structure immediately above endothelial cells on the aortic luminal surface by scanning electron microscopy. We were able to verify the presence of a membranous structure with this instrument because we used a combination of tissue fixation and observation methods, such as supravital perfusion fixation and scanning electron microscopy of paraffin-embedded thin-tissue sections. In conventional perfusion fixation for scanning electron microscopy any thin and soft membranous structure immediately shrinks or comes off, so that such a structure is absent or adheres to the surface of cell clusters, making it difficult to identify its presence. Transmission electron microscopy using Alcian blue stain and ruthenium red stain has shown that a complex of glycoprotein and glycosaminoglycan, called glycocalyx, is present on the arterial luminal surface (10–12). The glycocalyx is a single-layer membranous structure of about 10- to 100-nm thickness covering the luminal surface which closely resembles the membranous structure we observed. A membranous structure located on the luminal surface, including the glycocalyx, is thought to serve as a selective barrier for plasma components (13), so that not only endothelial function but also control by "a specifically differentiated membranous structure formed on the luminal surface" like we observed should be thoroughly considered to understand substance transmission through the tunica intima. Such a membranous structure is thus an important factor in determining the pathogenesis of atherosclerosis. Previous investigations (14, 15) reported that the arterial surfaces of the aorta consisted of longitudinal folds with ridges and valleys running in the longitudinal axis of the blood vessel. Folds and grooves on the aortic luminal surface covered with the membranous structure have a very useful and suitable morphology from the pulsatile hydrodynamic viewpoint. If the aortic luminal surface had neither a membranous structure nor a fold/groove structure, viscosity resistance would be increased, particularly during the diastolic phase in which pressure, flow volume, and flow rate all decrease, giving a high risk of the burst phenomenon which contributes to the appearance of turbulence, and of the hairpin eddy phenomenon which may cause longitudinal eddies (2–4). It was our observation that this membranous structure became swollen and its uneven surface became even more uneven when the aorta was damaged; there was also fibrosis and destruction, and the basic structure of folds and grooves had collapsed, further flattening the surface. Such a situation largely affects the viscoelasticity of both blood and the luminal surface, and may greatly decrease the rheological blood flow energy efficiency. Quantitative and morphological alteration and degradation of the membrane are therefore associated with decreased barrier function. It is more important to note that destruction of the membranous and fold/groove structures is thought to cause necrosis and destruction of endothelial cell clusters immediately below these structures.

As the parietal structure of the aorta was damaged, the membranous structure of the groove was also seriously damaged, easily demonstrating the tubular stomal structure. Each tubular stoma had an outlet (2–6 $f\hat{E}$m in diameter) sufficient to allow a material blood component (monocytes, *etc.*) to enter it (Fig. 6b). Although these stomas seemed to be independent entities, they could be defined as many holes separated by thin partitions or as numerous holes that continued in a sequence, when carefully observed under various photographic conditions (Fig. 7b). Some investigators have proposed that tubular holes on the aortic luminal surface are artifacts (16, 17), or that they are small discs and troughs formed by flaps of adjacent cells, these discs and troughs being misinterpreted for the so-called stomata (18), or that they are myo-endothelial herniation and endothelial cell pseudovacuoles (19). Bjorkrud et al. (20), however, reported the presence of tubular holes about 0.2 fÊm in diameter on the aortic luminal surface of the rabbit using supravital perfusion fixation. Previous reports on the aortic endothelial surface in the rat with experimental hypertension or diabetes (21), with experimental hypertension and hypercholesterolemia (22), or with aortic allografts (23), also showed the presence of the stomata structure. Since this stomatal structure showed no basic morphology of mature vessels, it is thought to be formed by the generation of a parietal membrane around intercellular spaces. The assumed significance of the presence of these stomas is that they are involved in the nutritional supply to the inner two-thirds of the aorta, corresponding to the nutritional supply of the vasa vasorum in the outer area (5, 6). Stress and distortion in the arterial wall, occurring by changes in aortic aperture from the systolic to the diastolic phase, alter parietal stress every moment. For example, stress in an artery of 3-cm diameter under blood pressure of 120/180 mmHg at the diastolic acme is estimated to reach a level as high as 239.9×103 dynes/cm^2. Needless to say, the elastic fibrous network, the collagen fibril network, and the contractile fiber cell layer in the inner two-thirds of the tunica media of the aorta have a thickness more than 30 times that of the endothelial cell layer. The amount of nutrition required by this large contractile layer cannot be ignored, because the arterial system is a dynamic organ. Fortunately, the aperture is enlarged, the groove is stretched, and the stoma is opened most widely during the systolic phase in the aorta.

During the diastolic phase, the aperture becomes small, and the stoma is diminished in size. A decrease/increase in parietal stress causes a relative difference from inner pressure, and plasma components are thought to stream into and out of the arterial wall according to negative and positive pressures. If the membranous structure on the luminal surface is degraded or destroyed, the stoma is deformed, stenosed, and closed, and the nutritional condition in the inner two-thirds of the arterial wall is markedly impaired. This may result in rapid progression of damage on the arterial wall by ischemia.

In conclusion, we identified various structures present in the thoracic aorta of the rabbit and human, including folds/grooves, a membranous structure on the luminal surface, and tubular stomas in grooves. We also discussed the nutritional supply to the inner area of the aortic wall, substance transmission, and hydrodynamic characteristics in relation to these structures.

SUMMARY

The luminal surface of the thoracic aorta was observed in human and rabbit specimens using a scanning electron microscope. The findings were as follows: (1) the luminal surface had folds and grooves of about 2 to 20 $f\hat{E}m$ width, running in the direction of blood flow; (2) there was a single-layer membranous structure about 0.2 to 1 $f\hat{E}m$ thick covering the entire layer of endothelial cells; (3) the grooves had tubular stomas, measuring 2-6 $f\hat{E}m$ in longer diameter and 1-5 $f\hat{E}m$ in shorter diameter, which were separated from one another by partitions; (4) tubular holes were readily observed in arteriosclerotic lesions, but they were markedly deformed and in some cases destroyed and monocyte-like cells had migrated into these holes. These microstructures appeared to play important roles in hydrodynamic movements, nutritional supply, and the potential function of substance transmission in the inner part of the aorta.

Acknowledgments

We are grateful to Dr. T. Ooyama of the Department of Clinical Physiology, Toho University School of Medicine and to Dr. Y. Tujita of Pharmacology and Molecular Research Laboratories, Sankyo Co., Ltd. for their suggestions and advice on the preparation of the manuscript.

REFERENCES

1. Folkow, B. and Neil, E. (eds.) (1971) In *Circulation*, pp. 57-72, Oxford University Press, New York.
2. Neuman, D. and Dinkelacker, A. (1991) *Appl. Sci. Res.* **48**, 105-114.
3. Suzuki, Y. and Kasagi, N. (1994) *J. Am. Inst. Aeronaut. Astronaut.* **32**, 1781-1790.
4. Rohr, J.J., Anderson, G.W., Reidy, L.W., and Hendricks, E.W. (1992) *Exp. Fluids* **13**, 361-368.
5. Clarke, J.A. (1964) *Z. Anat. Entwicklungsgesch.* **124**, 261-267.
6. Wollinsky, H. and Glagov, S. (1967) *Circ. Res.* **20**, 409-421.
7. Glagov, S., Zarinski, C.K., Giddens, D.P., and Ku, D.N. (1989) In *Diseases of the Arterial Wall* (Camilleri, J.-P. *et al.*, eds.), pp. 217-329, Springer-Verlag, London.
8. Haust, M.D. (1983) In *Cardiovascular Pathology* (Silver, M.D., ed.), vol. 1, pp. 191-315, Churchil Livingstone, New York.
9. Hasegawa, M., Saito, M., Fukunaga, Y. *et al.* (1984) *J. Jpn. Clin. Pathol. Soc.* (*Rinsho-byori*) **32**, 801-808 (in Japanese).
10. Ausprunk, D.H., Boudreau, C.L., and Nelson, D.A. (1981) *Am. J. Pathol.* **103**, 367-375.
11. Sims, D.E. and Horne, M.M. (1993) *Eur. J. Morphol.* **31**, 251-256.
12. Haldenby, K.A., Chappel, D.C., Winlobe, C.P., Parker, K.H., and Firth, J.A. (1994) *J. Vasc. Res.* **31**, 2-9.
13. Busch, P.C. (1984) In *Biology of Endothelial Cells* (Jaffe, E.A., ed.), pp. 178-188, Martinous Nijhoff Publishers, Boston.
14. Shimamoto, T., Yamashita, Y., and Sunaga, T. (1969) *Proc. Jpn. Acad.* **45**, 507-511.
15. Groniowski, J.W., Biezyskowa, W., and Walki, M. (1971) *Folia Histochem. Cytochem.* **9**, 243-249.
16. Garbarsch, C., Tranum, J.J., and van Deurs, B. (1982) *Acta Anat.* **112**, 79-91.
17. Duff, G.L., McMillan, G.C., and Ritchie, A.C. (1957) *Am. J. Pathol.* **33**, 845-873.
18. Albert, E.N. and Nayak, R.K. (1976) *Anat. Rec.* **185**, 223-234.

19. Stetz, E.M., Majno, G., and Joris, I. (1979) *Virchows Arch. Abt. Path. Anat. Histol.* **383**, 135–148.
20. Bjorkrud, S., Hansson, H.-A., and Bondjers, G. (1972) *Virchows Arch. Abt. B Zellpath.* **11**, 19–23.
21. Todd, M.E. (1991) *Can. J. Physiol. Pharmacol.* **70**, 536–551.
22. Sakata, N., Ida, T., Joshita, T., and Ooneda, G. (1983) *Acta Pathol. Jpn.* **33**, 1105–1113.
23. Bowyer, D.E. and Reidy, M.A. (1977) *J. Pathol.* **123**, 237–245.

Extracellular Matrix-Cellular Interaction: Molecules to Diseases (Y. Ninomiya et al., eds.), pp. 185–202,
Japan Sci. Soc. Press, Tokyo / S. Karger, Basel (1998)

A Novel Helical Filament Model of Hen Tendon Collagen Fibril

NOBUHIKO KATSURA AND TOSHIO ONO

Department of Oral Biochemistry, School of Dentistry, Nagasaki University, Nagasaki 852-8102, Japan

Collagen is the most abundant protein in the animal kingdom and is the major fibrous constituent of connective tissues. It has been widely accepted that the collagen fibril is made up of straight linear rods of the tropocollagen molecule arranged in a "quarter stagger array", where each rod overlaps its neighbor by about one quarter (D-period, 60–70 nm) of its length (300 nm). It would be difficult, however, to explain why a collagen fibril shows an identical lateral feature with deep wavy surface in all directions using this hypothesis. Piez and Miller (*1*) stated that "the collagen fibril has four structural levels of which at least three are coils. The polypeptide chain, the molecule, and the microfibril are helical structures; the fibril may consist of parallel or coiled microfibrils." A number of helical models of the third (microfibril) and of the fourth (fibril) level were also proposed (*2–8*). All of them, however, are the spring coil-like supra-helix in which there are inevitable discrepancies of molecular array between the center and the periphery of the cylindrical fibril.

Using advanced techniques of surface analysis such as the plasma polymerized film replica method (9–12), we found that the surface of collagen in tissue has a wave-like feature that appears to be formed of a lateral assembly of hollowless helical filaments (microfibril) with D-helical pitches (13). This new model well explains the characteristic features of tissue collagen, its mineral deposition, its tensile strength, its natural turnover, its fibrogenesis, and possible movability of the fibril.

The plasma polymerized film replica method applied here requires only drying or freezing with liquid nitrogen of a specimen but does not require chemical fixation, embedding, staining or ultra-thin sectioning. Chicken tarsus tendon and aged and partially mineralized hen tarsus tendon were used because they are mostly Type I collagen and well oriented.

I. PLASMA POLYMERIZED FILM REPLICA METHOD

Chicken and aged hen tarsi were obtained from a poultry slaughterhouse. Tendons were freeze-dried in 100% *t*-butanol after sequential substitution, or were air-dried after the treatment indicated in the respective figure. The dried tendon was split off longitudinally with a razor blade and coated with methane/ ethylene polymerized film in the negative glow phase domain on a cathode plate in a gas reactor (NL-OP30S, Nippon Laser & Electronics Lab., Nagoya, Japan). The mixing ratio of methane to ethylene was $3:1$ at $2\,KV$, 4–$6\,mA$, 6–8×10^{-2} Torr and discharge time was about 40 sec. The single-stage replica film of the specimen was released by dissolution of tissue in a commercial bleach, and was then scooped up on a 200 mesh copper grid for transmission electron microscopy (TEM). The replica film was visualized directly in a Hitachi H-800 electron microscope at $100\,KV$, where its assumed thickness was 6–7 nm under those conditions. Other details of the preparation of plasma polymerization replica film were described earlier (13).

II. WAVY SURFACE OF COLLAGEN FIBRIL

1. Pile of Sheets

Transmission electron micrograph of the plasma polymerized film replica of unmineralized chicken tendon which was freeze-dried under stretching shows a pile of sheets with D-periodic banding (Fig. 1A). Another sheet is visible underneath one of the broken sheets and the direction of the banding is well oriented. The edges in some broken sheets show fibril-like scrolls, but they arose during sample preparation. An unfixed specimen dipped in 1% Triton X-100 is shown in Fig. 1B. The peeled topmost sheet shows a rhombic pattern, indicating reflection of the sheet underneath.

2. Wavy Surface

The plasma film replica of air-dried chicken tendon soaked in Triton X-100 (without stretching) shows a continuous and identical wavy surface with a wave height of 6–10 nm and wavelength of 65 nm at the top- and side-views (Fig. 2). This indicates that the structural element must be a helical but not a linear element. It should be noted that phase discrepancies in some areas may indicate the possible existence of right (P)- and left (M)-handed helices.

3. Helical Filament

Aged hen tendon mineralizes and the onset of mineralization is observable at the boundary of the mineralized and unmineralized regions. This boundary portion shows many helical straight filaments about 23 nm in filament diameter, about 46 nm in helix diameter and about 65 nm in helix pitch, which is comparable to the reported banding intervals of both left-handed and right-handed helices as shown in Fig. 3 (bottom). Those assumed thicknesses include that of the plasma polymerized film of approximately 6–7 nm; therefore, the net thickness of the filament may be between 9–11 nm. The orienta-

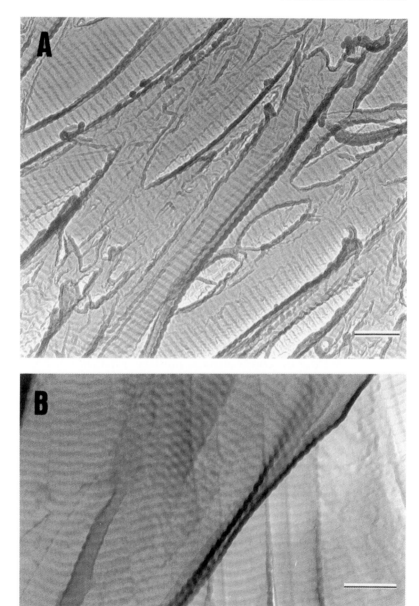

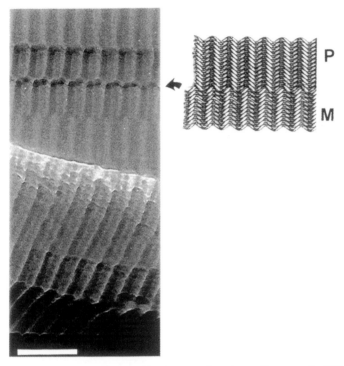

Fig. 2. Surface replica images of tendon collagen under TEM. Cracked edge of chicken tendon collagen fiber shows the identical wavy surfaces in the top- and side-views. Procedures identical to those in Fig. 1B. Note the phase discrepancies (arrow) and compare the wire model (right half) of right (P)- and left (M)-handed helices. See the section III "χ-Helical Filament", p. 190; bar: 200 nm.

←Fig. 1. Surface replica images of tendon collagen under TEM. (A) Fresh chicken tendon (approximately 55 days old) was fixed with 2% paraformaldehyde for 30 min at room temperature and then freeze-dried under stretching. Stacking of periodic banding sheets is clearly seen and the edges of broken sheet, which occurred during sample preparation, are rolled. This may be misinterpreted as a "fibril"; bar: 600 nm. (B) The unfixed specimen was dipped in 1% Triton X-100 for 10 min at room temp., washed with water and air-dried. The peeled topmost sheet shows a rhombic pattern indicating reflection of the sheet beneath; bar: 300 nm.

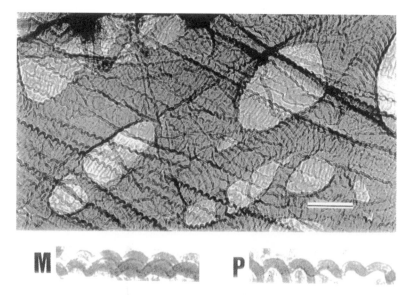

Fig. 3. Partially mineralized aged hen tarsus tendon surface replica, freeze-dried without fixation or stretching. Many straight helical filaments where helical pitches coincide with D-periodicities can be seen; bar: 500 nm. M: left-handed helix. P: right-handed helix.

tion of banding is disordered in Fig. 3 (compare with Fig. 1A). Plural helical filaments were shown at the dark edge of the fibril by tilting in the TEM (*13*). The wavy sheet must be constructed by lateral association of the helical filaments.

III. ϰ-HELICAL FILAMENT

1. Wavy Surface

The characteristic D-banding of collagen fibril seen in conventional TEM images has been thought to be due to staining differences between the "hole" and "overlap" zones and the fibril to be a thread-like column. Studies with atomic force microscopy (AFM) showed the rough surface of turkey leg (*14*) and rat tail tendon collagen fibril (*15, 16*). Our examination of these data and our own observations revealed that the collagen fibril has a sine-curved wavy surface 6–10 nm in wave height and 62–67 nm in wave length, rather than a saw-toothed surface. AFM excels in the measurement of vertical displacement, therefore the D-periodicity of collagen fiber shows its wavy roughness. As shown in Fig. 2, the identical wavy surfaces of the fiber in the top and side views indicate that the component of the fiber must be a helical element geometrically. This is shown clearly in Fig. 3. Enlargement by the plasma film coating and increased density with the mineral deposition of the filament clearly visualized such ultra-thin filaments.

An image reproduced from Fig. 3 is illustrated in Fig. 4A. This helix has no hollow in its center. It must be emphasized that this unique helix is hardly expansible and is less flexible against external force, so that it is hereafter designated as "κ-helix".

2. κ-Helical Filament Model

The identical κ-helical filaments (as shown in Fig. 4A) fit laterally to each other and form a wavy sheet. These sheets also stack on top of one another and form a closely packed structure with an isotropic cross section. As shown in Fig. 5, a stack of the wavy sheets can be transformed into various solid bodies: a

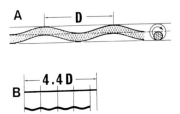

Fig. 4.   Schematic presentation of the helical filament and the relationships between linear and helical tropocollagen molecule. Scales are arbitrary. (A) The helical filament reproduced from Fig. 3. Note the absence of a hollow space at the center of the helix. (B) The 4.4-D linear length of tropocollagen molecule is shortened to 4-D in the helical filament.

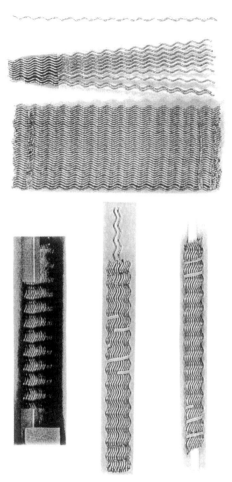

Fig. 5. Helical filament model of Type I collagen fibril. Wire models are shown in order of structural hierarchy. Top to bottom: a helical filament, periodic wavy sheet, periodic wavy-surfaced board, column and tube formed by lateral assembly of helical wires. The single helical wire in the column cannot be pulled out but it is easily screwed in or out. This hollowless helix was designated as "$\varkappa$-helix".

prism, a round column and a tube with a wavy surface in which the trough coils around them. The columnar fibril feature very often observed under TEM and scanning electron microscopy (SEM) is believed to be one of the features transformed under a particular environment.

The wire model explains satisfactorily the phase discrepancies in Fig. 2.

IV. FUNCTIONS OF COLLAGEN FIBRIL AND κ-HELICAL FILAMENT MODEL

1. Mineralization

The plasma polymerized film adheres to mineral firmly so that the mineral in tissue remains intact on the film after dissolution of the organic materials in tissue. Figures of both mineral and collagen fibrils, therefore, can be observed simultaneously in a single replica film. This is very useful for the study of biomineralization.

D-periodic and discontinuous image of mineral deposition was noted at the onset of mineralization of hen tendon as shown in Fig. 6A. This image became unclear when the replica film in a TEM was tilted around its fiber axis at an angle of 20° to a position in which the periodicity was clear and an almost even density image was obtained (Fig. 6B). High and low density areas in a D-period were interchanged after further tilting with an angle of 10–20° (picture not shown). A vertical view of wire model of a pile of ribbon of laterally associated M-helical wires and a P-helical ribbon shows D-periodic rows of spaces about 1/2D in width (Fig. 6C); these spaces are not seen when tilted at an angle of 17° (Fig. 6D). New periodic spaces appear in half phase discrepancy after additional tilting at an angle of 17° (Fig. 6F). Mineral platelet array should be declined about ±17° in a D when an even density position is assumed at 0° (Fig. 6E). This means that the crystallites tilt in the cross sectioned view. It is well known that the crystallographic c-axes of the hydroxy-apatite platelets and the collagen fiber axes are not perfectly parallel. Landis *et al.* observed the tilt of platelets within $\pm 20°$ in normally calcifying turkey tendon (*17*), and Mechanic *et al.* estimated that the average axial flex-tilt angle of collagen in skin was 16° (*18*). The above findings indicate that the crystallite array must be helical. Only the $\varkappa$-helical filament model offers possible helical room which can accommodate the crystallites of hydroxyapatite. Such nanometer spaces may provide the "critical space" in the "Nanospace Theory" for the mechanism of bio-mineralization (*19, 20*).

Dislocation of the collagen sheets and deployment into $\varkappa$-helical filaments can be seen in Fig. 3. The lateral association of the $\varkappa$-helical filaments must be maintained by hydrogen bond, electrostatic and hydrophobic interactions, and by encircling materials such as proteoglycans and minor collagens. Matrix vesicle proteinase (*21, 22*) might be involved in dislocation of the collagen sheet and dissociation to $\varkappa$-helical filaments in the process of mineralization. The three dimensional relationships between collagen sub-fibrils and mineral deposition, however, remain a riddle.

2. Tensile Strength and Metabolism

This helical filament model provides for a huge tensile strength along the longitudinal direction without any covalent bonds between those filament units laterally or between their head and tail, because it is an assembly of very long screw-in bolts. On the other hand, lateral addition or screw in of new

---

→Fig. 6. D-periodic images of mineral deposition at the onset of mineral-ization. This was taken at an accelerating voltage of 200 kV. A: D-periodic image. B: almost even density image when A was tilted at an angle of 20° from its fiber axis. C: vertical view of wire model of a pile of M-helical ribbon and P-helical ribbon. D-periodic rows of spaces of about 1/2D width can be seen. D: the periodic rows of spaces are not visible with a tilt of 17° from position C. E: declined mineral platelets with $\pm 17°$ in D-length. F: schematic presentation of the shadows cast by E. Position of even density is 0° and tiltings of $\pm 17°$ provided an alternative change of densities.

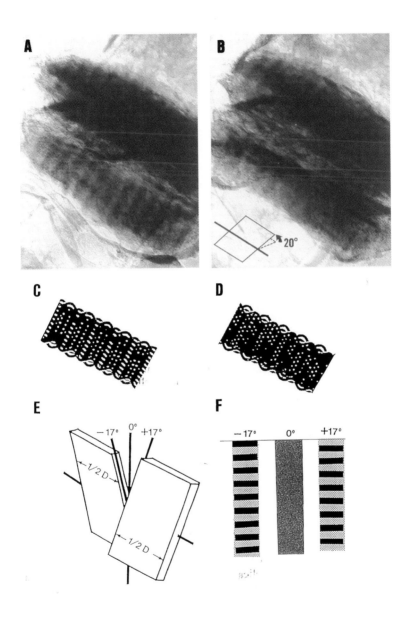

filament units to the existing assemblage is very easily performed. This model may explain the mysterious anti-parallel
assembly at the transition region of a bipolar fibril (*23-25*), and

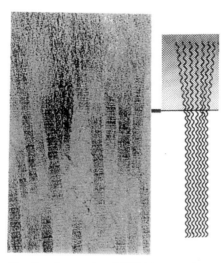

Fig. 7.   Conventional TEM image of rat molar cementum (upper half)
and periodontal ligament (lower half). Collagen fibrils of the ligament
deployed into filaments in mineralized cementum. Attached drawing
shows an interpretation according to the $\varkappa$-helical filament model.

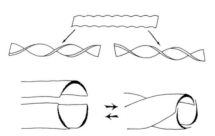

Fig. 8.   Possible twist movement of wavy ribbon of collagen. A wavy
ribbon composed of helical filaments twists when its helical pitches are
changed. Direction of rotation is reversed by unwinding of helices. A
tubing with the wavy ribbons varies in diameter.

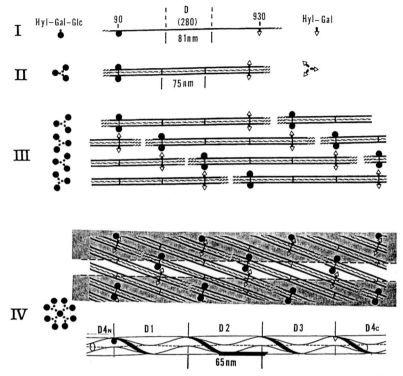

Fig. 9. Speculative explanation of molecular assemblage in "helical quarter stagger array". I. α-Chain with Hyl-Gal-Glc and Hyl-Gal. Glycosylated positions were estimated from the literature (33–36) and the distance between those positions was divided by 3. II. Tropocollagen with Hyl-glycoside residues. III. Current "linear quarter stagger array". IV. "Helical quarter stagger array". Upper is an image of radial projection in which the vertical width is determined by the circumference of the filament. The shaded areas are added for ease of understanding. Lower shows one tropocollagen molecule of a tetramer in a $\varkappa$-helical filament as a closed band; D4 is separated into two parts at both ends. Details in text. Small closed circles: cross section of peptide. Large closed circles: Hyl-Gal-Glc. Small open triangles: Hyl-Gal.

the collagen fibrogenesis and turnover without apparent change of the external form of the fiber. Collagen fibrils of periodontal ligament, for example, are deployed into filaments in cementum where the collagen is metabolized very quickly without apparent change in the mineral phase. An individual $\varkappa$-helical filament in cementum is easily exchangeable by rotation but it cannot be pulled out when filaments are tied in a bundle as shown in Fig. 7. This concept may be applicable to the biological connection of ligament and artificial implant (26).

3. Movability of Collagen

The diversified properties of collagen might be due to delicate changes of the micro-environment (e.g., pH, ionic strength, or temperature) which lead to changes in the helical pitch and helix diameter. In a model experiment, thermal deformation of the constituted helical plastic wire in a wavy ribbon causes it to roll around its long axis in a direction opposite to that of the rotation of the helix of the constituent filament (Fig. 8). This twist movement is quite different from the contraction of muscle. An elastic tubing or bag with such plastic ribbons acted as a wringer in hot water. This may explain the partial inside-out observation (27, 28) and the disappearance of D-bandings of the collagen fibril after dissociative treatment (8, 29), as well as collagen-involved movement and maintenance of posture in certain echinoderms (30) and in nematocysts of hydra (31, 32). If this possible twist movement of the collagen fibril is inherent in some mammalian tissues, cellular control of the diameter of the peripheral capillary could be expected as shown in Fig. 8, and collagen fibrils may play an active role in the homeostasis of the interstitial fluid as a dynamic structure rather than a static one. This movement is inhibited by the coexistence of the helices of P and M, and it is shown in tendon which has half wavelength discrepancies (Fig. 2). In dermal collagen fibrils, pairs of diagonal clefts in the opposite direction (2, 3, 5) were observed after dissociative treatment and this fact may support the above interpretation.

V. MOLECULAR ASSEMBLY IN $\varkappa$-HELIX

How does tropocollagen molecule assemble in the $\varkappa$-helix ? The method applied here provides only surface morphology but no information about the intra-filament. The length of the tropocollagen molecule in a linear state is 280–300 nm. According to the "quarter stagger model", this value was expressed as 4.4 D and the insufficient length of 0.6 D in 5 D was called the "hole or gap". The length required along this helical filament, however, is about 15% more than that of the helix. This means that the length of the above mentioned helix of tropocollagen (240–260 nm) corresponds to four-fold the D-period (60–65 nm) (Fig. 4B). Thus the 0.6 D gap is not significant.

The authors tentatively speculate the following (Fig. 9):

I) α-Chain: each α-chain has hydroxylysine glycosides near the N- and C-terminals when it is biosynthesized. One third of the fixed distance between those sites of glycosylation is about 81 nm (about 280 amino acid residues), and this length should be viewed as the pre-coiled length of D in α-chain.

II) Tropocollagen: triple helix of α-chain protrudes hydroxylysine glycosides in a triangle near both ends; the length of D is about 75 nm.

III) Current linear quarter stagger array: large hydroxylysine glycosides interfere with the close lateral assembly of the linear tropocollagen molecules.

IV) Helical quarter stagger array: possible array of tropocollagen in a helix could avoid such steric hindrance of hydroxylysine glycosides. Tetragonal assemblage of tropocollagen molecules in which one of the three hydroxylysine glycosides is positioned in the center and the other two are around it helically on the surface of the tetramer suggests a possible basic unit of the $\varkappa$-helical filament of collagen. The length of D is about 65 nm.

The above speculations must be examined by several different methods.

As mentioned in the introduction, a supra-helix model in

the third or the fourth level requires a continuous increase of the net length corresponding to the D-distance and of tilt of molecule (the angle of the spiral substructure (7)) from the center to the periphery of the structure, so that those models are unlikely. Twist movement of a pile of collagen sheets mentioned in section IV-3 may explain the dissociated features of fibrils (diagonally oriented clefts (5)) subsequent to exposure of the tissues to dissociative solvents which cause x-helix to unwind.

CONCLUSION

The wavy surface (6–10 nm in roughness) of collagen fibril is inexplicable using the current linear model of tropocollagen. Transmission electron micrographs of the plasma polymerized film replica of t-BuOH lyophilized mineralizing hen tendon showed: pile and dislocation of peeled, thin D-pitched wavy sheets, and D-pitched helical filaments at the onset of calcium-phosphate deposition. The first direct evidence of the constituting unit of helical element in collagen fiber leads us to postulate the following hierarchical structure: 1) a D-pitched helical filament approximately 10 nm in diameter composed of tropocollagen molecules, 2) a wavy sheet with D-periods composed of lateral association of the helical filaments, and 3) a wavy surfaced piled mass of the sheets which is transformable to board, rod, column and tube.

This helical model well explains the helical row of crystallite, the huge tensile strength, turnover without loss of strength and movability of collagen fibrils. The results obtained by the plasma polymerization film replica method, however, remain to be confirmed by other techniques based on different principles.

SUMMARY

Transmission electron micrographs of the plasma polymerized film replica of chicken and hen tendon collagen surface revealed: (i) 65 nm pitched (D-periodicity) hollowless helical thin filaments composed of tropocollagen molecules, (ii) a wavy sheet with D-wave length composed of lateral association of the helical filaments, and (iii) a wavy-surfaced piled mass of the sheets which is transformable to board, rod, column and tube. This helical filament model explains the characteristic features of tissue collagen, its mineral deposition, its tensile strength, its natural turnover, and its possible movability. The hollowless helix termed $\varkappa$-helix has unique characteristics of being less expansible and less flexible.

Acknowledgment

This study was supported by a scientific grant from the Japanese Ministry of Education, Science, Sports and Culture (No. 05454504, 1994).

REFERENCES

1. Piez, K.A. and Miller, A. (1974) *J. Supramol. Struct.* **2**, 121-137.
2. Bouteille, M. and Pease, D.C. (1971) *J. Ultrastruct. Res.* **35**, 314-338.
3. Pease, D.C. and Bouteille, M. (1971) *J. Ultrastruct. Res.* **35**, 339-358.
4. Belton, J.C., Michaeli, D., and Fudenberg, H.H. (1975) *Arthritis Rheum.* **18**, 443-450.
5. Lillie, J.H., MacCallum, D.K., Scaletta, L.J., and Occhino, J.C. (1977) *J. Ultrastruct. Res.* **58**, 134-143.
6. Ruggeri, A., Benazzo, F., and Reale, E. (1979) *J. Ultrastruct. Res.* **68**, 101-108.
7. Nimni, M.E. and Harkness, R.D. (1988) In *Collagen* (Nimni, M.E., ed.), vol. I, pp. 17-20, CRC Press, Boca Raton, Florida.
8. Katsura, N., Tanaka, O., and Yokoyama, M. (1991) *Connect. Tissue* **22**, 92-98.
9. Tanaka, A. and Sekiguchi, Y. (1986) Proc. Xlth Int. Cong. on Electron Microsc., Kyoto, pp. 2201-2202.
10. Tanaka, A., Yamaguchi, M., and Hirano, T. (1988) Proc. 46th Ann. Meeting Electron Microsc. Soc. America, pp. 414-415.
11. Tanaka, A. (1991) *J. Electron Microsc.* **26**, 141-150 (in Japanese).
12. Yamaguchi, M., Tanaka, A., and Suzuki, T. (1992) *J. Electron Microsc.* **41**, 7-13.
13. Katsura, N., Ono, T., and Tanaka, K. (1996) *Connect. Tiss.* **28**, 165-171 (in Japanese).

14. Lees, S., Prostak, K.S., Ingle, V.K., and Kjoller, K. (1994) *Calcif. Tiss. Int.* **55**, 180–189.
15. Baselt, D.R., Revel, J., and Baldeschwieler, J.D. (1993) *Biophys. J.* **65**, 2644–2655.
16. Revenko, I., Sommer, F., Minh, D.T., Garrone, R., and Franc, J.M. (1994) *Biol. Cell* **80**, 67–69.
17. Landis, W.J., Song, M.J., Leith, A., McEwen, L., and McEwen, B.F. (1993) *J. Struct. Biol.* **110**, 39–54.
18. Mechanic, G.L., Katz, E.P., Henmi, M., Noyes, C., and Yamauchi, M. (1987) *Biochemistry* **26**, 3500–3509.
19. Katsura, N. (1990) *Dentistry Jpn.* **27**, 57–63.
20. Katsura, N. (1991) In *Mechanisms and Phylogeny of Mineralization in Biological Systems* (Suga, S. and Nakahara, H., eds.), pp. 193–197, Springer-Verlag, Tokyo.
21. Katsura, N. and Yamada, K. (1986) *Bone* **7**, 137–143.
22. Katsura, N., Baba, T., Tanaka, O., and Sadamatsu, T. (1992) *Bone and Mineral* **17**, 303–304 (abstract).
23. Prockop, D.J. and Hulmes, D.J.S. (1994) In *Extracellular Matrix Assembly and Structure* (Yurchenco, P.D., Birk, D.E., and Mercham, R.P., eds.), pp. 61–66, Academic Press Inc., San Diego.
24. Kadler, K.E., Holmes, D.F., Trotter, J.A., and Chapman, J.A. (1996) *Biochem. J.* **316**, 1–11.
25. Birk, D.E., Hahn, R.A., Linsenmayer, C.Y., and Zycband, E.I. (1996) *Matrix Biol.* **15**, 111–118.
26. Katsura, N., Ono, T., and Tanaka, K. (1996) In *Bioceramics* (Kokubo, T., Nakamura, T., and Miyaji, F., eds.), vol. 9, pp. 85–88, Pergamon, Amsterdam.
27. Holmes, D.F., Mould, A.P., and Chapman, J.A. (1991) *J. Mol. Biol.* **220**, 111–123.
28. Romanic, A.M., Adachi, E., Hojima, Y., Engel, J., and Prockop, D.J. (1992) *J. Biol. Chem.* **267**, 22265–22271.
29. Petkov, R. (1978) *Anat. Anz.* **144**, 485–501.
30. Hidaka, M. and Takahashi, K. (1983) *J. Exp. Biol.* **103**, 1–14.
31. Holstein, T.W., Benoit, M., Herder, G.V., Wanner, G., David, C.N., and Gaub, H.E. (1994) *Science* **265**, 402–404.
32. Holstein, T.W. (1995) *Biologie in unserer Zeit* **25**, 161–169.
33. Butler, W.T. (1970) *Biochemistry* **9**, 44–50.
34. Isemura, M., Takahashi, K., and Ikenaka, T. (1976) *J. Biochem.* **80**, 653–658.
35. Light, N.D. and Bailey, A.J. (1980) *Biochem. J.* **189**, 111–124.
36. Tenni, R., Valli, M., Rossi, A., and Cetta, G. (1993) *Am. J. Med. Genet.* **45**, 252–256.

Extracellular Matrix-Cellular Interaction: Molecules to Diseases (Y. Ninomiya et al., eds.), pp. 203–215,
Japan Sci. Soc. Press, Tokyo/S. Karger, Basel (1998)

Collagen Gene Expression in Vascular Smooth Muscle Cells: Inhibition by B-Myb

KYRIAKOS E. KYPREOS AND GAIL E. SONENSHEIN

Department of Biochemistry, Boston University School of Medicine, Boston, Massachusetts 02118, U.S.A.

The extracellular matrix (ECM) within an artery consists primarily of collagens, elastin and proteoglycans, and functions as a structural framework while allowing for cell layering. The major fibrillar collagen species found in a blood vessel are the group I, interstitial collagen types I, III, and V/XI (*1, 2*). Types I and III collagen are the most abundant; a significant level of type V/XI collagen is also expressed. The development of an atherosclerotic lesion involves a fairly extensive increase in the amounts of all of the fibrillar collagens found in the artery. Types I and III collagen are elevated to a similar extent (*3–6*), while the relative ratio of type V/XI collagen to types I and III collagen increases as much as two-fold (*5–9*). During this process, smooth muscle cells (SMCs) undergo a phenotypic switch, losing their contractile phenotype and becoming more proliferative and highly synthetic (*10, 11*). Within the intima of the vessel wall, SMCs undergo some initial rounds of proliferation, which are followed by a synthetic period where they deposit ECM components (*12–16*). Thus, understanding the

203

pathways involved in the regulation of collagen gene expression by SMCs should provide valuable information on the molecular mechanisms that underlie the development of an atherosclerotic lesion.

Due to the complexity of the factors involved in atherogenesis, there is no perfect model for studying atherosclerosis. However, cultures of SMCs derived by explant or enzyme dispersion from the medial layer of normal arteries have been used extensively for studying SMC behavior. Vascular SMCs in culture display a synthetic, highly proliferative phenotype, similar to the intimal SMCs found in atherosclerotic lesions. For example, just as SMCs in atherosclerotic lesions produce more collagen than the normal contractile SMC, cultured vascular SMCs produce as much as 30 times more fibrillar collagens than contractile cells (*17, 18*). Several laboratories, including our own, have employed cultured vascular SMCs as a model system to study the regulation of collagen gene expression and have found an inverse correlation between proliferative rate and collagen production. Our recent studies indicate a major role for B-Myb, a member of the *myb* gene family expressed in a growth-related fashion, in control of matrix genes by the vascular SMC. In this paper, first we present a review of the literature on the relationship between SMC growth and collagen expression, then introduce B-Myb and finally summarize our recent finding on the role of this transcription factor in down-regulation of matrix gene expression.

I. COLLAGEN GENE EXPRESSION BY VASCULAR SMOOTH MUSCLE CELLS: INVERSE RELATIONSHIP TO GROWTH

During early stages of development of the vessel, as well as during growth in culture, SMCs are in the synthetic state. Synthetic SMCs synthesize large amounts of collagens and other extracellular matrix components, as well as enzymes involved in their deposition. Type I collagen, which is composed of a heterotrimer of two $\alpha 1(I)$ and one $\alpha 2(I)$ chains, is the most abundant collagen species produced by SMCs. Type III collagen, a homotrimer of $\alpha 1$(III) chains, is the second most abundant species. SMC cultures also make appreciable amounts of type V/XI collagen (5, 19–23). There are three type V collagen chains, $\alpha 1$, $\alpha 2$, and $\alpha 3$ (24). Type XI collagen was known to be closely related structurally to type V collagen and to share similar properties (25, 26). These two species were originally denoted as different types based on their apparent separate sites of synthesis. More recently, however, type XI chains were detected in bone in trimeric structures with type V (27), and mRNA for various type XI chains were found in the placenta (28) and in rhabdomyosarcoma cells (29). Only the $\alpha 1$ chain of type XI collagen is expressed in bovine vascular SMCs (23) and it can form heterotrimers with the $\alpha 1$ and $\alpha 2$ chains of type V (27, 29). Thus, it is now referred to as the type V/XI collagen family. Type V/XI collagen has been reported to be important in regulating the diameter of type I fibers (30–32), and our recent work indicates that type V/XI collagen is laid down within heterotypic fibrils by bovine aortic SMCs in culture (manuscript in preparation). Thus, increased expression of this collagen can have far-reaching effects on fibril formation and, by extension, the architecture of the vessel wall.

Synthesis of collagen protein by SMCs in culture occurs in an inverse pattern with respect to the proliferative rate of these cells. Work from several laboratories, including our own, has shown that types I, III, and V/XI collagen mRNA levels are relatively low when SMCs are at subconfluence and growing exponentially, and the levels increase as cells become more dense (17, 22, 23, 33, 34). Thus, after pulmonary artery or aortic SMCs were seeded at low density, only low collagen types I, III, and V/XI mRNA levels were detected. With time in culture as cells became more dense, their growth rate slowed, and the mRNA levels of these collagen chains increased significantly (22, 23, 34). Collagen protein synthesis reflected the increase in mRNA levels and accumulation of connective tissue also increased steadily as cells became more dense (33).

Serum-starvation has been shown to render SMCs quiescent (*35*). Within 12 hr after switching SMC cultures to media containing 0.5% serum, types I and III collagen mRNA levels increased (*36*). By 48 hr, a 2- to 15-fold induction was seen in quiescent compared to exponentially growing cells. Conversely, serum add-back resulted in re-entry into the cell cycle and a concomitant reduction in types I and III collagen mRNA levels between 12 and 24 hr following serum addition. These changes in mRNA levels were due, in part, to a drop in the rates of gene transcription (*36*). Similar trends were observed when SMC cultures were rendered quiescent by other means such as essential amino acid deprivation (*37*).

Basic fibroblast growth factor (bFGF), a member of the fibroblast growth factor family, is another growth factor implicated in atherogenesis. It is a potent inducer of SMC proliferation, both *in vivo* and *in vitro* and it is synthesized by a variety of cells in culture, including SMCs (*38-43*). Treatment of vascular SMCs with bFGF decreases expression of types I and III collagen (*44, 45*). Recently, we demonstrated a significant decrease in the levels of type V collagen steady-state mRNA and protein in vascular SMCs as a result of treatment with bFGF (*46*). The bFGF-induced decrease in $\alpha 2(V)$ collagen mRNA levels was mediated through a decrease in the rate of procollagen gene transcription.

In addition, expression of other matrix genes such as elastin, and enzymes involved in matrix deposition, such as lysyl oxidase, varies inversely with SMC growth state (*47*). A similar relationship has been seen with other cell types as well. For example, in density arrested nondividing human fetal lung fibroblasts, Miskulin *et al.* (*48*) found that $\alpha 1(I)$ and $\alpha 2(I)$ and $\alpha 1(III)$ collagen mRNA levels were significantly higher than those in logarithmically growing cells. Viral transformation of fibroblasts, which enhances their proliferative capacity, results in a substantial decrease in type I collagen mRNAs and protein synthesis (*49*). Overexpression of oncogenes such as *mos* or *ras* in fibroblastic cells has also been found to decrease collagen expression (*50, 51*). Furthermore, expression of matrix-degrading enzymes, such as transin/stromelysin and collagenase has been shown to relate directly to the proliferative state (*52, 53*). Thus, in many cells, including SMCs, overall regulation of connective tissue metabolism appears linked to the cellular growth state.

II. THE B-*myb* GENE: A MEMBER OF THE *myb* GENE FAMILY

In contrast to this inverse pattern of SMC growth and extracellular matrix production, expression of oncogenes often relates directly with the rate of proliferation. We earlier found that SMCs express B-*myb*, a member of the *myb* oncogene family (*54*). The B-*myb* gene was isolated based on its homology (approximately 90%) with the c-*myb* oncogene in its DNA binding region (*55*). The human B-*myb* mRNA is approximately 3.3-kilobases (kb) and codes for a B-Myb protein of 704 amino acids that has a molecular weight of approximately 93 kDa (*56*). B-Myb protein is capable of binding to the consensus c-Myb binding site (MBS) [YGRC(A/C/G)GTT(G/A)] (*57*), although R is preferably T/C for c-Myb and A/C for B-Myb. B-Myb was further found to regulate several reporter constructs that are known to be controlled by c-Myb independent of DNA binding, such as the DNA polymerase α promoter (*58, 59*). B-Myb has also been reported to recognize a second specific consensus sequence [PuPuAAANYG] (*58, 60*). Thus, B-*myb* is perhaps able to regulate a second subset of genes in addition to those in common with c-*myb*.

The expression pattern of B-*myb* appears to be broader than that of c-*myb* and of A-*myb*, another member of the *myb* family (*55*), suggesting B-*myb* has a more general functional role than other members (*54, 55, 61*). B-*myb* gene expression has always been found to correlate with cell proliferation. In synchronized 3T3 cells, stimulated lymphocytes, as well as SMCs, B-*myb* RNA expression was found to be specific to the G1/S phase transition (*54, 61–64*). For example, we found that B-*myb* RNA levels were high in actively proliferating cells and significantly lower in quiescent, serum deprived or confluent SMCs (*54*). Antisense experiments with fibroblasts and lymphocytes have demonstrated that inhibition of B-*myb* gene expression arrests cell cycle progression (*65, 66*). In contrast, our results indicate that B-*myb* does not play a similar role in control of SMC proliferation (*67*). For example, co-microinjection of quiescent SMCs with vectors expressing the competence factor c-*myc* (*68, 69*) and B-*myb* failed to push the cells into S phase, whereas up to 75% of cells entered S phase within 20 hr following co-microinjection with vectors expressing c-*myc* and c-*myb* or A-*myb* (*67*). Thus, B-*myb* does not appear to control progression of SMCs through the cycle, but is itself growth regulated.

Detailed analysis of the functional domains of B-Myb has been done. B-*myb* does not contain the acidic transactivation domain that is conserved between c-*myb* and A-*myb*. B-Myb has an acidic region (amino acids 207–273) that has been shown to be able to impart transactivation properties on the B-Myb protein; however, this region is not conserved with A- and c-*myb* and several groups have demonstrated that B-*myb* is a transcriptional repressor rather than an activator (58, 59, 70). Tashiro *et al.* (*71*) recently reported cell-type specific effects of B-*myb*. In transient transfection analysis, B-*myb* was shown to be a positive transactivator in several cell lines tested, including HeLa cells, whereas it was a transcriptional repressor of c-Myb-mediated transactivation in fibroblast and macrophage cell lines. The C-terminal conserved region has been postulated by this group to be responsible for binding to another protein, and it is this heterodimer that is then suggested to be a positive transactivator (*71*). In SMCs, we have found B-*myb* negatively affects multimerized MBS element driven heterologous promoter constructs, indicating it has a transcriptional repressor function in these cells (*54*).

III. B-*myb* REPRESSES MATRIX GENE TRANSCRIPTION

Given that B-Myb is a negative regulator of transcription in SMCs, and that it is expressed selectively in cells that are actively proliferating, we were intrigued by the results of a computer analysis indicating that several genes encoding ECM proteins contained putative MBS. In particular, analysis of the promoters of the $\alpha 1$ and $\alpha 2$ chains of type I collagen genes revealed several putative elements with homology to an MBS. Thus we set out to test the ability of B-Myb to down-regulate collagen gene transcription.

1. B-Myb Represses Type I Collagen Promoter Activity

Co-transfection analysis was performed to test the effects of B-*myb* expression on type I collagen promoter activity in SMCs. An $\alpha 2$(I) collagen promoter pMS-3.5/CAT construct, which contains 3.5 kb of sequence upstream of the promoter and 58 bp of exon 1 driving the chloramphenicol acetyltransferase (CAT) reporter gene (*72*) was used. The activity of the pMS-3.5/CAT vector displayed an average decrease in three experiments of $82 +/- 10.8\%$ upon co-transfection with a bovine B-*myb* vector, and similar down-regulation was seen with a human B-*myb* expression vector (*54*). To assess the effects of B-Myb expression on activity of the $\alpha 1$(I) promoter, the pOB3.6, which contains 3.6 kb of the $\alpha 1$(I) collagen promoter plus all of exon 1 and intron 1 upstream of the CAT reporter gene, was used in co-transfection analysis. The activity of pOB3.6 vector was down-regulated by approximately 92 and 79%, respectively, upon co-transfection with bovine or human B-*myb* expression vectors (*54*). Therefore, B-Myb expression leads to the down-regulation of activity of the promoters of the genes encoding both chains of type I collagen. More recently, using antisense and transfection analyses, we have demonstrated that B-Myb mediates the signals leading to down-regulation of type I collagen promoter activity seen in vascular SMCs upon bFGF treatment (manuscript in preparation).

2. B-Myb Down-regulates Type V Collagen Promoter Activity

Since expression of type V/XI collagen also varies inversely with growth rate of bovine aortic SMCs as influenced by cell density and serum growth factors (23), we have more recently performed similar co-transfection analysis using the promoter of the $\alpha 2$ chain of type V collagen. This promoter has been characterized independently by two groups, and the start site of transcription mapped (73, 74). A series of nested deletion upstream/promoter-CAT reporter constructs were kindly provided by F. Ramirez (Mt. Sinai Medical School, N.Y.). These constructs contain approximately 150 bp of exon 1 sequences, and upstream sequences ranging from 2,350 bp (termed pST-2.5COL5A2/CAT) to 100 bp (termed pST0.3COL5A2/CAT) (74). B-myb co-expression down-regulated the activity of all $\alpha 2$ type V collagen promoter constructs. Thus, the region of this promoter mediating negative regulation by B-Myb maps to a 265 bp fragment surrounding the start site of transcription. Furthermore, using lipofectamine to achieve higher levels of transfection efficiency, we have recently demonstrated that B-Myb expression decreases endogenous $\alpha 2(V)$ mRNA levels (manuscript in preparation).

3. B-Myb Negatively Regulates Elastin Promoter Activity

While previous studies employing bovine aortic SMCs failed to detect elastin deposition (75), more recently we have determined that these cells contain functional elastin mRNA, as judged by immunoprecipitation of in vitro translated proteins. Thus, in collaboration with P. Stone and S. Morris (Boston University Medical School), we re-examined the question of elastin deposition using a more sensitive technique of amino acid measurement and electron microscopy, and have demonstrated low levels of elastin protein deposition by bovine aortic SMCs. Since the level of elastin mRNA expression by aortic

SMCs varied inversely with SMC proliferative state, in collaboration with J. Foster (Boston University Medical School), we tested the ability of a B-*myb* expression vector to negatively regulate the elastin promoter. A 3.0-fold (67%) decrease in activity of the elastin promoter was observed upon expression of B-*myb*. This result suggests that B-Myb may be involved in the low level of elastin gene expression observed in actively proliferating, subconfluent SMCs.

Overall our findings demonstrate that B-*myb* gene expression has a negative regulatory effect on the activity of a number of matrix gene promoters in vascular SMCs, suggesting a major role for B-Myb in matrix metabolism.

SUMMARY

Collagen gene expression is strongly linked to the proliferative state of the vascular SMC. Agents that promote cell growth decrease collagen production, suggesting that a growth-related gene(s) mediates signals involved in the formation and remodeling of the extracellular matrix. One candidate is B-*myb*, a member of the *myb* gene family. In vascular SMCs, B-*myb* inhibits collagen and elastin gene expression without promoting progression through the cell-cycle. Thus, B-*myb* plays a significant role in the regulation of matrix metabolism by the SMC in culture. These observations suggest that B-*myb* is an important intracellular mediator of signals relating SMC growth state to the level of matrix gene expression, specifically in driving the down-regulation of matrix gene expression in proliferating SMCs. Of further interest, our more recent experiments suggest that B-*myb* mediates signals triggered by growth factors, such as bFGF. These factors have been shown to alter collagen synthesis *in vivo* as well as *in vitro*. Experiments are in progress using transgenic mouse models to test whether B-Myb plays a similar role within the vessel wall *in vivo*. If transgenic animals display greatly reduced levels of matrix proteins when high levels of B-Myb are achieved, such down-regulation would suggest that

B-Myb functions as an anti-fibrotic gene in vascular SMCs. As a potent inhibitor of collagen expression, B-*myb* would represent an important candidate gene for future gene therapy experiments. This finding would have important ramifications for our understanding of control and treatment of diseases involving excess matrix deposition by the SMC within the vessel wall, in particular the major diseases of atherosclerosis and restenosis. Additionally, future studies of interest might be directed to other very diverse fibrotic diseases involving SMCs, such as fibroids (76), the major cause of surgery (hysterectomy) among women in the United States, and impotence (77). Lastly, since B-*myb* is widely expressed in many cell types, the possibility that it similarly affects connective tissue expression in other cells deserves additional investigation.

Acknowledgment

This work was supported by National Institutes of Health Grants HL13262 and HL57326.

REFERENCES

1. Moss, M.S. and Benditt, E.P. (1970) *Lab. Invest.* **23**, 231–245.
2. Vuorio, E. and deCrombrugghe, B. (1990) *Annu. Rev. Biochem.* **59**, 837–872.
3. Park, H.-S., Kniep, A.C., Smith, S.C. *et al.* (1990) *Connect. Tiss. Res.* **25**, 67–76.
4. Shekhonin, B.V., Domogatsky, S.P., Idelson, G.L., Kotelianski, V.E., and Rukosuev, V.S. (1986) *Atherosclerosis* **67**, 9–16.
5. Murata, K., Motoyama, T., and Kotake, C. (1986) *Atherosclerosis* **60**, 251–262.
6. Morton, L.F. and Barnes, M.J. (1982) *Atherosclerosis* **42**, 41–51.
7. Ooshima, A. (1981) *Science* **213**, 666–668.
8. Ooshima, A. and Muragaki, Y. (1990) *Ann. N.Y. Acad. Sci.* **598**, 582–584.
9. Katsuda, S., Okada, Y., Minamoto, T., Oda, Y., Matsui, Y., and Nakanishi, I. (1992) *Arterioscler. Thromb.* **12**, 494–502.
10. Mosse, P.R.L., Campbell, G.R., Wang, Z.L., and Campbell, J.H. (1985) *Lab. Invest.* **53**, 556–562.
11. Mosse, P.R.L., Campbell, G.R., and Campbell, J.H. (1986) *Atherosclerosis*

6, 664–669.

12. Poole, J.C.F., Cromwell, S.B., and Benditt, E.P. (1971) *Am. J. Pathol.* **62**, 391–413.

13. Schwartz, S.M., Reidy, M., and Clowes, A. (1985) *Ann. N.Y. Acad. Sci.* **454**, 292–304.

14. Gordon, D., Reidy, M., Benditt, E., and Schwartz, S. (1990) *Proc. Natl. Acad. Sci. USA* **87**, 4600–4604.

15. Ross, R. (1993) *Nature* **362**, 801–809.

16. Strauss, B.H., Chisholm, R.J., Keeley, F.W., Gottlieb, A.I., Logan, R.A., and Armstrong, P.W. (1994) *Circ. Res.* **75**, 650–658.

17. Ang, A.H., Tachas, G., Campbell, J.H., Bateman, J.F., and Campbell, G.R. (1990) *Biochem. J.* **265**, 461–469.

18. Okada, Y., Katsuda, S., Matsui, Y., Minamoto, S., and Ishizake, R. (1989) *Acta Pathol. Jpn.* **39**, 15–22.

19. Mayne, R., Vail, M.S., and Miller, E.J. (1978) *Biochemistry* **258**, 446–452.

20. Fessler, L.I., Kumamoto, C.A., Meis, M.E., and Fessler, J.H. (1981) *J. Biol. Chem.* **256**, 9640–9645.

21. Furuto, D.K. and Miller, E.J. (1981) *Biochemistry* **20**, 1635–1640.

22. Liau, G. and Chan, L.M. (1989) *J. Biol. Chem.* **264**, 10315–10320.

23. Brown, K.B., Lawrence, R.L., and Sonenshein, G.E. (1991) *J. Biol. Chem.* **266**, 23268–23273.

24. Fessler, J.H. and Fessler, L.I. (1987) In *Structure and Function of Collagen Types* (Mayne, R. and Borgeson, R.E., eds.), pp. 81–103, Academic Press Inc., San Diego.

25. Eyre, D. and Wu, J.-J. (1987) In *Structure and Function of Collagen Types* (Mayne, R. and Borgeson, R.E., eds.), pp. 261–281, Academic Press Inc, San Diego.

26. Morris, N.P., Watt, S.L., Davis, J.M., and Bachinger, H.P. (1990) *J. Biol. Chem.* **265**, 10081–10087.

27. Niyibizi, C. and Eyre, D.R. (1989) *FEBS Lett.* **242**, 314–318.

28. Bernard, M., Yoshioka, H., Rodriguez, E. *et al.* (1988) *J. Biol. Chem.* **263**, 17159–17166.

29. Kleman, J.-P., Hartmann, D.J., Ramirez, F., and van der Rest, M. (1992) *Eur. J. Biochem.* **210**, 329–335.

30. Birk, D.E., Fitch, J.M., Babiarz, J.P., and Linsenmayer, T.F. (1988) *J. Cell Biol.* **106**, 999–1008.

31. Shea-McLaughlin, J., Linsenmayer, T.F., and Birk, D.E. (1989) *J. Cell. Sci.* **94**, 371–379.

32. Adachi, E. and Hayashi, T. (1986) *Connect. Tiss. Res.* **14**, 257–266.

33. Beldekas, J., Gerstenfeld, L., Sonenshein, G.E., and Franzblau, C. (1982) *J. Biol. Chem.* **257**, 12252–12256.

34. Stepp, M.A., Kindy, M., Franzblau, C., and Sonenshein, G.E. (1986) *J. Biol. Chem.* **261**, 6542–6547.

35. Kindy, M. and Sonenshein, G.E. (1986) *J. Biol. Chem.* **261**, 12865–12868.
36. Kindy, M.S., Chang, C.-J., and Sonenshein, G.E. (1988) *J. Biol. Chem.* **263**, 11426–11480.
37. Chang, C.J. and Sonenshein, G.E. (1991) *Matrix* **11**, 242–251.
38. Baird, A., Mormede, P., and Bohlem, P. (1985) *Biochem. Biophys. Res. Commun.* **126**, 358–364.
39. Vlodavsky, I., Folkman, J., Sullivan, R. *et al.* (1987) *Proc. Natl. Acad. Sci. USA* **84**, 2292–2296.
40. Vlodavsky, I., Fridman, R., Sullivan, R., Sasse, J., and Klagsbrun, M. (1987) *J. Cell. Physiol.* **131**, 402–408.
41. Schweigerer, L., Neufeld, G., Friedman, J., Abraham, J.A., Fiddes, J.C., and Gospodarowicz, D. (1987) *Nature* **325**, 257–259.
42. Mansson, E.P., Malark, M., Sawada, H., Kan, M., and McKeehan, W.L. (1990) *In Vitro* **26**, 209–212.
43. Speir, E., Sasse, J., Shrivastav, S., and Casscells, W. (1991) *J. Cell. Physiol.* **147**, 362–373.
44. Kennedy, S.H., Qin, H., Lin, L., and Tan, E.M.L. (1995) *Am. J. Pathol.* **146**, 764–771.
45. Majors, A. and Ehrhart, L.A. (1993) *Arterioscler. Thromb.* **13**, 680–686.
46. Kypreos, K.E. and Sonenshein, G.E. (1997) *J. Cell. Biochem.*, in press.
47. Toselli, P., Faris, B., Sassoon, D., Jackson, B.A., and Franzblau, C. (1992) *Matrix* **12**, 321–332.
48. Miskulin, M., Dalgleish, R., Kluve-Beckerman, B. *et al.* (1986) *Biochemistry* **25**, 1408–1413.
49. Adams, S.L., Boettiger, D., Focht, R., Holtzer, H., and Pacifici, M. (1982) *Cell* **30**, 373–384.
50. Setoyama, C., Liau, G., and de Crombrugghe, B. (1995) *Cell* **41**, 201–209.
51. Slack, J., Parker, M.I., Robinson, V., and Bornstein, P. (1992) *Mol. Cell. Biol.* **12**, 4714–4723.
52. Tolstoshev, P., Berg, R.A., Rennard, S.I., Bradley, K.H., Trapnell, B.C., and Crystal, R.G. (1981) *J. Biol. Chem.* **256**, 3135–3140.
53. Kerr, L.D., Holt, J.T., and Matrisian, L.M. (1990) *Science* **242**, 1424–1426.
54. Marhamati, D.J. and Sonenshein, G.E. (1996) *J. Biol. Chem.* **271**, 3359–3365.
55. Nomura, N., Takahashi, M., Matsui, M. *et al.* (1988) *Nucl. Acids Res.* **16**, 11075–11089.
56. Arsura, M., Luchetti, M.M., Erba, E., Golay, J., Rambaldi, A., and Introna, M. (1994) *Blood* **83**, 1778–1790.
57. Howe, K.M. and Watson, R.J. (1991) *Nucl. Acids Res.* **19**, 3913–3919.
58. Mizuguchi, G., Nakagoshi, H., Nagase, T. *et al.* (1990) *J. Biol. Chem.* **265**, 9280–9284.
59. Watson, R., Robinson, C., and Lam, E. (1993) *Nucl. Acids Res.* **21**, 267–272.

60. Nakagoshi, H., Kanei-Ishii, C., Sawazaki, T., Mizuguchi, G., and Ishii, S. (1992) *Oncogene* **7**, 1233–1239.

61. Lam, E.W.-F., Robinson, C., and Watson, R.J. (1992) *Oncogene* **7**, 1885–1890.

62. Golay, J., Capucci, A., Arsura, M., Castellano, M., Rizzo, V., and Introna, M. (1991) *Blood* **77**, 149–158.

63. Reiss, K., Ferber, A., Travali, S., Porcu, P., Phillips, P., and Baserga, R. (1991) *J. Cell. Physiol.* **51**, 5997–6000.

64. Lam, E. and Watson, R. (1993) *EMBO J.* **12**, 2705–2713.

65. Arsura, M., Introna, M., Passerini, F., Mantovani, A., and Golay, J. (1992) *Blood* **79**, 2708–2716.

66. Sala, A. and Calabretta, B. (1992) *Proc. Natl. Acad. Sci. USA* **89**, 19415–19419.

67. Marhamati, D.J., Bellas, R.E., Arsura, M., Kypreos, K.E., and Sonenshein, G.E. (1997) *Mol. Cell. Biol.* **17**, 2448–2457.

68. Kelly, K., Cochran, B., Stiles, C., and Leder, P. (1983) *Cell* **35**, 603–610.

69. Campisi, J., Gray, H., Pardee, A.B., Dean, M., and Sonenshein, G.E. (1984) *Cell* **36**, 241–247.

70. Foos, G., Grimm, S., and Klempnauer, K.H. (1992) *EMBO J.* **11**, 4619–4629.

71. Tashiro, S., Takemoto, Y., Handa, H., and Ishii, S. (1995) *Oncogene* **10**, 1699–1707.

72. Boast, S., Su, M.-W., Ramirez, F., Sanchez, M., and Avvedimento, E.V. (1990) *J. Biol. Chem.* **265**, 13351–13356.

73. Greenspan, D.S., Lee, S.-T., Lee, B.-S., and Hoffman, G.G. (1991) *Gene Expression* **1**, 29–39.

74. Truter, S., Di Liberto, M., Inagaki, Y., and Ramirez, F. (1992) *J. Biol. Chem.* **267**, 25389–25395.

75. Schwartz, E., Adamany, A.M., and Blumenfeld, O.O. (1980) *Biochim. Biophys. Acta* **624**, 531–544.

76. Rein, M.S. and Nowak, R.A. (1992) *Semin. Reproductive Endocr.* **10**, 310–319.

77. Luangkhot, R., Rutchick, S., Agarwal, V., Puglia, K., Bhargava, G., and Melman, A. (1992) *J. Urol.* **148**, 467–479.

Extracellular Matrix-Cellular Interaction: Molecules to Diseases (Y. Ninomiya et al., eds.), pp. 217-234,
Japan Sci. Soc. Press, Tokyo/S. Karger, Basel (1998)

Transcriptional Regulation of Type-I Collagen Genes

CHRISTIAN RABE, CHRISTOPHER P. DENTON, AND BENOIT DE CROMBRUGGHE

Department of Molecular Genetics, M.D. Anderson Cancer Center, Houston, Texas 77030, U.S.A.

Type-I collagen is an important structural protein that provides strength to a number of tissues. In fibrotic diseases, diverse mechanisms can lead to activation of collagen producing cells resulting in an overproduction and increased deposition of this protein, thereby interfering with the organisms' normal function.

Type-I collagen consists of two alpha1 chains and one $\alpha2$ chain. These chains result from posttranslational processing of precursor molecules. After cleavage of the signal peptide in each of these precursor polypeptides, two proα1(I)-chains associate with one proα2(I)-chain to form a triple helical molecule, which in turn undergoes posttranslational modification such as hydroxylation and, after secretion into the extracellular space, is further processed and assembled into larger fibrillar structures (for review see ref. *1*).

This review focuses on the transcriptional regulation of the two type-I procollagen chains. Although there are convincing data on other mechanisms of collagen regulation, such as

217

modulation of mRNA stability (2) or the regulation of proteases acting on type-I collagen (3), a major proportion of collagen-deposition is likely to be determined by transcriptional control mechanisms.

These mechanisms may be grouped for practical purposes according to the location of the regulatory regions in the collagen genes: the proximal promoter; regions further upstream of the transcription start site; and regions located in introns of the transcribed portion of the gene itself. These regions will be separately discussed in the following sections. A general model involving several mechanisms concerning tissue-specific expression, the organization of tissue-specific elements, a functional redundancy in the regulation of type–I collagen transcription, and mechanisms governing the coordinate expression of both chains can be developed from the available data and is summarized at the end of this review.

I.   METHODS USED TO INVESTIGATE THE TRANSCRIPTIONAL REGULATION OF TYPE-I COLLAGEN

Approaches to identify *cis*-regulatory elements of any gene include a deletional analysis to roughly locate elements regulating transcription in expression systems such as transfection experiments or *in vitro* transcription, *in vitro* footprinting to further locate potential protein binding sites with a precision of a few base pairs, a functional mutational analysis of these sites and experiments such as gel-shift assays to identify proteins binding to these sites.

Methods to investigate the cell-specificity of transcriptional mechanisms *in vivo* include DNase I or restriction enzyme hypersensitivity experiments with intact nuclei, and *in vivo* footprinting experiments that can demonstrate cell-specific changes in chromatin conformation and cell-specific *in vivo* DNA-occupancy. The most physiological experimental system for investigating transcriptional regulation *in vivo* is the transgenic mouse assay: using reporter genes such as β-galactosidase this experimental system allows study of the cell-specificity and strength of transcriptional mechanisms in living organisms over time. All methods outlined above have been used in the determination of transcriptional mechanisms of the two type-I procollagen genes.

II. PROXIMAL PROMOTERS

Much interest has focused on the proximal promoters of the proα1(I)- and proα2(I)-collagen genes. Definition of the proximal promoters is somewhat arbitrary and is often guided by the presence of convenient restriction sites for the isolation of these

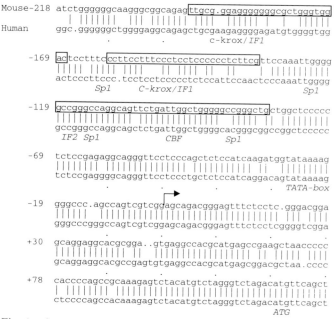

Fig. 1. Sequence comparison and known footprints (boxes) of the human and murine proα1(I)-collagen promoters. Sequences compiled from Genebank and the authors' own sequencing data. Factors binding *in vitro* are indicated in italics. Arrow denotes mRNA start.

elements. For the purposes of this review the proximal promoter
of the mouse proα1(I)-collagen gene includes 220 bp upstream
of the transcription start, while the proximal promoter of the
mouse proα2(I)-collagen gene includes 350 bp upstream of the
transcription start site. These proximal promoters show a high

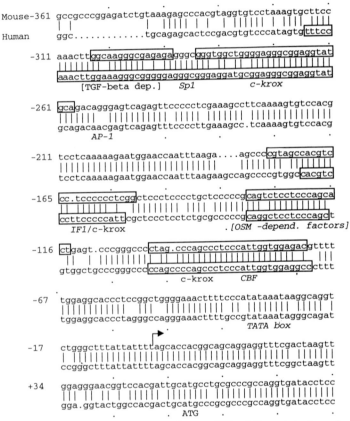

Fig. 2. Sequence comparison and known DNaseI-footprints (boxes) of
the human and mouse proximal proα2(I)-collagen promoters. Sequences
compiled from Genebank and the authors' own sequencing data. Factors
binding *in vitro* are indicated in italics. Arrow denotes mRNA start. OSM:
oncostatin M.

Transcription-factor	Binding to the alpha1(I)-promoter	alpha2(I)-promoter	Effect on transcription (*)
CBF	yes	yes	increased
c-krox	yes	yes	increased
BFCOL1	yes	yes	complex role
IF-1	yes	yes	decreased
IF-2	yes	N.D.	decreased
Sp1	yes	yes	increased
Sp3	N.D.	yes	complex role
AP-1	N.D.	yes	increased

Fig. 3. Factors potentially acting on the type-I procollagen promoters —compilation of factos implicated in the transcriptional regulation. (*) The effects on transcription were either tested directly with the use of recombinant proteins or deduced from mutations in the promoters that abolish binding of these factors. N.D.: not determined.

degree of homology between species (mouse, rat, human, and to a lesser degree chicken), but less between each other (see ref. 4, and Figs. 1 and 2). Since production of the two chains must follow a fixed 2:1 molar ratio, it is plausible that the mechanisms acting on these promoters are similar. Indeed, evidence described below shows that known or putative transcription factors such as CBF, c-krox, or IF1 could act on both promoters to regulate transcription (see Fig. 3 and ref. 49). Furthermore, both promoters contain sequences that mediate the effects of cytokines like transforming growth factor (TGF)-β and tumor necrosis factor (TNF)-α on collagen transcription.

1. Proα1(I)-collagen Gene

Figure 1 compares human and mouse proximal promoters between -220 bp and $+110$ bp (mouse). There is a close homology between the promoters in these species suggesting the presence of physiologically important mechanisms acting on these conserved sequences.

1) Localization of cis-regulatory sequences by deletional/ mutational analysis and in vitro footprints
A) Human gene: Jimenez et al. (5) showed in transient transfection experiments that while various promoter constructs with a 5′ end between −804 and −174 bp upstream of the transcription start site had roughly equal expression in NIH3T3 cells, the activity decreased when larger constructs were used.

Even though these data may also reflect decreased transfection efficiency for the larger constructs, mouse data also show similar results (see below).

Mori et al. (6) provided evidence that the TNF-α-mediated suppression of proα1(I)-collagen transcription in a transfection system is mediated by sequences between −101 and −97 bp and between −46 and −38 bp.

B) Mouse gene: While longer promoter constructs exhibit lower activity, the 220 bp mouse proα1(I)-collagen promoter has very high activity in transient transfection of fibroblasts (7). Brenner et al. (8) demonstrated by progressively deleting the mouse proα1(I)-collagen promoter from its 5′-end that sequences downstream of −181 are needed for high level activity of the promoter in transient transfection. Karsenty and de Crombrugghe (7) found three DNase I-footprints in the mouse α1(I) promoter using NIH3T3-nuclear extracts (−120 to −80; −160 to −130; −190 to −170). Mutations in the proximal footprint, affecting a CCAAT-motif between −96 and −100, decreased promoter activity to 15% compared to the 220 bp wild-type promoter. Mutations in the same footprint but upstream of this CCAAT-motif increased promoter activity significantly, as did mutations in the two footprints between −160 and −130 and −190 and −170.

2) Identification of DNA-binding proteins
A) Human gene: Li et al. (9) showed that Sp1 could stimulate the activity of the human proα1(I)-collagen promoter in transient transfection experiments employing cotransfections of an Sp1 expression vector in a background devoid of Sp1.

B) Mouse gene: Based on functional studies (above) and gelshift experiments Karsenty and de Crombrugghe (7) proposed that inhibiting factors (called IF1 and IF2) were binding to the footprinted sequences between -190 and -170 and between -160 and -130 (IF1) and in the -120 to -80 footprint bracketing the CCAAT-motif (IF2).

Using the footprint between -190 and -170 as a probe to screen an expression library, Galera et al. (10) identified a zinc-finger transcription factor, c-krox, that can bind to two GC-rich sequences between -190 and -170 and between -160 and -130 of the mouse proα1(I)-collagen promoter and stimulates transcription.

The CCAAT-binding factor (CBF) can bind to the CCAAT-binding motif of the mouse-proα1(I)-collagen promoter and stimulates transcription in an in vitro transcription assay. A binding site between -339 to -361 bp binds a negative regulatory factor that appears to be present selectively in collagen I cell lines, but not in several cell lines that do not produce collagen, possibly indicating an involvement in the tissue specific expression of the proα1(I)-collagen chain (11).

2. Proα2(I)-collagen Gene

Figure 2 comparing the proximal proα2(I)-collagen promoter sequence in mice and humans shows again a very high degree of sequence homology, suggesting the conservation of mechanisms acting on these sequences.

1) Localization of cis-regulatory sequences by deletional/ mutational analysis and in vitro footprints

A) Human gene: Progressive 5'-deletion of the human promoter in transient transfection experiments in primary human fibroblasts showed that promoter activity was decreased by shortening the promoter length from -289 to -264 bp. Footprinting analysis with fibroblast nuclear extracts showed protection of a segment between -319 and -267 (coding strand). Mutations of GC-rich regions in this footprinted area strongly decreased promoter activity (12).

Additional in vitro footprints (see also Fig. 2) with nuclear extracts from primary fibroblasts showed three footprinted regions (-173 to -155; -133 to -119; -101 to -72) in the human promoter. Mutations in a TCCCCC-motif between -164 and -159 increased expression of the reporter gene in transient transfection. Mutations in a TCCTCC-motif between -128 and -123 decreased promoter activity more than five fold. Mutating the ATTGG ($=$CCAAT)-motif to TTTGG decreased promoter activity to less than 10% of the wild-type promoter (*13*).

B) Mouse gene: Transient expression of various promoter constructs transfected into NIH3T3-fibroblasts showed that a segment between -346 and -104 of the mouse proα2(I)-collagen promoter was essential for strong expression of the promoter in transfection experiments (*14*). Further deletion analysis revealed that a promoter segment between -346 and -300 contained a regulatory element that appears to participate in conferring transcriptional activation by TGF-β of this promoter (*14, 15*). An extensive analysis of internal deletion constructs of the proα2(I)-collagen promoter using both *in vitro* transcription and transfection into Balb3T3-fibroblasts by Hasegawa *et al.* (*16*) described a segment between -170 and -40 that was essential for high level expression of reporter genes driven by the promoter.

In vitro footprinting with fibroblast nuclear extracts demonstrated the presence of three footprints (-98 to -75 bp; -131 to -114 bp; -176 to -152 bp) in the functionally important sequence located between -170 and -40. Each of the footprinted sequences contributed to the promoter's strength in transient transfection experiments of internally deleted constructs (*16*). In this study two additional footprints (-285 to -258; -304 to -290) were also identified. Figure 2 shows that the footprints in the proximal promoter of the human and mouse proα2(I) collagen genes are very similar, suggesting very similar if not identical mechanisms of regulation.

Footprinting using recombinant c-krox, preferentially expressed in skin, showed that this protein can bind to three sequences of the murine $\alpha2$(I)-promoter *in vitro*: a region between -264 and -277, a region between -112 and -189, and a region between -72 and -109 (*17*). This factor has been shown to activate the $\alpha1$(I)-promoter in DNA transfection experiments (*10*).

Mutagenesis of a CCAAT-motif between -84 and -80 in one of the footprints decreased promoter activity in transient transfection 8 fold. In the same study mutation in an element located at -250 and some mutations in the putative nuclear factor 1(NF1)-binding site between -315 and -295 bp that was partially footprinted (*16, 18*) could also decrease promoter activity in transfection (*19*).

2) *Identification of DNA-binding proteins*

The data on factors binding to the regulatory elements are mostly based on gel-shift experiments but the precise functional role of specific DNA-binding proteins in promoter activity is not well understood.

A) Human gene: Gel-shift competition and supershift experiments show that Sp1 binds to an oligonucleotide representing part of the footprinted region between -319 and -267. A TCCTCC motif located between -128 and -123 also binds Sp1 and Sp3, as does a repressor site between -164 and -159. The TCCTCC motif between -128 and -123 mediates the transcriptional effects of the cytokine oncostatin M (*12, 13, 20, 21*). In cells devoid of a background of Sp1 and Sp3, both factors can activate transcription of the human promoter, a somewhat surprising result for Sp3, since Sp3 often acts as an inhibitor of transcription. Binding of Sp1/Sp3 was demonstrated for the regulatory elements between -303 and -271, between -128 and -123, and between -164 and -159 (*20*).

B) Mouse gene: One of the proteins binding in the region of the most proximal footprint at the CCAAT-motif between -84 to -80 bp has been identified as CBF using supershifting antibodies to this factor in gel-shift experiments. This transcription factor consists of three different subunits whose presence is necessary for its physiological activity (*16, 22-24*). Also binding in gel-shift experiments to sequences corresponding to the footprint are Sp1, other proteins with an Sp1-like binding specificity, and proteins that bind to a krox consensus oligo. Binding of Sp1, other proteins with a similar binding specificity, and proteins binding to a krox consensus oligo could also be demonstrated in gel-shift experiments employing oligos corresponding to the footprints between -176 and -152 and between -131 and -114 (16).

Hasegawa et al. (25) used the yeast one-hybrid system to clone a new transcription factor designated BFCOL1 that can bind to a segment between -180 and -152 bp of the mouse $\alpha 2(I)$-promoter; this factor is also expressed in non-collagen producing cells and its role in the transcriptional regulation of the procollagen genes is still undefined.

3. Mechanisms of TGF-β Responsiveness of the Proximal Promoter

Jimenez et al. (5) described a sequence between -164 and -142 bp of the human pro$\alpha 1(I)$-collagen promoter that acts as a TGF-β responsive positive element.

Deletion analysis of the mouse $\alpha 2(I)$ revealed that a promoter segment between -350 and -300 contained a positive regulatory element that could mediate transcriptional activation by TGF-β of this promoter (15). This segment contains a motif similar to an NF1-binding site (18). However, it is doubtful that NF1 plays a major role in TGF-β signal transduction, mainly because an NF1-binding site is not conserved in the human promoter.

In the human pro$\alpha 2(I)$-collagen gene Inagaki et al. (26) and Greenwel et al. (27) located a TGF-β responsive region, also mediating effects of TNF-α, between -376 bp and -255 bp, the activation of which may be mediated through tyrosine dephosphorylation of nuclear proteins (28). Others have located a TGF-β responsive element between -241 and -265 bp of the human promoter in the region of an AP-1 binding site (29). This view is supported by data from Chang and Goldberg (30), who found that collagen promoter activity could be increased by cotransfection with junB. Additional studies are needed to further elucidate the mechanisms by which TGF-β mediates transcription of the type I collagen genes.

4. Chromatin Structure and Cell-specificity
1) Proα1(I)-collagen gene

DNase I hypersensitivity studies of the proα1(I)-collagen gene in intact nuclei of collagen type-I expressing cells, but not of other cells, revealed a relatively open configuration of chromatin indicated by DNase I hypersensitivity in the region of the proximal promoter (31, 32).

In other experiments it was shown by *in vivo* footprint methods that proteins bound to the promoter of intact cells only in cells which expressed type-I collagen. *In vivo* footprints describe the physiological situation much more accurately than *in vitro* methods. These data contrast with the results of *in vitro* footprints which show identical patterns of protection of both type-I collagen producing cells and of cells that do not produce type-I collagen (33). Overall, these experiments suggest that changes in chromatin configuration allow access of ubiquitous transcription factors such as CBF to regulatory elements of the collagen promoter, implying a model whereby still unknown tissue-specific factors would be involved in the opening of chromatin so that ubiquitous factors can interact with the promoter and regulate transcription. The participation of coactivators with histone acetyl-transferase activity in this process is an attractive hypothesis that will need to be tested (34, 35).

Rossert *et al.* (36) demonstrated that the 220 bp minimal α1(I)-promoter was essentially silent in transgenic mice. Sokolov *et al.* (37) demonstrated that tissue specific expression of either COL1A1-minigenes or COL2A1-minigenes could be achieved in transgenic mice using just 476 base pairs of the proα1(I)-collagen promoter. In this study bone expression was also observed—this is in contrast to other authors' findings (36, 38, 39) and may possibly be explained by different sensitivities of the assays used to detect the reporter genes and differences in constructions (minigenes *vs.* β-galactosidase).

2) *Proα2(I)-collagen gene*

DNase I hypersensitivity studies of the proximal proα2(I)-collagen promoter showed that in cells expressing type-I collagen but not in other cell types the whole promoter-region between -550 and $+150$ bp relative to the transcription start was in a relatively open configuration. A much stronger DNase I hypersensitive site was, however, found at approximately -100 bp (*40*). A recent survey of sequences up to 28 kb upstream of the murine transcription start reconfirmed the presence of this DNase I hypersensitive site in the proximal promoter (*41*).

In vivo protection of regulatory sequences of the proα2(I)-collagen promoter was shown by Chen *et al.* (*33*) to occur only in collagen type I expressing cells as judged by *in vivo* footprinting. While in many instances the opening of the chromatin may be facilitated by sequences located more upstream of the promoter, the 350 bp mouse proα2(I)-collagen promoter by itself shows only weak activity in subsets of type I collagen producing cells in transgenic mice: mice harboring the 350 bp proximal proα2(I)-collagen promoter also showed some weak tissue-specific expression of reporter genes (*42, 43*). Multimerized sequences between -284 and -315 of the murine proximal proα2(I)-collagen promoter were able to confer a low level of tissue specific expression of a Luciferase reporter gene to a 40 bp proα2(I)-collagen promoter (basically only consisting of a TATA-box). Since this sequence between -284 bp and -315 bp can also mediate the effects of TGF-β on the murine proα2(I)-collagen promoter, it is possible that TGF-β or factors acting through the same pathway determine this tissue-specific activity of the isolated proximal promoter. A similar weak expression was also observed in transgenic mice harboring the 378bp human proα2(I)-collagen promoter (Rabe, Denton, and de Crombrugghe, unpublished data).

III. UPSTREAM SEQUENCES

1. Proα1(I)-collagen Gene

A 900 bp murine proα1(I)-collagen promoter construct shows a relatively low level of expression predominantly in skin of transgenic mice, while a 2.3 kb promoter showed almost exclusive expression in osteo- and odontoblasts, but not in other type-I-collagen producing cells. A 3.2 kb promoter, in addition, showed high-level expression in tendon and fascial fibroblasts (*38*). These results favor a modular model of separate DNA elements that direct expression of reporter genes to different type-I collagen producing cells. One possible hypothesis is that tissue specific transcription factors would interact with such upstream elements and lead to an opening of the chromatin at the proximal promoter, allowing ubiquitous transcription factors to interact with the proximal promoter. Once the DNA in the proximal promoter were accessible to transcription factors, it could serve for the quantitative fine-regulation of transcription.

The sequences directing osteoblast specific expression were later reduced to a 117 bp enhancer segment between -1656 and -1540. This element directed expression to osteoblasts when fused to the proximal proα1(I)-collagen promoter. Gel shift assays showed that part of this sequence binds osteoblast-specific proteins while other parts bind ubiquitous proteins, thus being another example of the interaction between ubiquitous and tissue-specific transcription factors in the regulation of the type I-collagen genes (*36*). Pavlin *et al.* (*44*) and Dodig *et al.* (*39*) identified a homologous element in the rat proα1(I)-collagen gene and demonstrated binding of Msx2 to this element. The physiological significance of Msx2 binding is not well understood since Msx2 is not expressed in all osteoblasts.

In view of recent data (*32*) showing further DNAse hypersensitive sites at approximately -8 kb of the α1(I)-collagen gene, other regulatory sites may exist.

2. Proα2(I)-collagen gene

Some data (*41*) suggest an inhibitory role of sequences between -2 kb and -6.5 kb of the murine proα2(I)-collagen gene, but this region has not been further characterized.

A region between 17 kb and 13.5 kb upstream of the transcription start site of the mouse proα2(I)-collagen gene confers high level fibroblast-specific expression when linked either to the homologous proximal proα2(I)-collagen or to the proximal proα1(I)-collagen promoter in transgenic mice. Unpublished experiments show that this region is also able to confer a high degree of tissue specific expression of marker genes in transgenic mice when linked to a heterologous herpes simplex virus thymidine-kinase promoter which does not by itself confer tissue specific expression of marker-genes in transgenic mice (Rabe, Denton, and de Crombrugghe, unpublished results).

Recent results showing a cluster of DNase I-hypersensitive sites in a region approximately 20 kb upstream of the transcription start of the proα1(I)-collagen gene (*32*) suggest the possible existence of far-upstream regulatory elements in this gene.

IV. THE REGULATORY FUNCTION OF THE INTRONS *IN VIVO*

While data suggest that the introns of the type-I genes may harbor important regulatory regions just as in the proα1(II)-collagen gene, these elements have not been clearly defined.

An important role of the first intron of the mouse proα1(I)-collagen gene was first suggested by the (embryonally lethal) phenotype of the homozygous Mov-13 mice that were derived by the insertion of DNA correseponding to the Moloney murine leukemia virus genome into the germ line of mice. This Mov-13 strain carries the DNA corresponding to the viral genome inserted just downstream of exon1 of the proα1(I)-collagen gene (*45*). The presence of DNase I-hypersensitive sites in the first and fifth intron of the proα1(I)-collagen gene (*31, 32*) also suggests that these introns may play a regulatory role. Studies using various human promoter fragments with or without the first intron present linked to a human growth hormone minigene established a role for the first intron in directing tissue-specific collagen expression to the dermis (*46*). In contrast, other studies in transgenic mice harboring human collagen α1(I)-minigenes with deletions of the first intron did not show any influence of the first intron with respect to tissue-specific expression (*37*). The apparent differences in these results cannot be reconciled at the moment.

The first intron of the mouse α2(I)-collagen gene acts as a tissue-specific enhancer in cell culture experiments (*47*). Experiments in transgenic mice, however, failed to show an effect of these intron sequences on the expression-pattern of a CAT-transgene driven by -2000 bp collagen α2(*I*)-promoter (*42*).

V. A GENERAL MODEL FOR THE TRANSCRIPTIONAL REGULATION OF TYPE-I COLLAGEN

The following model combines the available data concerning the transcriptional regulation of type-I procollagen:

1.   Ubiquitous transcription factors work in conjunction with tissue-specific factors. Tissue-specific factors could help open the chromatin to allow the ubiquitous transcription factors to interact with promoter elements. This view is supported by the presence in cells that express type-I collagen of tissue-specific conformational changes of chromatin detected by DNase I hypersensitivity and also by the presence of *in vivo* footprints in areas of the promoters where ubiquitous proteins bind *in vitro*, and by the absence of these hypersensitive sites and *in vivo* footprints in cells that do not express type-I collagen.

2.   Tissue-specific elements show a modular organization. Several upstream elements—possibly interacting with putative tissue-specific factors—may have a role in the opening of chromatin only in specific cell types. This model is supported by the patterns of expression in transgenic mice harboring different lengths of the distal proα1(I)-collagen promoter and by transgenic mice containing the far-upstream proα2(I)-collagen element.

3. There is functional redundancy in the quantitative regulation of type-I collagen transcripts. The high level of transcription is the net result of multiple mechanisms acting on the type-I procollagen promoters, including ubiquitous factors such as CBF, Sp1/Sp3 and more specific stimulatory signals such as those activated by TGF-β.

4. The coordinate regulation of the two different chains of type I is probably the result of similar mechanisms acting on both promoters. The existence of several known or putative transcription factors acting on the two promoters and the fact that sequences mediating cytokine-responsiveness are present in both of them support the view that there is common control of the coordinate expression of both type-I collagen genes.

Acknowledgments

C. Rabe was supported by a fellowship from the Deutsche Forschungsgemeinschaft (Ra 750/1-1). C.P. Denton was supported by an advanced training fellowship from The Wellcome Trust.

REFERENCES

1. Rossert, J. and de Crombrugghe, B. (1996) In *Principles of Bone Biology* (Bilezikian, J.P. *et al.*), pp. 127–142, Academic Press Inc., San Diego.
2. Eckes, B., Mauch, C., Huppe, G., and Krieg, T. (1996) *Biochem. J.* **315**, 549–554.
3. Langholz O., Rockel, D., Mauch, C. *et al.* (1995) *J. Cell Biol.* **131**, 1903–1915.
4. Garrett, L.A. (1996) Dissertation, The University of Texas Health Science Center, Houston.
5. Jimenez, S.A., Varga, J., Olsen, A. *et al.* (1994) *J. Biol. Chem.* **269**, 12684–12691.
6. Mori, K., Hatamochi, A., Ueki, H., Olsen, A., and Jimenez, S.A. (1996) *Biochem. J.* **319**, 811–816.
7. Karsenty, G. and de Crombrugghe, B. (1990) *J. Biol. Chem.* **265**, 9934–9942.
8. Brenner, D.A., Rippe, R.A., and Veloz, L. (1989) *Nucl. Acids Res* **17**, 6055–6064.
9. Li, L., Artlett, C.M., Jimenez, S.A., Hall, D.J., and Varga, J. (1995) *Gene*

164, 229–234.

10. Galera, P., Musso, M., Ducy, P., and Karsenty, G. (1994) *Proc. Natl. Acad. Sci. USA* **91**, 9372–9376.

11. Ravazzolo, R., Karsenty, G., and de Crombrugghe, B. (1991) *J. Biol. Chem.* **266**, 7382–7387.

12. Tamaki, T., Ohnishi, K., Hartl, C., LeRoy, E.C., and Trojanowska, M. (1995) *J. Biol. Chem.* **270**, 4299–4304.

13. Ihn, H., Ohnishi, K., Tamaki, T., LeRoy, E.C., and Trojanowska, M. (1996) *J. Biol. Chem.* **271**, 26717–26723.

14. Schmidt, A., Rossi, P., and de Crombrugghe, B. (1986) *Mol. Cell. Biol.* **6**, 347–354.

15. Rossi, P., Karsenty, G., Roberts, A.B., Roche N.S., Sporn, M.B., and de Crombrugghe, B. (1988) *Cell* **52**, 405–414.

16. Hasegawa, T., Zhou, X., Garrett, L.A., Ruteshouser, E.C., Maity, S.N., and de Crombrugghe, B. (1996) *Nucl. Acids Res.* **24**, 3253–3260.

17. Galera, P, Park, R.-W., Ducy, P., Mattei, M.-G., and Karsenty, G. (1996) *J. Biol. Chem.* **271**, 21331–21339.

18. Oikarinen, J., Hatamochi, A.,and de Crombrugghe, B. (1987) *J. Biol. Chem.* **262**, 11064–11070.

19. Karsenty, G., Golumbek, P., and de Crombrugghe, B. (1988) *J. Biol. Chem.* **263**, 13909–13915.

20. Ihn, H. and Trojanowska, M. (1997) *Nucl. Acids Res.* **25**, 3712–3717.

21. Ihn, H., LeRoy E.C., and Trojanowska, M. (1997) *J. Biol. Chem.* **272**, 24666–24672.

22. Sinha, S., Maity, S., Lu, J.,and de Crombrugghe, B. (1995) *Proc. Natl. Acad. Sci. USA* **92**, 1624–1628.

23. Coustry, F., Maity, S.N., and de Crombrugghe, B. (1995) *J. Biol. Chem.* **270**, 468–475.

24. Maity, S.N., Sinha, S., Ruteshouser, E.C., and de Crombrugghe, B. (1992) *J. Biol. Chem.* **267**, 16574–16580.

25. Hasegawa, T., Takeuchi, A., Miyaishi, O., Isobe, K.,and de Crombrugghe, B. (1997) *J. Biol. Chem.* **272**, 4915–4923.

26. Inagaki, Y., Truter, S., and Ramirez, F. (1994) *J. Biol. Chem.* **269**, 14828–14834.

27. Greenwel, P., Inagaki, Y., Hu, W., Walsh, M., and Ramirez, F. (1997) *J. Biol. Chem.* **272**, 19738–19745.

28. Greenwel, P., Hu, W., Kohanski, R.A., and Ramirez, F. (1995) *Mol. Cell. Biol.* **15**, 6813–6819.

29. Chung, K.-Y., Agarwal, A., Uitto, J., and Mauviel, A. (1996) *J. Biol. Chem.* **271**, 3272–3278.

30. Chang, E. and Goldberg, H. (1995) *J. Biol. Chem.* **270**, 4473–4477.

31. Breindl, M., Harbers, K., and Jaenisch, R. (1984) *Cell* **38**, 9–16.

32. Salim-Tari, P., Cheung, M. Safar, C.A. *et al.* (1997) *Gene* **198**, 61–72.

33. Chen, S.S., Ruteshouser, E.C., Maity, S.N., and de Crombrugghe, B. (1997) *Nucl. Acids Res.* **25**, 3261–3268.
34. Wolffe, A.P. (1997) *Nature* **387**, 16–17.
35. Wu, C. (1997) *J. Biol. Chem.* **272**, 28171–28174.
36. Rossert, J., Chen S.S., Eberspaecher, H., Smith, C.N., and de Crombrugghe, B. (1996) *Proc. Natl. Acad. Sci. USA* **93**, 1027–1031.
37. Sokolov, B.P., Mays, P.K., Khillan, J.S., and Prockop, D.J. (1993) *Biochemistry* **32**, 9242–9249.
38. Rossert, J., Eberspaecher, H., and de Crombrugghe, B. (1995) *J. Cell Biol.* **129**, 1421–1432.
39. Dodig, M., Kronenberg, M.S., Bedalov, A. *et al.* (1996) *J. Biol. Chem.* **271**, 16422–16429.
40. Liau, G., Szapary, D., Setoyama, C., and de Crombrugghe, B. (1986) *J. Biol. Chem.* **261**, 11362–11368.
41. Bou-Gharios, G., Garrett, L.A., Rossert, J. *et al.* (1996) *J. Cell. Biol.* **134**, 1333–1344.
42. Goldberg H., Helaakoski, T. Garrett, L.A. *et al.* (1992) *J. Biol. Chem.* **267**, 19622–19630.
43. Niederreither, K., D'Souza, R., and de Crombrugghe, B. (1992) *J. Cell Biol.* **119**, 1361–1370.
44. Pavlin, D., Lichtler, A.C., Bedalov, A. *et al.* (1992) *J. Cell Biol.* **116**, 227–236.
45. Harbers, K., Kuehn, M., Delius, H., and Jaenisch R. (1984) *Proc. Natl. Acad. Sci. USA* **81**, 1504–1508.
46. Liska, D.J., Reed, M.J., Sage, E.H., and Bornstein, P. (1994) *J. Cell Biol.* **125**, 695–704.
47. Rossi, P. and de Crombrugghe, B. (1987) *Proc. Natl. Acad. Sci. USA* **84**, 5590–5594.
48. Maity, S.N., Golumbek, P.T., Karsenty, G., and de Crombrugghe, B. (1988) *Science* **241**, 582–585.
49. Karsenty, G. and de Crombrugghe, B. (1991) *Biochem. Biophys. Res. Commun.* **177**, 538–544.

Extracellular Matrix-Cellular Interaction: Molecules to Diseases (Y. Ninomiya et al., eds.), pp. 235-260, Japan Sci. Soc. Press, Tokyo / S. Karger, Basel (1998)

Expression of Collagen IV Genes and Goodpasture Antigens

YOSHIKAZU SADO,[*1] ICHIRO NAITO,[*2] AND YOSHIFUMI NINOMIYA[*3]

*Divisions of Immunology[*1] and Ultrastructural Biology,[*2] Shigei Medical Research Institute, Okayama 701-0202 and Department of Molecular Biology and Biochemistry, Okayama University Medical School, Okayama 700-8558,[*3] Japan*

Evolutionary change of extracellular matrices is indispensable for the development of multicellular organisms. Such matrices have evolved into a large variety of specialized structures including basement membranes, which are always closely associated with epithelial cells, endothelial cells, muscle cells, adipocytes, and Schwann cells. This close association is instrumental in defining biological functions of the basement membranes. In fact, many basement membranes separate epithelia from underlying extracellular matrices, while some specialized membranes such as glomerular basement membranes in kidney sit between epithelia and endothelia or other types of cells. Glomerular basement membranes have been investigated extensively in terms of ultrastructure and molecular characteristics. Collagen IV molecules appear to assemble into a network through various types of interactions. Assembly of collagen IV triple-helical molecules can be described as a three-step process: dimerization through NC1 domains, tetramerization via the 7S domains, and the lateral association by hydrophobic interac-

235

tions between antiparallel molecules. Recent reports dealing with the new α(IV) chains stimulated investigators to reconsider a hypothesis explaining the mechanism of how these new collagen IV molecules participate in the assembly of the polygonal basement membrane network. This chapter summarizes updated information about the special organization of genes encoding α(IV) chains, structural features and distribution of type IV collagen molecules in various basement membranes, and the nephritogenicity of fragments of type IV collagen molecules.

I. DISCOVERY OF THE SIX GENETICALLY DISTINCT α(IV) CHAINS

Collagens are a distinct family of multifunctional proteins that are found in extracellular space. Their molecules are composed of three α chains that contain the repeating Gly-X-Y sequence and form collagen triple-helices. Type IV collagen is the fourth genetically distinct molecule that was identified from basement membranes from Engelbreth-Holm-Swarm (EHS) tumors (1). They form a thin sheet structure that separates the epithelia and extracellular matrices. Each monomeric molecule was thought to consist of two α1 chains and one α2 chain. Unlike the interstitial collagen molecules, each α chain is longer and more flexible and the triple-helix is interrupted by a number of imperfections in the Gly-X-Y sequence. In fact, the α1(IV) chain contains 21 interruptions and the α2(IV) chain contains 23, superimposing to produce 26 irregular flexibilities of various lengths (2, 7-9, Fig. 1).

The presence of the other four α(IV) chains, α3(IV)–α6(IV) was identified by protein chemistry and cDNA cloning (3-6). The complete cDNA sequences for the six α(IV) chains in human are available now (6-13). In other species, the mouse (14), the sea urchin (15), Drosophila (16), and Caenorhabditis elegans (17) α1(IV) chains, and the mouse (18), Ascaris suum (19) and C. elegans (20) α2(IV) chains have been reported to date. Each α chain polypeptide can be divided into three functional domains: the amino-terminal 7S domain which is rich in cysteine and lysine residues, the central COL domain which forms a triple-helix, and the carboxy-terminal NC1 domain (*21–23*). The triple-helices make the molecules rod-like structure for interstitial collagens, however, the type IV collagen molecules look like a thread about 400 nm long due to the multiple interruptions of triple-helices. The NC1 domain forms a large globule at the carboxyl end. The central COL domain consists of multiple Gly-X-Y triplets with many interruptions of relatively short noncollagenous sequences.

Sequence alignment between the human six α(IV) chains is shown in Fig. 1. The α chains in general consist of a 21–38 residue signal peptide, a short stretch of amino-terminal noncollagenous sequence, a 1,410–1,449 residue COL domain with 21–26 interruptions of 2–21 residues, and a 227–232-residue carboxy-terminal NC1 domain. The locations for potential carbohydrate attachment sites, and interruptions of Gly-X-Y repeat are well conserved in the six α(IV) chains (see Fig. 2). Locations of the cysteinyl residues are well conserved in the amino-terminal noncollagenous sequence, in some interruption regions, and in the NC1 domain. Especially, all 12 cysteinyl residues are conserved, forming six disulfide bridges. Interestingly, the NC1 domain appears to be duplicated from a gene encoding half of the NC1 domain of about 115 residues (*12*). Each half NC1 domain contains six cysteinyl residues, which have an almost identical disulfide-bond arrangement as the other half. Comparison of the amino acid sequence in the COL domain reveals that the amino acid residues at the X or Y positions of the Gly-X-Y triplets are not well conserved, resulting in a relatively low homology. However, the amino acid sequences within the NC1 domains are highly conserved among the six α chains. More careful comparison of the primary structure among the human six α(IV) chains showed them to be very similar in structure and classifiable into two groups: group A consisting of α1, α3, and α5 chains and group B, with α2, α4, and α6 chains. Comparing the structure of the genes encoding the six α(IV) chains, as

```
α1    1:-----------MGPR-LSVWLLLLPAALLLHEEHSRAA--AKGG-CAGSGC-GKCDCHG
α2    1:----MGRDQR-AVAGPALRRWLLLGTVTVGFLAQSVLAGVKKFDVPCGGRDCSGGCQCYP
α3    1:M-SARTA-PR-P-QV---LLL-P---LLLVLLAAA-P-AASKGCV--CKDKG--Q-CFCDG
α4    1:MWSLHIVLMRCSFRLTKSLATGPWSLILILFSVQYVYGSGKKYIGPCGGRDC-SVCHCVP
α5    1:-----------MKLRGVSLAAGLFLLALSLWGQPAEAA--ACYG-C-SPG--SKCDCSG
α6    1:-----------MHPGL--WLLL--VTLCLTEELAAAGEKSYGKPCGGQDCSGSCQCFP
                                              *        * *      *  *
```

```
α1   61:VKGQKGERGLPGLQGVIGFPGMQGPEGPQGPPGQKGDTGEPGLPGTKGTRGPPGASGYPG
α2   61:EKGGRGQPGPVGPQGYNGPPGLQGFPGLQGRKGDKGERGAPGVTGPKGDVGARGVSGFPG
α3   61:AKGEKGEKGFPGPPGSPGQKGFTGPEGLTGPQGPKGFPGLPGLTGSKGVRGISGLPGFSG
α4   61:EKGSRGPPGPPGPQGFPGLNGRKGDKGERGAPGVTGPKGDKGPTGVPGFPG
α5   61:IKGEKGERGFPGLEGHPGLPGFPGPEGPPGPRGQKGDDGIPGPPGPKGIRGPPGLPGFPG
α6   61:EKGARGRPGPIGIQGPTGPQGFTGSTGLSGLKGERGFPGLLGPYGPKGDKGPMGVPGFLG
        * *    *    *    * *    *    *   *    *    *  *  *     *    *
```

```
α1  121:NPGLPGIPGQDGPPGPPGIPGCNGTKGERGPLPGPGLPGFAGNPGPPGLPGMKGDPGE-I
α2  121:ADGIPGHPGQGGPRGRPGYDGCNGTQGDSGPQGPPGSEGFTGPPGPQGPKGQKGEP-YAL
α3  121:SPGLPGTPGNTGPYGLVGVPGCSGSKGEQGFPGLPGTLGYPGIP---GAAGLKGQKGAPA
α4  121:LDGIPGHPGPPGRGKPGMSGHNGSRGDPGFPGGRGALG-PGGP--LGHPGEKGEKGNSV
α5  121:TPGLPGMPGHDGAPGPQGIPGGNGTKGERGFPGSPGFPGLQGPPGPPGIPGMKGEPGSII
α6  121:INGIPGHPGQPGPRGPPGLDGCNGTQGAVGFPGPDGYPGLLGPPGLPGQKGSKGDP--VL
        * **  *       *    *   *    *     *      *   *    *    *
```

```
                   7S
α1  181:LGHVPGMLLKGERGFPGIPGTPGPPGLPGLQGPVGPPGFTGPPGPPGPPGPPGEKGQMGL
α2  181:-PKEERDRYRGEPGEPGLVGFQGPPGRPGHVGQMGPVGAPGRPGPPGPPGPKGQQGNRGL
α3  181:-KEEDI-ELDA-KGDPGLPGAPGPQGLPGPPGFFGFPGAMGPRGPKGH
α4  181:FILGAVKGIQGDRGDPGLPGLPGSWG-AGGP--AGPTGYPGEPGLVGPGQPGRPGLKGN
α5  181:MSSLPGP--KGNPGYPGPPGIQGLPGPTGIPGPIGPPGPPGLMGPPGPPGLPGPKGNMGL
α6  181:-APGSFKGMKGDPGLPGLDGITGPQGAPGFPGAVGPAGPPGLQGPPGPPGPLGPDGNMGL
                 * **     *  *    *    *       **     *   * *    *
```

```
α1  241:SFQGPKGDKGDQGVSGPPGVPGQ-A-Q---VQ-E--KGDFATKGEKGQKGEPGFQ---G-
α2  241:GFYGVKGEKGDVGQPGPNGIPSDTLH-PIIA-PTG--VT-FHPD-Q-YKGEKGSEGEPGI
α3  241:MGERVIGHKGERGVKGLTGPP--GPPGTVIVTLTGPDNRTDLKGEKGDKGAMGEPGPPGF
α4  241:PGVGVKGQMGDPGEVGQQGSP--G-PTLLVEPPDFCLYKGE-KGIKGIPGMVGLPGPPGR
α5  241:NFQGPKGEKGDGLQGPPGPPGQISEQKRPIDVEFQKGDQGLPGDRGPPGPPGIRGPPG-
α6  241:GFQGEKGVKGDVGLPGPAGPPPSTGELEFMGFPKGKKGSKGEPGPKGFPGISGPPGFPGL
        * *    *   *    * *                      *    *         *  *
```

```
α1  301:MPGV--GEKGEPGKPGPRGKPGKDGDKGEKGSPGFPGEPGYPGLIGRQGPQGEKGEAGPP
α2  301:RGISLKGEEGIMGFPGLRGYPG-LSGE--KGSPGQKGSRGLDGYQGPDGPRGPKGEAGDP
α3  301:SGLPGESYGSEKGAPGDPGLQGKPGKDGVPGFPGSEGVKGNRGFPGLMGEDGIKGQKGDI
α4  301:KGESGIGAKGEKGIPGFPGRGDPGSYGSPGFPGLKGELGLVGDPGLFGLIGPKGDPGNR
α5  301:PPGGEKGEKGEAGEPGKRGKPGKDGENGQPGIPGLPGDPGYPGEPGRDGEKGQKGDTGPP
α6  301:GTTGEKGEKGEKGIPGLPGRGPMGSEGVQGPPGQQGKKGTLGFPGLNGFQGIEGQKGDI
        *   *  *   *   **  *   *    *    *  *   *    *    *  *    *
```

```
α1  361:GPPGIVI---GTG-PLGEKGERGYPGTPGPREGPKGFPGLPGQPGPPGLPVPGQAGAP
α2  361:GEMGPKGF--I-GDPGIPALYG-GPPGPDGKRGPGPPGPGLPGPPGPDGFLFGLKGAKGRA
α3  361:GPPG-FRGPTEYYDTYQEKGDEGTPGPPGPRGARGPQGPSGPPGVPGSPGSSRPGLRG--
α4  361:GHPG-PPG-VLVTPPLPLKGPPGDGPGFPGRYGETGDVGPPGEPGLLGRPGEACAGMIGPP
α5  361:GPPGLVIPRPGTGITIGEKGNIGLPGLPGEKGERGFPGIQGPPGLPGPPGAAVMGPPGPP
α6  361:GLPGPDVFI-DID-GAVISGNPGDPGVPGLPGLKGDEGIQGLRGPSGVP-GLPAL-SGVP
        * **    *    *   *    * *    *     *   *    *  *  *     *
```

```
α1  421:GFPGERGEKGDRGFPGTSLPGPSGRDGLPGPPGSPGPPGQPG-YTNGIVECQPGPP-GDQ
α2  421:GEMGPKGF--I-GDPGIPALYG-GPPGPDGKRGPGPPGLPGPPGDGFLFGLKGAKGRA
α3  421:AP-GWPGLIGSKGERGRPGKDAMGTPGSPGCAGSPGLPGSPG-PPGPPG-DIGF-RKGPP
α4  421:GPQGFPGLPGLPGEAGIPGRPD-SAPGKPGPGLPGLPGPAPG-LQGLPGSSVIYCSVGNP
α5  421:GFPGERGQKGDEGPPGISIPGPPGLDGQPGAPGLPGPPG-PG-SPHIPPSDEICEP-GPP
α6  421:GALGPQGFPGLKGDQGNPGRTTIGAAGLPGRDGLPGPPGPPGPPSPE-FETETLHNKE-S
        * *     *    *    *    *    *    *     * *    *   **  *
```

```
α1  481:GPPGIPGQPGFIGEIGEKGQKGESCLICDIDGYRGPPGPQG-P---PGEIGFPGQPGAKGD
α2  481:GFPGLPG--SP-GAPGPKGWKGDAGECRGTEGDEAIKGLPGLPGPKGKFAGINGEPGRKGD
α3  481:GDHGLPGYLGSPGIPGVDGPKGEPGLLCTQC--PYIPGPPGLPGLPGLHGVKGIPGRQGA
α4  481:GPQGIKGKVGPPGGRGPKGEKGNEG-LCA-C-EPGPMGPGPPGLPGRQGSKGDLGLPGW
α5  481:GPPGSPGDKGLQGEQGVKGDKGDTCFNGIGTGISGPPGQPGLPGLPGPPGSLGFPGQKGE
α6  481:GFPGLRGEQGPKGNLGLKGIKGDSGFCAC-DGGVPNTGPPGEPGPPGPWGLIGLPGLKGA
        *  *    *  *    *    *    *  *  *    *    *     *    *
```

```
α1  541:RGLPGRDGVAGVPGPQGTPGLIGQPGAKGEPGEFY-F-DLRLKGDKGDPGFPGQPGMPGR
α2  541:KGDPGQHGLPGFPGLKGVPGLNGIAPGPKGAKGDSRTITTKGERGQPGVPGVPGMKGDDGS
α3  541:AGLKGSPGSPGNTGLPGFPGFPGAQGDPGLKGEKGE-TL-QPEGQVGVPGDPGLRGQPGR
α4  541:LGTKGDPGPPGAEGPPGLPGKHGASGPPGNKGAKGDMVVSRVKGHKGERGPDGPPGFPGQ
α5  541:KGQAGATGPKGLPGIPGAPGAPGFPGSKGPGGDILTF-P-GMKGDKGRELGSGPAPGLPGL
α6  541:RGDRGSGGAQGPAGAPGLVGPLGPSGPKGKKGEPILSTIQGMPGDRGDSGSQGFRGVIGE
        *  *    *    *    *    *    *    *      *    *    *    *
```

```
α1  601:AGSPGRDGHPGLPG-P----KGSPG--SVGLK-GERGPPGGVGFP--GSRGDTGPPGPPG
α2  601:PGRDGLDGFPGLPGPPGDGIKGPPGDPGYPGIPGTKGTPGEMGPPGLGLPGLKGQRGFPG
α3  601:KGLDGIPGTPGVKGLP-----GPKGELALSGEKGDQGPGDPGSP--GSPGPAGPAG--PP
α4  601:PGSHGRDGHAGEKGDP-----GPPGDHE-DATPGGKGFPGPLGPP--GKAGPVGPPG--L
α5  601:PGTPGQDGLPGLPG-P----KGEPG--GITFK-GERGPPGNPGLP--GLPGNIGPMGPPG
α6  601:PGKDGVPGLPGLPGLPGDGGQGFPGEKGLPGLPGEKGHPGPPGLPGNGLPGLPGPRGLPG
        *    *     *        *       *   *  * ** ***  * * * *    *

α1  661:-YGPAGPIGDKGQAGFPGGPGSPG-LPGPK-GEPG--KIVPLPGPP-GAEGLPGSPGFPG
α2  661:DAGLPGPPGFLGPPGPAGTPGQID©TDVKRAVGGDRQEAIQPG©IGGPKGLPGLPGPPG
α3  661:G--Y-GPQGEPGLQGTQGVPG--APGPPGEAGPRG-E-LSVSTPVP-GPPGPPGPPGHPG
α4  661:G--FPGPPGERGHPGVPGHPG--VRGPDGLKGQKG-DTIS©NVTYP-GRHGPPGFDGPPG
α5  661:-LALQGPVGEKGIQGVAGNPGQPG-IPGPK-GDPG--QTITQPGKP-GFRGNPGRDGDVG
α6  661:DKG-KD--G-L--PGQQGLPG-----S--K-GI--T-LP©IIPGSY-GPSGFPGTPGFPG
         *    *  **     *          *       *   ****  *

α1  721:PQGDRGFPGTPGRPGLPGEKGAVGQPG-IG---FPGPPGPKGVDGLPGDMGPPGTPGRPG
α2  721:PTGAKGLRGIPGFAGADGGPGPRGLPGDAGREGFPGPPGFIGPRGSKGAVGLPGPDGSPG
α3  721:PQGPPGIPGSLGK©GDPGLPGPDGEP---GIPGI-GFPGPPGPKGDQGFPGTKGSLG©PG
α4  721:PKGFPGPQGAPGLSGSDGHKGRPGTP---GTAEIPGPPGFRGDMGDPGFGGEKGS-SPVG
α5  721:LPGDPGLPGQPGLPGIPGSKGEPGIPG-IG---LPGPPGPKGFPGIPGPPGAPGTPGRIG
α6  721:PKGSRGLPGTPG-Q-P-GSSGSKGEPGSPGLVHLPELPGFPGPRGEKGLPGFPGLPGKDG
        *    *     *  *         *       *     **  *   *     *   *

α1  781:FNGLPGNPGVQGQKGEPGVGLPGLKGLPGLPGIPGTPGEKGSIGVPGVPGEHGAIGPPGL
α2  781:PIGLPGPDGPPGERGLPGEVLGAQPGPRGDAGVPGQPGLKGLPGDRGPPGFRGSQGMPGM
α3  781:KMGEPGLPGKPGLPGAKGEPAVAMPGGPGTPGFPGERGNSGEHGEIGLPGLPGLPGTPGN
α4  781:PPGPPGGSPGVNGQKGIPGDPAFGHLGPPGKRGLSGVPGIKGPRGDPG©PGAEGPAGIPGF
α5  781:LEGPPGPGFPGPKGEPGFALPGPPGPPGLPGFKGALGPKGDRGFPGPPGPPGRTGLDGL
α6  781:LPGMIGSPGLPGSKGATGDIFGAENGAPGEQGLQGLTGHKGFLGDSGLPGLKGVHGKPGL
        *    *     *     *     *    *        *   ** *    **   *   *

α1  841:QGIRGEPGPPGLPGSVGSPGVPG-IG-PPGARGPPGGQGPPGLSGPPGIKGEKGFPGFPG
α2  841:PGLKGQPGLPGPSGQPGLYGPPGLHG-FPGAPGQEGPLGLPGITPGREGLPGDRGDPGDTG
α3  841:EGLDGPRGDPGQPGPPGEQGPPG-R-©IEGPRGAQGLPGLNGLKGQQGRRGKTGPKGDPG
α4  841:LGLKGPKGREGHAGFPGVPGPPG-HS©ERGAPGIPGQPGLPGYPGSPGAPGGKGQPGDVG
α5  841:PGPKGDVGPNGQPGPMGPPGLPG-IG-VQGPPGPPGIPGPIGQPGLHGIPGEKGDPGPPG
α6  841:LGPKGERGSPGTPGQVGQPGTPGSSGPY-GIKGKSGLPGAPGFPGISGHPGKKGTRGKKG
        * *     *    ** *     *     *     *    *  *   *   *   *  *

α1  901:L-DMPGPKGDKGAQGLPGITGQSGLPGLPGQQGAPGIPGFPGSKGEMGVMGTPGQPGSPG
α2  901:APGPVGMKGLSGDRGDAGFTGEQGHPGSPGFKGIDGMPGTPGLKGDRGSPGMDGFQGMPG
α3  901:IPG-LDRSGFPGETGSPGIPGHQGEMGPLGQRGYPG---NPGILGPPGEDGVIGMMGFPG
α4  901:PPGPAGMKGLPGEPGPGAHGPPG-LPGIPGF-F-G---DDGLPGPRGPKGFPGFPGFPG
α5  901:L-DVPGPPGERGSPGIPGAPGPIGPPGSPGLPGKAGRSGFPGTKGEMGMMGPPGPPGPLG
α6  901:PPGSIVKKGLPGLKGLPGNPGLVGLKGSPGSPGVAGLPALSGPKGEKGSVGFVGFPGIPG
        *   *    *    *    *     *       *     *   *      *   *  *

α1  961:PWGAPGLPGEKGDHGFPGSSGPRGDPGLKGDKGDVGLPGKPG-SMDKVDMGSMKGQKGDQ
α2  961:LKGRPGFPGSKGEAGFFGIPGLKGLAGEPGFKGSRGDPGPPG-PPP-VILPGMKDIKGEK
α3  961:AIGPPGPGNPGTPGHRGSPGIPGVKGQRGTPGAKGEQGDKG-NPGPSEISHVIGDKGEP
α4  961:FPGERGKPGAEG©PGAKGEPGEKGMSGLPGDRGLRGAKGAIG-PPGDEGEMAIISQKGTP
α5  961:IPGRSGVPGLKGDDGLQGGQPGLPGPTGEKGSKGEPGLPGPPG-PMDPNLLGS-KGEKGEP
α6  961:LPGISGTRGLKGIPGSTGKMGPSGRAGTPGEKGDRGNPGPVGIPSPRRPMSNLW-LKGDK
        *   *    *    *      *    *    *      *           *   **

α1 1021:GEKGQIGPIGEKGSRGDPGTPGVPGKDGQAGPGPGQPGP-KGDPGISGTPGAPGLPGPKGS
α2 1021:GDEGPMGLKGYLGAKGIQGMPGIPGLSGIPGLPGRPGHIKGVKGDIGVPGIPGLPGFPGV
α3 1021:GLKGFAGNPGKGNRGVPGMPGLKGLKGLPGPAGPPG-PRGDLGSTGNPGEPGLLGIPGS
α4 1021:GEPGPPGDDGFPGERGDKGTPGMQGRRGELGRYGPPGFHRGEPGEKGQPGPPGPPGPPGS
α5 1021:GLPGIPGVSGPKGYQGLPGDPGQPGLSGQPGLPGPPGP-KGNPGLPGQPGLIGPPGLKGT
α6 1021:GSQGSAGSNGFPGPRGDKGEAGRPGPPGLPGAPGLPGIIKGVSGKPGPPGFMGIRGLPGL
        *    *    *     *    *   *     *     *   *    *    *    *

α1 1081:VGGMGLPGTPGEKGVPGIPGPQGSPGLPGDKGAKGEKGQAG---PPGIGIPGLRGEKGDQ
α2 1081:AGPPGITGFPGFIGSRGDKGAPGRAGLYGEIGATGDFGDIG-DTINLPGRPGLKGERGTT
α3 1081:MGNMGMPGSKGKRGTLGFPGRAGRPGLPGIHGLQGDKGEPGYSEGTRPGPPGPTGDPGLP
α4 1081:TGLRGFIGFPGLPGDQGEPGSPGPPGFSGIDGARGPKGNK G-DPASHFGPPGPKGEPGSP
α5 1081:IGDMGFPGLPGVEGPPGPSGVPGQPGSPGLPGQKGDKGDPG---ISSIGLPGLPGQKGEP
α6 1081:KGSSGITGFPGPMPGESGSQGIRGSPGLPGASGLPGLKGDNG-QTVEISGSPGPKGQPGES
        *    *    *   *     *     *   * *        * ** *   *

α1 1141:GIAGFPGSPGEKGEKGSIGIPGMPGSPGLKGSPGSVGYPGSPGLPGEKGDKGLPGLDGIP
α2 1141:GIPGLKGFFGEKGTEGDIGFPGITGVTGVQGPGLKGQTGFPGLTGPPGSQGELGRIGLP
α3 1141:GDMGKKGEMGQPGPPGHLGPAGPEGAPGSPGSPGLPGKPGPHGDLGFKGIKGLLG-P-P-
α4 1141:G©PGHFGASGEQGLPGIQGPRGSPGRPGPPGSSGPPG©PGDHGMPGLRGQPGEMGDPGPR
α5 1141:GLPGYPGNPGIKGSVGDPGLPGIFGITGVTGPGFPGTPGPPGPKGISGPPGNPGLP
α6 1141:GFKGTKGRDGLIGNIGFPGNKGEDGKVGVSGDVGLPGAPGFPGVAGMRGEPGLPGSSGHQ
        *    *    *    *    *    *    *     *    *    *    *     *
```

```
α1 1201:GVKGEAGLPGTPGPTGPAGQKGEPGSDGIPGSAGEKGEPGLPGRGFPGFPGAKGDKGSKG
α2 1201:GGKGDDGWPGAPGLPGFPGLRGIRGLHGLPGTKGFPGSPGSDIHGDPGFPGPPGERGDPG
α3 1201:GIRGPPGLPGFPGSPGPMGIRGDQGRDGIPGPAGEKGETGLLR-APPGPRGNPGAQGAKG
α4 1201:GLQGDPGIPGPPGIKGPSGSPGLNGLHGLKGQKGTKGASGLHDVGPPGPVGIPGLKGERG
α5 1201:GEPGPVGGGGHPGQPGPPGEKGKPGQDGIPGPAGQKGEPGQPGFGNPGPPGLPGLSGQKG
α6 1201:GAIGPLGSPGLIGPKGFPGFPGLHGLNGLPGTKGTHGTPGPSITGVPGPAGLPGPKGEKG
        *   *   *   *   *   *   *   *

α1 1261:-EVGFPGLAGSPGIPGSKGEQGFMGPPGPQGQPGLPG-SPGHA-TEGPKGDRGPQGQPGL
α2 1261:EANTLPGPVGVPGQKGDQGAPGERGPPGSPGLQGFPG-ITPPSNISGAPGDKGAPGIFGL
α3 1261:-DRGAPGFPGLPGRKGAMGDAGPRGPTGIEGFPGPPG-LPGAI-IPGQTGNRGPPGSRGS
α4 1261:-DPGSPGISP-PGPRGKKGPPGPPGSSGPPGPAGATGRAPKDIPDPGPPGDQGPPGPDGP
α5 1261:-DGGLPGIPGNPGLPGPKGEPGFHGFPGVQGPPGPPG-SPGPA-LEGPKGNPGPQGPPGR
α6 1261:YPGIGIGAPGKPGLRGQKGDRGFPGLQGPAGLPGAPG-ISLPSLIAGQPGDPGPRPGLDGE
        *        * **      *     *          *       *   *     *

α1 1321:PGLPGPMGPPG-LPGIDGVKGDKGNPGWPGAPGVPGPKGDPGFQGMPGIGGSPGITGSKG
α2 1321:KGYRGPPGPPG-SAALPGSKGDTGNPGAPGTPGTKGWAGDSGPQGRPGVFGLPGEKGPRG
α3 1321:PGAPGPPGPPGS-HVI-GIKGDKGSMGHPGPKGPPGTAGDMGPPGRLGAPGTPGLPGPRG
α4 1321:RGAPGPPGLPGSVDLLRGEPGDGGLPGPPGPPGPPGPPGPPGYKGFPGGDGKDGQKGPMGFPG
α5 1321:PGLPGPGPEGPPG-LPGNGGIKGEKGNPGQPGLPGLPGLKGDQGPPGLQGNPGRPGLNGMKG
α6 1321:RGRPGPAGPPG-PPGPSSNQGDTGDPGFPGIPGPKGPKGDQGIPGFSGLPGELGLKGMRG
        *   * **  * **             *      *       *      *    *    *

α1 1381:DMGPPGVPGFQGPKGLPGLQGIKGDQGDQGVPGAKGLPGPPGPPGPYDIIKGEPGLPGPE
α2 1381:EQGFMGNTGPTGAVGDRGPKGPKGDPGFPGAPGTVGAPGIAGIPQKIAIQPGTVGPQGRR
α3 1381:DPGFQGFPGVKGEKGNPGFLGSIGPPGPIGPKGPPGPVGGDPGT-LKIISLPGSPGPPGTP
α4 1381:PQGPHGFPGPPGEKGLPGPPGRKGPTGLPGPRGEPGPPADVDDGPRIPGLPGAPGMRGPE
α5 1381:DPGLPGVPGFPGMKGPSGVPGSAGPEGEPGLIGPPGPPGLPGPSGQSIIKGDAGPPGIP
α6 1381:EPGFMGTPGKVGPPGDPGFPGMKGKAGARGSSGLQGDPGQTPTAEAVQVPPGPLGLPGID
        *    *   *   *   *   *   *   *         *   *      *   *  *

α1 1441:GPPGLKGLQGLPGPKGQQGVTGLVGIPGPPGIPGFDGAPGQKGEMGPAGPTGPRGFPGPP
α2 1441:GPPGAPGEIGPQGPGEPGFRGAPGKAGPQGRGGVSAVPGFRGDEGPIGHQGPIGQEGAP
α3 1441:GEPGMQGEPGPPGPPGNLGPGGPRGKPGKDGKPGTPGPAGEKGNKGSKGEPGPAGSDGLP
α4 1441:GAMGLPGMRGPPG-PGGKGEPGLDGRRGVDGVPGSPGPPGRKGDTGEDGYPGGPGPPG-P
α5 1441:GQPGLKGLPGPQGPQGLPGPTGPPGDPGRNGLPGFDGAGGRKGDPGLPGQPGTRGLDGPP
α6 1441:GIPGLTGDPGAQGPVGLQGSKGLPGIPGKDGPSGLPGPPGALGDPGLPGLQGPPGFGEAP
        *   *   *   *   *   *   *   *   *   *   *   *     *   *

α1 1501:GPDGLPGSMGPPGTPSVDHGFLVTRHSQTIDDPQGPSGTKILYHGYSLLYVQGNERAHGQ
α2 1501:GRPGSPGLPGMPG-RSVSIGYLLVKHSQTDQEPMGPVGMNKLWSGYSLLYFEGQEKAHNQ
α3 1501:GLKGKRGDSG-SPATWTTRGFVFTRHSQTTAIPSGPEGTVPLYSGFSFLFVQGNQRAHGQ
α4 1501:-I-DGPGPGKG-FGPGYLG-GFLLVLHSQTDQEPTGPLGMPRLWTGYSLLYLEGQEKAHNQ
α5 1501:GPDGLQGPPGPPGTSSVAHGFLITRHSQTTDAPQGPQGTLQVYEGFSLLYVQGNKRAHGQ
α6 1501:GQQGPFGMPGMPG-QSMRVGYTLVKHSQSEQVPPGPIGMSQLWVGYSLLFVEGQEKAHNQ
        *   *   *          *        *

α1 1561:DLGTAGSGLRKFSTMPFLFGNINNVGNFASRNDYSYWLSTPEPMPMSMAPITGENIRPFI
α2 1561:DLGLAGSGLARFSTMPFLYGNPGDVGYYASRNDKSYWLSTTAPLP-IM-PVAEDEIKPYI
α3 1561:DLGTLGSGLQRFTTMPFLFGNVNDVGNFASRNDYSYWLSTPALMPMNNMAPITGRALEPYI
α4 1561:DLGLAGSGLPVFSTLPFAYGNIHQVGHYAQRNDRSYWLASAAPLPM-M-PLSEEAIRPYV
α5 1561:DLGTAGSGLRRFSTMPFMFGNINNVGNFASRNDYSYWLSTPEPMPMSMQPLKGQSIQPFI
α6 1561:DLGFAGSGLPRFSTMPFIYGNINEVGHYARRNDKSYWLSTTAPIP-MM-PVSQTQIPQYI
        ***     *   *  ** *    *   **  *  ****  **

α1 1621:SRGAVGEAPAMVMAVHSQTIQIPPGPSGWSSLWIGYSFVMHTSAGAEGSGQALASPGSGL
α2 1621:SRGSVGEAPAIAIAVHSQDVSIPHGPAGWRSLWIGYSFLMHTAAGDEGGGQSLVSPGSGL
α3 1621:SRGTVGEAPAIAIAVHSQTDIPPGPHAGVKGFSFIMFTSAGSEGTGQALASPGSGL
α4 1621:SRGAVGEAPAQAVAVHSQDQSIPPGPQTWRSLWIGYSFLMHTGAGDQGGGQALMSPGSGL
α5 1621:SRGAVGEAPAVVIAVHSQTIQIPHGPQGWDSLWIGYSFMMHTSAGAEGSGQALASPGSGL
α6 1621:SRGSVGEAPSQAIAVHSQDITIPQGPLGWRSLWIGYSFLMHTAAGAEGGGQSLVSPGSGL
        ***  ***      *****     **   *  *  *** *** ***  **      ******

α1 1681:EEFRSAPFIEGHG-RGTGNYYANAYSFWLATIERSEMF-KKPTPSTLKAGELRT-HVSRG
α2 1681:EDFRATPFIEGNGGRGTGHYYANKYSFWLTTIPE-QSFQGSPSADTLKAGLIRT-HISRG
α3 1681:EEFRASPFLEGHGR-GTGNYYSNSYSFWLASLNPERMFRK-PIPSTVKAGELEK-IISRG
α4 1681:EDFRAAPFLEGQGRQGTGHFFANKYSFWLTTVKADLQFSSAPAPDTLKESQAQRQKISRG
α5 1681:EEFRSAPFIEGHG-RGTGNYYANSYFWLATVDVSDMF-SKPQSETLKAGDLRT-RISRG
α6 1681:EDFRATPFIEGSGARGTGHYFANKYSFWLTTVEERQQFGELPVSETLKAGQLHT-RVSRG
        * **  ** ** *** *     *  *****          *   * *      ***

α1 1741:QVGMRRT-
α2 1741:QVGMKNL-
α3 1741:QVGMKKRH
α4 1741:QVGVKYS-
α5 1741:QVGMKRT-
α6 1741:QVGMKSL-
        ***
```

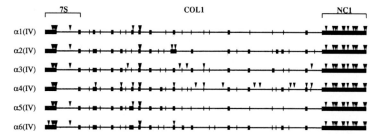

Fig. 2. Overall distribution of the cysteinyl residues and interruption of the Gly-X-Y triplets within the COL domain among α(IV) chains. Locations of the cysteinyl residues are indicated by triangles. Note that these residues are well conserved in the 7s and NC1 domains, but α4(IV) chains contain more cysteinyl residues than others. The NC1 domains and interruption of the Gly-X-Y triplets are indicated by bars.

discussed below, makes it possible to imagine that the six current genes diverged from genes that were evolved through duplication from an ancestral gene (see below and Fig. 3).

A certain degree of homology is observed between the nucleotide sequences of the cDNA encoding the six α(IV) collagen chains and their amino acid sequences. We speculate that the type IV collagen genes and their proteins evolved as shown in Fig. 3. Presumptive ancestral type IV collagen α chain could have formed a homotrimeric molecule. Gene duplication events gave rise to a common ancestral gene for α1(IV), α3(IV), and α5(IV) chains and another common ancestral gene for

←Fig. 1. Comparison of the amino acid sequences of the six α(IV) chains. Amino acid sequences of the six α(IV) chains had high homology and were aligned. To get high homology we introduced gaps with short bars. Amino acid residues at the Gly-X-Y regions are underlined. When the amino acid residues are all conserved in the six α(IV) chains, stars are shown under the residues. Cysteinyl residues are all highlighted. Note that the positions of most of the cysteinyl residues were well conserved, especially for the 12 residues at the NC1 domain.

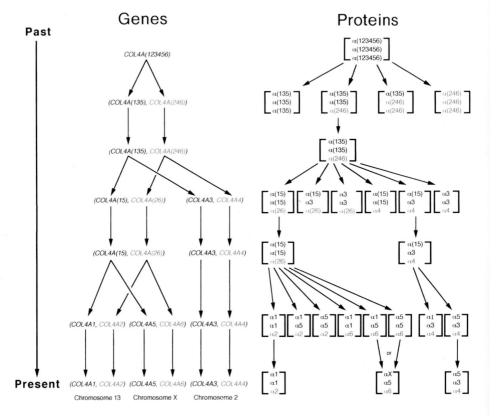

Fig. 3. Evolution of type IV collagen genes and their proteins by gene
duplication. New combinations of triple-helical structure of α chains were
produced by gene duplication and certain combinations were fixed as time
passed. Triple-helix structure was presumed to be two odd-number chains
and one even-number chain. The ancestor gene of the α1–α6 chains is
indicated as *COL4A(123456)*, and its proteins as α(123456). The other
ancestor genes and their proteins are shown in a similar way. Combination of α chains is indicated in brackets.

α2(IV), α4(IV), and α6(IV) chains. These two genes differentiated as time passed into individual genes with similar but
different sequences. Consequently, trimeric collagen molecules
have been formed with two even-number α chains and one

odd-number α chain. Later, we theorize that the ancestral odd-number gene gave rise to a common ancestral gene for $\alpha 1(IV)$ and $\alpha 5(IV)$ chains and a gene for $\alpha 3(IV)$ chain by gene duplication. Probably at about the same time another gene duplication might have happened in the ancestral even-number gene and a common ancestral gene for $\alpha 2(IV)$ and $\alpha 6(IV)$ chains and a gene for $\alpha 4(IV)$ chain were generated, because the even-number ancestral gene and the odd-number ancestral gene were already located together at the same chromosome. As time passed, the genes differentiated and the ancestral chain for $\alpha 1(IV)$ and $\alpha 5(IV)$ formed trimeric molecules with the ancestral chain for $\alpha 2(IV)$ and $\alpha 6(IV)$ together with $\alpha 3(IV)$ and $\alpha 4(IV)$ chains. The gene duplication event then occurred again. Duplication at this time happened in the ancestral gene for $\alpha 1(IV)$ and $\alpha 5(IV)$ chains and in the ancestral gene for $\alpha 2(IV)$ and $\alpha 6(IV)$ chains together, which brought about the four independent genes for $\alpha 1(IV)$, $\alpha 5(IV)$, $\alpha 2(IV)$, and $\alpha 6(IV)$ chains. During this differentiation, a new $\alpha 5(IV)$ chain was selected to form a trimeric molecule with $\alpha 3(IV)$ and $\alpha 4(IV)$ chains and new $\alpha 1(IV)$ chain was abandoned. $\alpha 1(IV)$ chain formed a heterotrimeric molecule with $\alpha 2(IV)$ chain, and $\alpha 5(IV)$ and $\alpha 6(IV)$ chains together formed a molecule. These three combinations are the trimeric forms that we detected in the glomerular basement membranes by immunohistochemistry using specific monoclonal antibodies.

Speculation from comparison of the structure of the six collagen IV genes can explain the presence of a collagen heterotrimer composed of $\alpha 3(IV)$, $\alpha 4(IV)$, and $\alpha 5(IV)$ chains in glomerular basement membranes. A heterotrimer with $\alpha 5(IV)$ and $\alpha 6(IV)$ chains can also be explained by this hypothesis without contradiction. The characteristic features of the evolution of collagen IV genes are that duplication events happened twice in a set of neighboring ancestral collagen IV genes, these events gave rise to the six different genes, and the six gene products are functioning in basement membranes in various tissues.

II. SUPRAMOLECULAR AGGREGATES OF TYPE IV COLLAGENS

Type IV collagen molecules appear to assemble into a complex network through noncovalent hydrophobic interactions. The network becomes rigid and cross-linked through disulfides and lysyl oxidase-derived bridges. Reconstitution studies (24) and analysis of tissue basement membranes (25) have revealed the two binding interactions that contribute to network formation: 1) C-terminal dimerization of collagen molecules, an interaction where a disulfide exchange occurs, and 2) 7S tetramer formation of four collagen molecules at their amino-termini. The tetramer formation prevented the 7S domains from proteolytic degradation and the 7S domain was originated from its sedimentation coefficient. This 7S domain is associated with asparagine-linked carbohydrates, which may define the correct axial and azimuthal orientation of the molecules for proper matching of disulfide pairs and lysine-derived cross-links (26-28). Further interactions of the tetramers and dimers form a complex network that includes lateral interactions. Other components such as laminin, nidogen, and heparan sulfate proteoglycans participate in building the three dimensional basement membrane meshwork. These biologically active aggregates attract cells to interact in different ways depending upon their differentiation states.

In concert with other constituents of basement membranes, collagen IV provides mechanical strength and serves as a scaffold; it is also involved in cell binding. The cell-binding site for the human $[\alpha1(IV)]_2\alpha2$ molecule is located approximately 100 nm away from the amino-terminus where the two integrin receptors, $\alpha1\beta1$ and $\alpha2\beta1$, are recognized (29, 30). The amino acid residues on the $[\alpha1(IV)]_2\alpha2$ molecule essential for $\alpha1\beta1$ recognition have been pinpointed at the Asp[461] on $\alpha1$ and Arg[461] on $\alpha2$ chain. Asp and Arg also seem to be essential for $\alpha2\beta1$ recognition (31).

III.   CHROMOSOMAL LOCATION AND TRANSCRIPTION UNIT OF TYPE IV COLLAGEN GENES

The COL4A1 gene encoding the human α1(IV) chain was first isolated and characterized in detail (32). Its chromosomal localization was 13q34 (33–36) which was, surprisingly, exactly the same as that of the COL4A2 gene for α2(IV) chain (37–39). Soon after these reports, a quite intriguing demonstration of close linkage of the two genes was described: the COL4A1 and COL4A2 on human chromosome 13 are oriented head-to-head and connected by a short promoter region (40, 41). The two genes share a common bidirectional promoter, therefore, the direction of each transcription is the opposite, as shown in Fig. 4A. The transcription start sites for the COL4A1 and COL4A2 are separated by only 130 bp. Similar structure and transcription of the genes for the mouse col4a1 and col4a2 were also reported (42–47). Detailed investigation of the promoter activity for divergent transcription was done only for COL4A1 and COL4A2. Interestingly, the promoter by itself does not possess the transcription activity for the two directions, but activator regions for both genes that are located around the first exon/ intron boundaries activate the promoters for each direction (40). SP1 binding site in the middle of a potential palindrome symmetry within the bidirectional promoter region is characteristic as are other cis-acting elements such as CCAAT motif and CTC box (48–51).

The gene encoding the human α5(IV) chain, COL4A5, was assigned to chromosome Xq22 (4). When this was reported, one expected that there might be another gene that belonged to type IV collagen gene on the opposite side of the COL4A5. It took some time for the predicted gene to be discovered, and it is now known as COL4A6 (5, 6). As expected, the COL4A6 is located not far from the COL4A5 (Fig. 4C), however, it harbors some characteristic features (52): 1) the two alternative promoters that drive the two distinct transcriptions encode α6(IV) chains with different signal peptides; 2) the two transcripts are expressed in a tissue-specific fashion (*53*); 3) analysis of the sequence immediately upstream of the transcription start sites revealed some features of housekeeping genes. To investigate if the translation products of the two genes are co-localized in various tissues, we raised $\alpha5(IV)$ chain- and $\alpha6(IV)$ chain-specific rat monoclonal antibodies against synthetic peptides reflecting sequences near the carboxy terminus of each NC1 domain using our new technique, the rat lymph node method (*54*) as described below (*53*). We demonstrated that the COL4A5 and COL4A6 genes are expressed in a tissue-specific manner, presumably due to the unique function of the bidirectional promoter for both genes

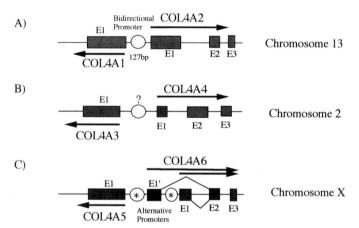

Fig. 4. Schematic representation of the three pairs of bidirectional promoters and COL4A1-6. Expression of the COL4A1 and COL4A2 genes is well characterized among the three pairs. It seems to be controlled by a bidirectional promoter located in the 127 bp-intergenic region between the two genes (*47–51*). The COL4A3 and COL4A4 genes are not yet characterized but both are known to be present within a 500 kb genomic fragment on chromosome 2. The genes COL4A5 and COL4A6 are aligned in opposite directions in the chromosome Xq22 region. There are two alternative promoters that drive two transcripts for the COL4A6 gene (*52*).

(*52*), which is different from that for COL4A1 and COL4A2 (*48–51*).

The gene coding for the human α3(IV) chain, COL4A3, was assigned to chromosome 2q35–37 (*55*). Similarly, the gene for the human α4(IV) chain, COL4A4, was localized at the same spot (*56, 57*). Although it is expected that the two genes are arranged head-to-head as seen in COL4A1 and COL4A2, no detailed structure dealing with the close relation of the two genes has yet been reported (see Fig. 4B). Our preliminary result showed that these genes are located much closer together than the COL4A1 and COL4A2, suggesting that the promoter regions as well as *cis*-acting elements necessary for the transcriptional regulation for both are overlapped onto the coding regions for the opposite genes (*58*).

IV. ESTABLISHMENT OF PEPTIDE DERIVED α(IV) CHAIN-SPECIFIC MONOCLONAL ANTIBODIES

During the course of our investigation into nephritogenic activity from basement membrane fraction, we came across a situation requiring monoclonal antibodies specific for the α(IV) chains. Sado and his colleagues developed a novel technique for monoclonal antibodies using enlarged rat lymph nodes (*54*), with which ten times more target hybridoma cells can be obtained than by conventional technique using mouse spleen cells (Table I).

Since one of the objectives of our experiments was to raise

TABLE I
Number of Positive Cells Using Rat Lymph Node Method

Exp.	Mouse spleen method[a]	Rat spleen method[a]	Rat lymph node method[a]
1	10	10	50
2	11	8	90
3	9	7	80

[a]Number of positive cells per total 384 cells (*54*).

α(IV) chain specific antibodies, we synthesized peptides from various parts of the human α(IV) sequences derived from the nucleotide sequences of cDNAs. Many of them are from the 27 residue-long peptides that are located close to the carboxy

```
α1 :  SVDHGFLVTRHSQTIDDPQCPSGTKILYHGYSLLYVQGNERAHGQDLGTAGSCLRKFSTMPF
α2 :   VSIGYLLVKHSQTDQEPMCPVGMNKLWSGYSLLYFEGQEKAHNQDLGLAGSCLARFSTMPF
α3 : ATWTTRGFVFTRHSQTTAIPSCPEGTVPLYSGFSFLFVQGNQRAHGQDLGTLGSCLQRFTTMPF
α4 : PGYLG-GFLLVLHSQTDQEPTCPLGMPRLWTGYSLLYLEGQEKAHNQDLGLAGSCLPVFSTLPF
α5 :  SVAHGFLITRHSQTTDAPQCPQGTLQVYEGFSLLYVQGNKRAHGQDLGTAGSCLRRFSTMPF
α6 :  MRVGYTLVKHSQSEQVPPCPIGMSQLWVGYSLLFVEGQEKAHNQDLGFAGSCLPRFSTMPF


      LFCNINNVCNFASRNDYSYWLSTPEPMPMSMAPITGENIRPFISRCAVCEAPAMVMAVHSQ
      LYCNPGDVCYYASRNDKSYWLSTTAPLP-IM-PVAEDEIKPYISRCSVCEAPAIAIAVHSQ
      LFCNVNDVCNFASRNDYSYWLSTPALMPMNMAPITGRALEPYISRCTVCEGPAIAIAVHSQ
      AYCNIHQVCHYAQRNDRSYWLASAAPLPM-M-PLSEEAIRPYVSRCAVCEAPAQAVAVHSQ
      MFCNINNVCNFASRNDYSYWLSTPEPMPMSMQPLKGQSIQPFISRCAVCEAPAVVIAVHSQ
      IYCNINEVCHYARRNDKSYWLSTTAPIP-MM-PVSQTQIPQYISRCSVCEAPSQAIAVHSQ


      TIQIPPCPSGWSSLWIGYSFVMHTSAGAEGSGQALASPGSCLEEFRSAPFIECHG-RGTCN
      DVSIPHCPAGWRSLWIGYSFLMHTAAGDEGGGQSLVSPGSCLEDFRATPFIECNGGRGTCH
      TTDIPPCPHGWISLWKGFSFIMFTSAGSEGTGQALASPGSCLEEFRASPFLECHGR-GTCN
      DQSIPPCPQTWRSLWIGYSFLMHTGAGDQGGGQALMSPGSCLEDFRAAPFLECQGRQGTCH
      TIQIPHCPQGWDSLWIGYSFMMHTSAGAEGSGQALASPGSCLEEFRSAPFIECHG-RGTCN
      DITIPQCPLGWRSLWIGYSFLMHTAAGAEGGGQSLVSPGSCLEDFRATPFIECSGARGTCH
```

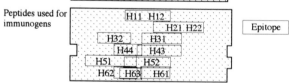

```
Peptides used for
immunogens
                                    H11  H12
                                         H21  H22
                         H32         H31
                            H44      H43
                     H51             H52
                        H62   H63    H61
```

                                                              Epitope

Fig. 5.   Comparison of the six α(IV)NC1 domains and epitope regions for the 13 monoclonal antibodies specific for each α(IV) chain. Amino acid sequences of the six α(IV) chains within the large box are the peptides used for antigens for monoclonal antibodies. Small boxes within the large box indicate the epitopes for the individual monoclonal antibodies.

termini of the NC1 domains for each chain (see Fig. 5), to which a cysteinyl residue was added at the amino-terminus to link to hemocyanin. We were able to screen hybridoma cells that produce antibodies that specifically recognize antigens harboring the amino acid sequences within the NC1 domains of 54 to 83% homology. Thirteen different antibodies, at least two per α chain for the six α(IV) chains, were analyzed further to define

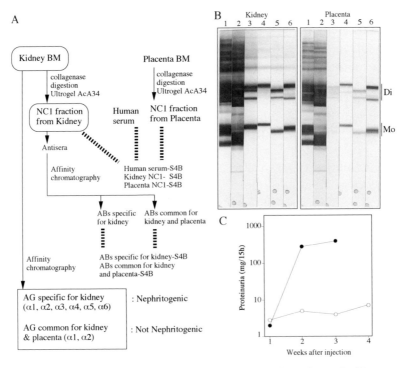

Fig. 6.   A: purification method for nephritogenic antigens. B: Western blot analysis using specific antibodies. Bacterial collagenase digested extracts from kidney and placenta were analyzed by the monoclonal antibodies specific for the six α chains. C: nephritogenic fraction extracted from kidney (●) and non-nephritogenic fraction common for kidney and placenta (○) were injected individually into rats. Proteinuria was observed 2 weeks after injection of the nephritogenic fraction.

epitopes for each using the multi-pin method developed by Geysen *et al.* (*59*). The results showed that even if two typical antibodies against the same antigen recognized different epitopes as shown in Fig. 5, they demonstrated the same staining pattern by immunohistochemistry and Western blot (*53, 54,* see Fig. 6B). The antibodies were utilized to examine the expression and distribution of the six different α(IV) genes. Differential distribution pattern for the six α chains was obtained by immunohistochemistry: the $\alpha 1$(IV) and $\alpha 2$(IV) chains are present in all basement membranes, whereas $\alpha 3$ and $\alpha 4$ chains are mainly restricted to the basement membranes in glomeruli and lung alveoli and were found in lesser amounts in some other basement membranes (*53, 54*). The $\alpha 5$ and $\alpha 6$ chains are colocalized in basement membranes in the skin and smooth muscle cells. Interestingly, in human adult kidney the $\alpha 6$(IV) chain was never detected in the glomerular basement membrane, where the $\alpha 5$(IV) chainn was strongly positive.

V. CHAIN COMPOSITION OF TYPE IV COLLAGEN MOLECULES

The most abundant and ubiquitously distributed composition of α chains for type IV collagen in basement membranes entails a molecule of $[\alpha 1(IV)]_2 \alpha 2(IV)$ and probably a molecule of $[\alpha 1(IV)]_3$ as well (*61-63*). After the discovery of the four other α chains, which are less abundant than the major two, the three large questions raised about them were: 1) biological function, 2) molecular forms, and 3) their distribution in tissues. We attempted to analyze dimerized forms of the α chains via NC1 domains because it is relatively easy to extract NC1 domains if the COL1 domain is removed by bacterial collagenase digestion. Extraction was performed under non-denaturing and non-reducing conditions. Extracted materials were applied on SDS-PAGE, blotted onto polyvinylidene membranes and reacted with α chain specific monoclonal antibodies. However, we detected only homodimerized forms by this method. If heterodimeric forms are there, we should have detected the same size dimers with two different antibodies. The result did not indicate this to be the case, suggesting that C-terminal dimerization occurs only between the same α chains. In turn, 7S tetramer could probably be formed at N-termini by the same or different collagen molecules.

Further, we tried to figure out molecular forms and α chain selection by immunohistochemistry using α chain specific monoclonal antibodies (53, 54). Indirect immunofluorescence staining pattern suggested that distribution of the α3, α4, and α5 chains was quite identical in the human kidney. We then tried to conjugate FITC and Texas red to antibodies specific for α chains individually by different combinations. Tissue sections were stained with the two labeled antibodies. When antigens colocalized on sections, it was seen in yellow, whereas where only one antigen existed it was seen in green or red, depending on the labels. In this way we checked all the combinations of the six antibodies with two colors. Some representative results are shown in Fig. 7. The glomerular and some part of the tubular basement membranes were always stained with α3, α4, and α5 antibodies. In epidermal basement membranes and most of Bowman's capsules, α5 and α6 chains colocalized. These results suggest that there may be three molecular forms: 1) $[\alpha 1(IV)]_2$ $\alpha 2(IV)$, 2) $\alpha 3(IV)\alpha 4(IV)\alpha 5(IV)$, and 3) $\alpha 5(IV)$ and $\alpha 6(IV)$. The first form was already identified by biochemical method (64) and we do not yet know if the third chain of $\alpha 1(IV)$ or the $\alpha 5(IV)$ chain participates in the last form of collagen IV molecules.

The other piece of evidence suggesting that the $\alpha 3(IV)$, $\alpha 4(IV)$, and $\alpha 5(IV)$ chains form a molecule is that in the Alport syndrome case, the mutation of which was identified to cause a premature translation product by introducing a termination codon, the glomerular and tubular basement membranes were unstained by not only $\alpha 5(IV)$ antibody but $\alpha 3(IV)$ and $\alpha 4(IV)$ antibodies (65). This result suggested that $\alpha 5(IV)$ chain was degraded during biosynthesis due to the mutation, and this

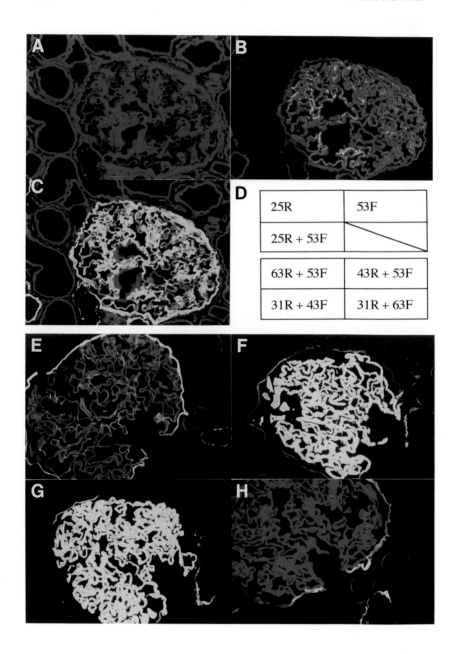

degradation may have caused degradation of the other two chains without forming a complete triple helix.

VI. GOODPASTURE ANTIGEN AND NEPHRITOGENICITY

Goodpasture reported two autopsy cases of nephritis accompanied by pulmonary bleeding after recovery from severe influenza infection (66). This type of nephritis was assumed to be an autoimmune disease caused by autoantibodies against some components of the kidney and was designated Goodpasture syndrome (67). Thereafter, these autoantibodies were found to be against glomerular basement membranes. Further, Goodpasture antigen was identified to be in the NC1 domain of the α3(IV) collagen chain, because autoantibodies present in the patient's serum reacted with the domain (68, 69). In the 1980s, animal models for this human nephritis were created by us and other investigators which developed quite similar symptoms as those observed in the Goodpasture syndrome (70–73).

During the course of our research using animal models, we were able to directly prove that Goodpasture antigens were derived from short stretches of amino acid sequences from

---

←Fig. 7.   Distribution of the six α(IV) chains in kidney by immunohistochemistry. A: α2(IV) chain specific monoclonal antibody, H25 was labeled with Texas red and used to stain the kidney. Basically all the basement membranes in the glomerulus, the mesangial region, and tubules are stained heavily. B: fluorescence labeled-monoclonal antibody, H53, specific for α5(IV) chain was used to stain the kidney. Basement membranes in glomeruli, Bowman's capsule, and specific areas of tubular basement membranes were stained. C: double stained with Texas red-labeled H25 and fluorescence labeled-H53. Both α2(IV) and α5(IV) chains seem to colocalize in the yellow staining region: in glomeruli, Bowman's capsule, and specific areas of tubular basement membranes. D: keys for the individual panels are indicated. E: double staining with Texas red-labeled H63 and fluorescence labeled-H53. F: Texas red-labeled H43 and fluorescence labeled-H53. G: Texas red-labeled H31 and fluorescence labeled-H43. H: Texas red-labeled H31 and fluorescence labeled-H63.

α3(IV) chain, α4(IV) chain, and even α5(IV) chain from several experiments to induce glomerulonephritis using 1) native α(IV)NC1 fractions, 2) oligopeptides themselves, and 3) monoclonal antibodies against oligopeptides from α(IV)NC1 domains. Further, we have some results that at least several Goodpasture sera reacted not only with α3NC1 domain but also with α4(IV)NC1 domain (data not shown).

VII. NEPHRITOGENICITY OF BASEMENT MEMBRANE TYPE IV COLLAGEN CHAINS

Bovine α(IV)NC1 domains were extracted by bacterial collagenase digestion and column chromatography (see Fig. 6). A fraction injected into the rat footpad induced severe glomerulonephritis, which was similar to that seen in human Goodpasture syndrome (72). Western blot analysis demonstrated that the fraction contained mainly the NC1 domains of α1(IV) through α5(IV) chains. However, further purified fraction using antibody affinity chromatography that contains the α1(IV)NC1 and the α2(IV)NC1 did not induce nephritis nor bleeding in lung, suggesting that the nephritogenicity was derived from one of the NC1 domains from α3, α4, and α5 or their combination as shown in Fig. 6.

The nephritogenic activity of the α(IV)NC1 fractions from basement membrane of the human kidney was assayed in animal experiments as described and compared with that of the bovine lung and placenta (74). The kidney and lung fractions showed strong nephritogenicity but the placenta fraction did not have such activity. We further purified kidney-specific fraction and kidney-placenta-lung common fraction using antibody-coupled affinity chromatography. The former fraction showing strong nephritogenic activity contained all the NC1 domains of α1 to α5(IV) collagen chains, but the latter and common fraction containing only α1(IV)NC1 and α2(IV)NC1 was poorly nephritogenic.

VIII. ISOLOGOUS MONOCLONAL ANTIBODIES INDUCE GLO-MERULONEPHRITIS

Experimental models for anti-glomerular basement membrane-nephritis can be classified into two groups. The first is an active model in which basement membrane constituents are injected into animals and induce nephritis. We successfully separated the primary target antigen from the bovine renal basement membrane and identified it as $\alpha 3-\alpha 5$(IV)NC1 domain as described above. The second is a passive one in which antibodies against glomerular basement membrane are injected into animals. When heterologous antibody was used, Masugi nephritis nephrotoxic serum nephritis has been reported in the past (75). Both heterologous and homologous antibodies can induce anti-glomerular basement membrane-nephritis.

Here, we tried to demonstrate whether isologous monoclonal antibodies can indeed induce anti-glomerular basement membrane-nephritis. Injection of the two isologous monoclonal antibodies, SR2 and SR3, caused nephritis in rats (76, 77). The antibodies were obtained from hybridoma cells by fusion of the spleen of a nephritic WKY/NCrj rat injected with rat basement membrane extract and myeloma cells, and stained the glomerular and tubular basement membranes in a linear pattern. His-tological changes were detected within a day of injection. Proteinuria and haematuria appeared on day 2 and became severe. The result suggested that nephritis can be induced by an isologous antibody and that the rat IgG2a subclass is at least nephritogenic.

IX. SHORT STRETCH OF AMINO ACID RESIDUES FROM THE $\alpha 3$NC1 AND THE $\alpha 4$NC1 CAUSED NEPHRITOGENICITY

Sugihara et al. tested the nephritogenic activities of 27 amino acid residue long synthetic peptides derived from each α(IV)NC1 domain (77). The location within each domain and the amino acid sequence are shown in Figs. 1 and 5. Each peptide was coupled to keyhole limpet hemocyanin and injected with adjuvant into the footpads of female WKY/NCrj rats. From 20 rats the number showing proteinuria and haematuria was 2 with $\alpha3$, 8 with $\alpha4$, and 1 with $\alpha5$ peptides. Histological changes including linear deposition of IgG along the glomerular basement membranes were characteristic of those observed in anti-glomerular basement membrane-nephritis. The results indicated the presence of fairly strong nephritogenicity in $\alpha3$, $\alpha4$, and $\alpha5$ but not in $\alpha1$, $\alpha2$, or $\alpha6$. The response from $\alpha5$ was quite low, indicating that nephritogenic activity was mainly from the sequence from $\alpha3$ and $\alpha4$ chains. An experiment to learn whether recombinant proteins produced by eukaryotic cells have nephritogenic activity is now in progress in our laboratory in collaboration with Hudson's laboratory. The advantage of this experiment is that we can test the activity using pure antigens. Our preliminary result indicated that the recombinant $\alpha3(IV)NC1$ and $\alpha4(IV)NC1$ has strong activity.

SUMMARY

Research on the anti-glomerular basement membrane nephritis has made considerable progress in the past ten years and the antigens that cause the nephritis have been narrowed down to a short stretch of amino acid sequences of the NC1 domains of the $\alpha3(IV)$ and $\alpha4(IV)$ chains. It is possible now to detect autoantibodies in the sera for use in diagnostic purposes when you use the ELISA using the NC1 fractions of the $\alpha3-\alpha5$ chains (78, 79). In this chapter we described the structure and expression of the collagen IV genes, predicted some molecular forms of collagen IV molecules, and summarized the current state of the anti-glomerular basement membrane nephritis antigens. Further direction of the research will focus on minimum requirements as an epitope for the anti-glomerular basement membrane nephritis, pathogenesis of glomerulonephritis in inflammatory progression, and development of therapeutic procedures.

Acknowledgments

The original research that provides the basis for this review was supported in part by Grants-in Aid for Scientific Research 04454564, 06454250, 08457154, from the Ministry of Education, Science, Sports and Culture of the Japan International Scientific Research Program: Joint Research 04044124, 06044154, 07044268, and grants from the Yamanouchi Foundation for Research on Metabolic Disorders, the Ryoubi-Teien Memorial Foundation and the Terumo Life Science Foundation.

REFERENCES

1. Kefalides, N.A. (1971) *Biochem. Biophys. Res. Commun.* **45**, 226–234.
2. Brazel, D., Pollner, R., Oberbaumer, I., and Kuhn, K. (1988) *Eur. J. Biochem.* **172**, 35–42.
3. Butkowski, R.J., Langveld, J.P.M., Mieslander, J., Hamilton, J., and Hudson, B.G. (1987) *J. Biol. Chem.* **262**, 7874–7877.
4. Hostikka, S.L., Eddy, R.L., Byer, M.G., Hoyhtya, M., Shows, T.B., and Tryggvason, K. (1990) *Proc. Natl. Acad. Sci. USA* **87**, 1606–1610.
5. Zhou, J., Mochizuki, T., Smeets, H. *et al.* (1993) *Science* **261**, 1167–1169.
6. Oohashi, T., Sugimoto, M., Mattei, M.-G., and Ninomiya, Y. (1994) *J. Biol. Chem.* **269**, 7520–7526.
7. Brazel, D., Oberbaumer, I., Dieringer, H. *et al.* (1987) *Eur. J. Biochem.* **168**, 529–536.
8. Soininen, R., Haka-Risku, T., Prockop, D.J., and Tryggvason, K. (1987) *FEBS Lett.* **225**, 188–194.
9. Hostikka, S.L. and Tryggvason, K. (1988) *J. Biol. Chem.* **263**, 19488–19493.
10. Mariyama, M., Leinonen, A., Mochizuki, T., Tryggvason, K., and Reeders, S. (1994) *J. Biol. Chem.* **269**, 23013–23017.
11. Leinonen, A., Mariyama, M., Mochizuki, T., Tryggvason, K., and Reeders, S.T. (1994) *J. Biol. Chem.* **269**, 26172–26177.
12. Pihlajaniemi, T., Pohjalainen, E.-R., and Myers, J. (1990) *J. Biol. Chem.* **265**, 13758–13766.
13. Zhou, J., Herz, J.M., Leinonen, A., and Tryggvason, K. (1992) *J. Biol. Chem.* **267**, 12475–12481.
14. Muthukumaran, G., Blumberg, B., and Kurkinen, M. (1989) *J. Biol. Chem.* **264**, 6310–6317.

15. Exposito, J.-Y., D'Alessio, M., DiLiberto, M., and Ramirez, F. (1993) *J. Biol. Chem.* **268**, 5249-5254.

16. Blumberg, B., MacKrell, A.J., and Fessler, J.H. (1988) *J. Biol. Chem.* **263**, 18328-18337.

17. Guo, X., Johnson, J.J., and Kramer, J.M. (1991) *Nature* **349**, 707-709.

18. Saus, J., Quinones, S., MacKrell, A. *et al.* (1989) *J. Biol. Chem.* **264**, 6318-6324.

19. Pettitt, J. and Kingston, I.P. (1991) *J. Biol. Chem.* **266**, 16149-16156.

20. Sibley, M.H., Johnson, J.J., Mello, C.C., and Kramer, J.M. (1993) *J. Cell Biol.* **123**, 255-264.

21. Timpl, R. and Dziadek, M. (1986) *Int. Rev. Exp. Pathol.* **29**, 1-112.

22. Timpl, R. (1989) *Eur. J. Biochem.* **180**, 487-502.

23. Yurchenco, P.D. (1994) In *Extracellular Matrix Assembly and Structure* (Yurchenco, P.D., Birk, D.E., and Mecham, R.P., eds.), pp. 351-388, Academic Press, San Diego.

24. Yurchenco, P.D. and Furthmayer, H. (1984) *Biochemistry* **23**, 1839-1850.

25. Yurchenco, P.D. and Ruben, G.C. (1987) *J. Cell Biol.* **105**, 2559-2568.

26. Glanville, R.W., Quian, R.-Q., Siebold, B., Risteli, J., and Kuhn, K. (1985) *Eur. J. Biochem.* **152**, 213-219.

27. Siebold, B., Quian, R.-Q., Glanville, R.W., Hoffman, H., Deutzmann, R., and Kuhn, K. (1987) *Eur. J. Biochem.* **168**, 569-575.

28. Langeveld, J.P.M., Noelken, N.E., Hard, K. *et al.* (1991) *J. Biol. Chem.* **266**, 2622-2631.

29. Vandenberg, P., Kern, A., Ries, A., Luckenbill-Edds, L., Mann, K., and Kuhn, K. (1991) *J. Cell Biol.* **113**, 1475-1483.

30. Kern, A., Eble, J., Golbik, R., and Kuhn, K. (1993) *Eur. J. Biochem.* **215**, 151-159.

31. Eble, J.A., Golbik, R., Mann, K., and Kuhn, K. (1993) *EMBO J.* **12**, 4795-4802.

32. Soininen, R., Huotari, M., Ganguly, A., Prockop, D.J., and Tryggvason, K. (1989) *J. Biol. Chem.* **264**, 13565-13571.

33. Pihlajaniemi, T., Tryggvason, K., Myers, J.C. *et al.* (1985) *J. Biol. Chem.* **260**, 7681-7687.

34. Solomon, E., Hiorns, L.R., Spurr, N. *et al.* (1985) *Proc. Natl. Acad. Sci. USA* **82**, 3330-3334.

35. Boyd, C.D., Weliky, K., Toth-Fejel, S.E. *et al.* (1986) *Hum. Genet.* **74**, 121-125.

36. Emanuel, B.S., Sellinger, B.T., Gudas, L.J., and Myers, J.C. (1986) *Am. J. Hum. Genet.* **38**, 38-44.

37. Griffin, C.A., Emanuel, B.S., Hansen, J.R., Cavenee, W.K., and Myers, J.C. (1987) *Proc. Natl. Acad. Sci. USA* **84**, 512-516.

38. Killen, P.D., Francomano, C.A., Yamada, Y., Modi, W.S., and O'Brien, S.J. (1987) *Hum. Genet.* **77**, 318-324.

39. Boyd, C.D., Toth-Fejel, S., Gadi, I.K. *et al*. (1988) *Am. J. Hum. Genet.* **42**, 309–314.
40. Poschl, E., Pollner, R., and Kuhn, K. (1988) *EMBO J.* **7**, 2687–2695.
41. Soininen, R., Huotari, M., Hostikka, S.L., Prockop, D.J., and Tryggvason, K. (1988) *J. Biol. Chem.* **263**, 17217–17220.
42. Burbelo, P.D., Martin, G.R., and Yamada, Y. (1988) *Proc. Natl. Acad. Sci. USA* **85**, 9679–9682.
43. Kaytes, P., Wood, L., Theriault, N., Kurkinen, M., and Vogeli, G. (1988) *J. Biol. Chem.* **263**, 19274–19277.
44. Killen, P.D., Burbelo, P.D., Martin, G.R., and Yamada, Y. (1988) *J. Biol. Chem.* **263**, 12310–12314.
45. Burbelo, P.D., Bruggeman, L.A., Gabriel, G.C., Klotman, P.E., and Yamada, Y. (1991) *J. Biol. Chem.* **266**, 22297–22302.
46. Tanaka, S., Kaytes, P., and Kurkinen, M. (1993) *J. Biol. Chem.* **268**, 8862–8870.
47. Yamada, Y. and Kuhn, K. (1993) In *Molecular and Cellular Aspects of Basement Membranes* (Rohrbach, D.H. and Timpl, R., eds.), pp. 121–146, Academic Press, San Diego.
48. Heikkila, P., Soininen, R., and Tryggvason, K. (1993) *J. Biol. Chem.* **268**, 24677–24682.
49. Schmidt, C., Fischer, G., Kadner, H., Genersch, E., Kuhn, K., and Poschl, E. (1993) *Biochim. Biophys. Acta* **1174**, 1–10.
50. Fischer, G., Schmidt, C., Opitz, J., Gully, Z., Kuhn, K., and Poschl, E. (1993) *Biochem. J.* **292**, 687–695.
51. Genersch, E. Eckerskorn, C., Lottspeich, F., Herzog, C., Kuhn, K., and Poschl, E., (1995) *EMBO J.* **14**, 791–800.
52. Sugimoto, M., Oohashi, T., and Ninomiya, Y. (1994) *Proc. Natl. Acad. Sci. USA* **91**, 11679–11683.
53. Ninomiya, Y., Kagawa, M., Iyama, K.-I. *et al*. (1995) *J. Cell Biol.* **130**, 1219–1229.
54. Kishiro, Y., Kagawa, M., Naito, I., and Sado, Y. (1995) *Cell Struct. Funct.* **20**, 151–156.
55. Morrison, K.E., Mariyama, M., Yang-Feng, T.L., and Reeders, S.T. (1991) *Am. J. Hum. Genet.* **49**, 545–554.
56. Kamagata, Y., Mattei, M.-G., and Ninomiya, Y. (1992) *J. Biol. Chem.* **267**, 23753–23758.
57. Mariyama, M., Zheng, K., Yang-Feng, T.L., and Reeders, S.T. (1992) *Genomics* **13**, 809–813.
58. Momota, R., Sugimoto, M., Oohashi, T., Kigasawa, K., Yoshioka, H., and Ninomiya, Y. (1998) *FEBS Lett.* **424**, 11–16.
59. Geysen, H.M., Rodda, S.J., Mason, T.J., Tribbick, G., and Schoofs, P.G. (1987) *J. Immunol. Methods* **102**, 259–274.
60. Sado, Y., Kagawa, M., Kishiro, Y. *et al*. (1995) *Histochem. Cell Biol.* **104**,

267-275.

61. Timpl, R. and Dziadek, M. (1986) *Int. Rev. Exp. Pathol.* **29**, 1-112.

62. Timpl, R. (1989) *Eur. J. Biochem.* **180**, 487-502.

63. Hudson, G.B., Reeders, S.T., and Tryggvason, K. (1993) *J. Biol. Chem.* **268**, 26033-26036.

64. Dolz, R., Engel, J., and Kuhn, K. (1988) *Eur. J. Biochem.* **178**, 357-366.

65. Naito, I. Kawai, S., Nomura, S. *et al.* (1996) *Kidney Int.* **50**, 304-311.

66. Goodpasture, E..W. (1919) *Am. J. Med. Sci.* **158**, 863-870.

67. Maddock, R.K. Jr., Stevens, L.E., Reemtsma, K, and Bloomer, H.A. (1967) *Ann. Intern. Med.* **67**, 1259-1264.

68. Wieslander, J., Bygren, P., and Heinegard, D. (1984) *Proc. Natl. Acad. Sci. USA* **81**, 1544-1548.

69. Butkowski, R.J., Wieslander, J., Wisdom, B.J. *et al.* (1985) *J. Biol. Chem.* **260**, 3739-3747.

70. Sado, Y., Okigaki, T., Takamiya, Y., and Seno, S. (1984) *J. Clin. Lab. Immunol.* **15**, 199-204.

71. Naito, I. and Sado, Y. (1989) *J. Clin. Lab. Immunol.* **28**, 187-193.

72. Sado, Y., Kagawa, M., Naito, I., and Okigaki, T. (1991) *Virchows Archiv. B Cell Pathol.* **60**, 345-351.

73. Sado, Y., Naito, I., and Okigaki, T. (1989) *J. Pathol.* **158**, 325-332.

74. Rauf, S., Kagawa, M., Kishiro, Y. *et al.* (1996) *Virchows Archiv.* **428**, 281-288.

75. Masugi, M. (1934) *Beitr. Pathol. Anat.* **92**, 429-466.

76. Sado, Y., Kagawa, M., Rauf, S. *et al.* (1992) *J. Pathol.* **168**, 221-227.

77. Sugihara, K., Sado, Y., Ninomiya, Y., and Wada, H. (1996) *J. Pathol.* **178**, 352-358.

78. Neilson, E.G., Kalluri, R., Sun, M.J. *et al.* (1993) *J. Biol. Chem.* **268**, 8402-8405.

79. Saxena, R., Isaksson, B., Bygren, P. *et al.* (1989) *J. Immunol. Methods* **118**, 73-78.

Extracellular Matrix-Cellular Interaction: Molecules to Diseases (Y. Ninomiya et al., eds.), pp. 261–269, Japan Sci. Soc. Press, Tokyo/S. Karger, Basel (1998)

Genetic Determinants of Extracellular Matrix Assembly and Function

FRANCESCO RAMIREZ, KOSTANTINOS ANDRIKOPOULOS, MAURIZIO DI LIBERTO,[*1] YUTAKA INAGAKI,[*2] PATRICIA GREENWEL, AND SHIZUKO TANAKA

Brookdale Center for Developmental and Molecular Biology, Mount Sinai School of Medicine, New York, New York 10029, U.S.A.

The term connective tissue is traditionally used to define the rather acellular tissue of mesodermal origin which provides structural support to the organism. Unfortunately, the term is both limiting and misleading. It is limiting because it fails to convey the full range of functions which include storage, transport, defense and repair, among others. It is misleading because it implies homogeneity and immutability where heterogeneity and plasticity are the very basis for the physiological diversity of the connective tissue. Since no single term can encapsulate such a complexity, the use of the plural form may at least alleviate the problem of emphasizing the functional diversity of connective tissues. Functional diversification of connective tissues is, in

[*1] Current address: Cellular Biochemistry and Biophysics Program, Memorial Sloan-Kettering Cancer Center, New York, New York 10021, U.S.A.
[*2] Current address: Department of Internal Medicine, National Kanazawa Hospital, Kanazawa 920-8650, Japan

turn, based on the assembly of tissue and stage-specific extracellular matrices, and thus on the regulation of and the interplay between matrix-specific gene products. The present article reviews studies relevant to these topics focusing on the major matrix components, the collagens.

I. GENETIC DETERMINANTS OF EXTRACELLULAR MATRIX ASSEMBLY

The collagens are a large family of glycoproteins which participate in the formation of higher-order macromolecular structures (*1*). Among them are the fibrillar networks whose structural components include types I, II, III, V, and XI collagen (*1*). The biosynthesis of a fibrillar trimer can be generalized as a three-step process which occurs part intracellularly and part extracellularly (*1*). During the first step—the cytosolic step— three procollagen subunits assemble at the carboxy-termini, and nucleate triple-helix formation in an amino-terminal direction with the concomitant cessation of post-translational modifications. The next two steps occur extracellularly. In the first of them, procollagen trimers are processed to the mature form with the removal of the propeptides. This renders the trimers highly insoluble and initiates lateral self-assembly into quarter-staggered aggregates. In the second and less understood extracellular step, the fibrils grow to form tri-dimensional aggregates that display tissue-specific morphologies indicative of distinct functions (*2*). For example, type I collagen fibrils appear as a woven wickerwork in skin, unidirectional cables in tendon, and an orthogonal network in the cornea (*2*).

It was widely believed that the primary structure of the procollagen subunit contains in itself all of the information necessary to direct fibrillogenesis. It was also thought that tissue-specific differences in fibril morphology were mostly due to variations in the degree of post-translational modifications of the procollagen trimer. This view is gradually changing with the increasing appreciation that the assembly of a fibrillar aggregate is a much more complex process than previously thought. Among other factors, it is hypothesized to depend on the integration within the fibrils of more than one collagen type, on the extent and rate of processing of the procollagen precursors, on the association of the fibrils with non-fibrillar collagens, and on direct or indirect interactions between the fibrils and other matrix components (2). The first of these determinants—i.e.: the co-polymerization of different collagens to give rise to heterotypic fibrils—has recently received substantial experimental scrutiny.

Types V and XI are quantitatively minor collagens which are found within the major fibrils of types I and II collagen, respectively (1). This and in vitro reconstitution experiments have suggested that the minor collagens regulate the growth and orientation of the fibrillar networks in cartilaginous and non-cartilaginous tissues, respectively (2). We have explored the significance of heterotypic fibrils in reference to collagen I fibrillogenesis using the technique of gene targeting by homologous recombination in mouse embryonic stem cells (3). To this end, we have targeted one exon within the gene coding for the $\alpha 2$ subunit of type V collagen and thus created a mutant allele which produces a shortened $\alpha 2(V)$ chain (4). Homozygous mice for the $\alpha 2(V)$ mutation survive poorly past the weaning stage. They are smaller and weaker than the wild-type littermates, they move with difficulty and have extremely fragile skin. Upon closer examination, the animals were found to exhibit severe kyphosis probably caused by loss of tensile strength in the supporting ligaments. Such a severe distortion of the vertebral column is the major cause of death, since it interferes with the animal's ability to breath by restricting the volume of the thoracic cavity. Most importantly, electron microscopy revealed profound disorganization of collagenous fibrils in several tissues, including dermis, perichondrium, and cornea. Thus, the genetic data provided conclusive proof for the proposed role of type V collagen in the tri-dimensional organization of the collagen I network in non-cartilaginous matrices. Interestingly, no abnormalities were seen in lungs, vascular tissue, or kidneys where heterotypic collagen I/V fibrils were hypothesized to maintain tissue integrity. Conceivably, other matrix components may substitute for the loss of normal collagen I/V fibrils or they may themselves play a more prominent role in these particular tissues.

Aside from establishing the function of type V collagen, the phenotype of the targeted mice has allowed us to predict the human counterpart (4). Accordingly, a type V collagenopathy was expected to belong to the Ehlers Danlos group of disorders and, more precisely, to one with the combined features of types I and II (4). Such a prediction was eventually proven to be correct. Human type V collagen mutations have, in fact, been associated with pleiotropic manifestations in the skin, eye, skeleton, and vascular system, as well as with disorganized collagen I fibrillar networks (5–7). The same combination of genetic evidence in humans and in mice has firmly established the role of the other minor collagen, type XI, as a regulator of collagen II fibrillogenesis in cartilage (8, 9).

II. GENETIC DETERMINANTS OF EXTRACELLULAR MATRIX FUNCTION

The biosynthesis and degradation of collagen molecules represent the opposite ends of the balance that regulates extracellular matrix remodeling and thus the physiology of tissue repair (10). This balance is, in turn, the product of the dynamic interplay between mesenchyme-specific genes and cytokines released by inflammatory cells (10). Environmental insults which interfere with this delicate equilibrium result in either decreased tissue integrity or excessive collagen deposition. The latter is the central hallmark of fibrosis, a pathological process that irreversibly damages organ function. Thus, fibrosis can be generalized as a chronic and uncontrolled inflammatory/repair process that alters the normal interplay between cellular stimuli and genes coding for extracellular matrix proteins. Collagen synthesis relies upon the utilization of diversified regulatory programs that employ time and cell type specific promoters and enhancers (*11*). As a result, collagen transcription appears to be orchestrated by complex interactions between a variety of *cis*-acting regulatory elements and *trans*-acting nuclear factors. In the case of type I collagen, transcription of the $\alpha 1$ and $\alpha 2$ genes is tightly controlled in that steady-state mRNA levels parallel the relative ratio of the protein chains (*11*). Relevant to pathology, altered production of type I collagen constitutes a focal problem in numerous fibrotic disorders that affect the function of organs like liver, skin, lung, and kidney. These biosynthetic changes are believed to be elicited by cytokines released by the cells in response to as yet unidentified stimuli (*10*).

To gain insight into the modulation of type I collagen synthesis, we have examined the factors that participate in the transcriptional response of the human $\alpha 2(I)$ collagen gene (COL1A2) to the antagonistic stimuli of transforming growth factor-β (TGF-β) and tumor necrosis factor-α (TNF-α). The rationale for choosing this model is based on the critical role that these effectors play in both the physiology and the pathology of connective tissues. TGF-β is a potent stimulator of fibroblast proliferation and of type I collagen production (*12*). TNF-α, on the other hand, inhibits fibroblast type I collagen synthesis and thus exerts an antagonistic action on TGF-β stimulation (*13*). Early work has shown that both TGF-β and TNF-α affect collagen I synthesis by acting mostly at the transcriptional level, but with different kinetics (*12, 13*). Collagen I stimulation by TGF-β is, in fact, an early response that requires no *de novo* protein synthesis; in contrast, collagen I repression by TNF-α is a relatively late event that depends on new protein synthesis. The obvious meaning of these early data is that TGF-β stimulates collagen transcription by post-translational activation of pre-existing factors, whereas TNF-α downregulates it by inducing production of one or more intermediate products. Our recent work has confirmed this hypothesis (*14–17*).

We have shown that the TGF-β and TNF-α signaling pathways converge on the same transcriptional complex (called CYRC) bound to the -330 to -255 region (called CYRE) of the human COL1A2 promoter (14, 15). The sequence consists of two footprinted areas, Box A and Box B. Based on the independent binding of the *trans*-factors and on the distinct contributions of the *cis*-elements to promoter activity, Box A can be further sub-divided into Boxes 5A and 3A. Deletion of Box 5A leads to increased COL1A2 transcription, whereas mutations in Box 3A result in loss of activity. The TGF-β responsive element (TbRE) includes Boxes 3A and B, whereas the TNF-α responsive element (TaRE) encompasses the TbRE and Box 5A. Box 3A binds the ubiquitous activator Sp1, and Box 5A binds a repressor termed C1R. It is unclear what binds to Box B but this sequence is nevertheless critical to the activity of the whole TGF-β responsive complex (TbRC).

In TGF-β activated fibroblasts there is a rapid increase of the *in vitro* binding of TbRC which parallels the stimulation of the endogenous COL1A2 gene (14). The increase includes Sp1 but only through the transactivation of an ill-defined factor of the TbRC (Cx). This conclusion is based on the observation that binding of the oligonucleotide with only the Sp1 site of Box 3A does not increase as does that of the probe containing both Boxes 3A and B. The proposed model of TGF-β stimulation can also be extended to explain the inhibitory effect of TNF-α. This cytokine induces similar visual changes in the binding pattern of the TbRC; it also stimulates the interaction of the C1R repressor at the Box 5A site (15). It is our belief that the antagonistic stimuli of the cytokines at the TbRE site are elaborated by changing either the identity of the Sp1-interacting factor, the nature of the protein modification or by a combination of both. In addition to reducing the positive action of the TbRC by one of the above mechanisms, TNF-α increases the effectiveness of the negative C1R with the net result of CYRC downregulating COL1A2 transcription.

The involvement of Sp1 has been confirmed by a more recent study which has analyzed the effects of inhibiting Sp1 activity on the basal activity of the COL1A2 gene and its ability to be stimulated by TGF-β (*16*). There has also been some progress in evaluating the nature of the modification that modulates the activity of the transcriptional complex (*17*). We have, in fact, correlated tyrosine dephosphorylation of nuclear proteins to increased binding of the TbRC and collagen gene stimulation. The conclusion was based on several lines of evidence. First, pre-incubation of nuclear extracts with PTPase—a tyrosine phosphatase—but not with PP2A—a serine/threonine phosphatase—enhanced binding of the TbRC to the same degree as culturing the cells in the presence of TGF-β. Second, addition of genistein—a tyrosine kinase inhibitor—to cultured fibroblasts led to markedly increased COL1A2 expression, whereas addition of sodium orthovanadate—a tyrosine phosphatase inhibitor —substantially decreased it. Similarly, transient and stable transfection experiments showed that genistein and sodium orthovanadate have opposite effects on TbRE-mediated transcription. Third, nuclear proteins isolated from genistein-treated cells interacted with the TbRE significantly more strongly than those from untreated cells. Finally, pre-treatment of cells with sodium orthovanadate abrogated the binding of nuclear proteins to the TbRE. Incidentally, pre-incubation of the nuclear extract with phosphatases had no effect on C1R binding to Box 5A. Thus, increased C1R binding is not regulated by changes in phosphorylation. In view of this evidence and the protein synthesis-dependence of the collagen response, TNF-α may either alter the activity of the C1R protein or stimulate the expression of the C1R gene.

SUMMARY

The study of the regulation and the function of the collagen genes is contributing greatly to our understanding of the genetic determinants responsible for the assembly of extracellular matrices and the physiology of connective tissues. The ultrastructural consequences of mutating minor collagens on the function of the major collagens have clearly demonstrated that the role of matrix proteins should be viewed within the context of the end-products, the tissue-specific macroaggregates. The analysis of collagen gene expression in response to environmental stimuli has indicated that tissue diversification is the product of combinatorial interactions of ubiquitous and cell-specific factors, as well as of the convergence of diverse intracellular signaling pathways. As this work progresses, we will undoubtedly learn more about the molecular and cellular mechanisms which underlay the ever-changing interplay between the cell and its extracellular environment. Ultimately, this information will enable us to superimpose these genetic determinants upon the cellular phenotype of the developing organism and thus, to finally acquire full appreciation of the physiology and pathology of the connective tissues.

Acknowledgments

This is article # 246 from the Brookdale Center for Developmental and Molecular Biology. The authors wish to thank Ms. Karen Frith for typing the manuscript.

REFERENCES

1. van der Rest, M. and Garrone, R. (1991) *FASEB J.* **5**, 2814–2823.
2. Linsenmayer, T.F. (1991) In *Cell Biology of Extracellular Matrix* (Hay, E.D., ed.), 2nd Ed., pp. 7–44, Plenum Press, New York.
3. Capecchi, M.R. (1989) *Trends Genet.* **51**, 70–76.
4. Andrikopoulos, K., Liu, X., Keene, D.R., Jaenisch, R., and Ramirez, F. (1995) *Nature Genet.* **9**, 31–36.
5. Toriello, H.V., Glover, T.W., Takahara, K. *et al.* (1996) *Nature Genet.* **13**, 361–365.
6. Wenstrup, R.J., Langland, G.T., Willing, M.C., D'Souza, V.N., and Cole, W.G. (1996) *Human Mol. Genet.* **5**, 1733–1736.
7. Da Paepe, A., Nuytinck, L., Hausser, I., Anton-Lamprecht, I., and Naeyaert, J.M. (1997) *Am. J. Hum. Genet.* **60**, 547–554.
8. Li, Y., Lacerda, D.A., Warman, M.L. *et al.* (1995) *Cell* **80**, 423–430.
9. Vikkula, M., Mariman, E.C.M., Lui, Y.C.H. *et al.* (1995) *Cell* **80**, 431–437.
10. Clark, R.A.F. (1995) In *The Molecular and Cellular Biology of Wound*

Repair (Clark, R.A.F., ed.), 2nd Ed., pp. 3–50, Plenum Press, New York.

11. Slack, J.L., Liska, D., and Bornstein, P. (1993) *Am. J. Med. Genet.* **45**, 140–151.

12. Ignotz, R.A., Endo, T., and Massagué, J. (1987) *J. Biol. Chem.* **262**, 6443–6446.

13. Solis-Herruzo, J.A., Brenner, D.A., and Chojkier, M. (1988) *J. Biol. Chem.* **263**, 5841–5845.

14. Inagaki, Y., Truter, S., and Ramirez, F. (1994) *J. Biol. Chem.* **269**, 14828–14834.

15. Inagaki, Y., Truter, S., Tanaka, S., Di Liberto, M., and Ramirez, F. (1995) *J. Biol. Chem.* **270**, 3353–3358.

16. Greenwel, P., Inagaki, Y., Hu, W., Welsh, M., and Ramirez, F. (1997) *J. Biol. Chem.* **272**, 19738–19745.

17. Greenwel, P., Hu, W., Kohanski, R.A., and Ramirez, F. (1995) *Mol. Cell. Biol.* **15**, 6813–6819.

Extracellular Matrix-Cellular Interaction: Molecules to Diseases (Y. Ninomiya et al., eds.), pp. 271–288, Japan Sci. Soc. Press, Tokyo/S. Karger, Basel (1998)

Structural Organization and Pathological Changes of the Vitreous Humor

RONG LU,[*1] ZHAO-XIA REN,[*2] AND RICHARD MAYNE[*2]

*Department of Physiological Optics, School of Optometry[*1] and Department of Cell Biology,[*2] University of Alabama at Birmingham, Birmingham, Alabama 35294, U.S.A.*

In the eye of most mammals (including homo sapiens) the vitreous cavity is filled with a loosely formed gel of connective tissue that can be easily dissected away from the surrounding tissues and its structure analyzed. The vitreous gel is largely sequestered from the rest of the body by the basement membrane of the inner surface of the retina (called the inner limiting membrane) and the basement membrane of the lens (called the posterior lens capsule). Normally, the vitreous contains only a few cells called hyalocytes which appear to be a type of macrophage (*1*). The vitreous gel is composed of two separate components of unique organization (*2–4*). The collagen fibrils are long, do not branch and are of small but probably constant diameter (*5*). Between the fibrils is a loosely formed network of hyaluronan molecules which is best observed in the electron microscope after rotary shadowing (*6*).

Previous studies showed that the collagen fibrils of the vitreous are closely related to the collagen fibrils of hyaline cartilage containing type II, type IX, and type V/XI collagen

271

(7–9). Although preparations of collagen fibrils can be obtained from embryonic cartilages (10, 11), our initial objective was to use the vitreous as a source of collagen fibrils that can be easily investigated by rotary shadowing and their structure analyzed by electron microscopy (5). Subsequently, we have also analyzed the vitreous fibrils for different collagen types (9), and for novel non-collagenous proteins that are associated with the collagen fibrils (12).

Normally, with aging, the vitreous undergoes local liquefaction or syneresis which can eventually lead to separation of the collapsed vitreous from the surface of the retina. Premature syneresis can also occur as the result of injury or any disease process affecting vitreous structure (13). Syneresis and vitreous separation can potentially lead to tears or detachment of the retina at sites where the collagen fibrils remain firmly attached to the inner limiting membrane (14). Such detachments can be repaired with a variety of surgical techniques (15). In proliferative diabetic retinopathy and proliferative vitreoretinopathy cells enter the vitreous cavity and their proliferation results in remodeling of the vitreous which can cause retinal detachment (16). Proliferating cells in the vitreous are often associated with the specialized region of the retina known as the macula, and it is proposed that tractional forces generated from cell-matrix interactions at the vitreoretinal border are responsible for macular hole formation (17).

Genetic diseases such as Stickler, Marshall, and Wagner syndromes are also associated with a poorly formed vitreous and this can lead to retinal detachment (18). Recent results describing the genes and mutations that lead to these conditions will be described below. In general, this article will not attempt to review all of the information describing the structure and biochemistry of the vitreous for which many comprehensive reviews are already available (2, 4, 13, 19, 20).

I. STRUCTURE OF THE VITREOUS

1. Collagen Fibrils

Several groups have analyzed the collagen types that can be solubilized from the vitreous by pepsin digestion (see Table I). It appears that the fibrils are assembled from type II collagen (7), type IX collagen (8), and molecules that contain the $\alpha 1(XI)$ and $\alpha 2(V)$ chains (collectively called type V/XI collagen) (9). The major collagen is type II (approximately 80–90% of total collagen), whereas type IX collagen appears to be associated with the surface of the fibrils (see Fig. 1) as occurs in hyaline cartilage (10). However, in the vitreous, the N-terminal knob of the NC4 domain is missing presumably due to the use of the downstream promoter for the transcription of the $\alpha 1(IX)$ chain (21, 22). Less is known about the location of type V/XI collagen in the vitreous fibrils. Attempts to demonstrate that the short N-terminal collagenous domain called COL2 projects from the surface of the fibrils, as proposed for type V collagen of tendon (23) and for type XI collagen in the smaller fibrils of hyaline cartilage (24, 25), have so far been unsuccessful. However, the importance of type V/XI collagen in vitreous structure is clearly shown by mutations in the $\alpha 1(XI)$ chain that give rise to both Stickler syndrome (26, 27), and Marshall syndrome (28). It seems likely that type V/XI plays a key role in the formation and overall organization of vitreous collagen to give

TABLE I
Collagen Types Present in the Collagen Fibrils of Bovine Vitreous

Type	Chain	Chain organization in triple helix
II	$\alpha 1(II)$	$[\alpha 1(II)]_3$
IX	$\alpha 1(IX)$	$\alpha 1(IX)\ \alpha 2(IX)\ \alpha 3(IX)$
	$\alpha 2(IX)$	
	$\alpha 3(IX)$	
V/XI	$\alpha 1(XI)$	$[\alpha 1(XI)]_2 \alpha 2(V) +$ other
	$\alpha 2(V)$	chain organizations

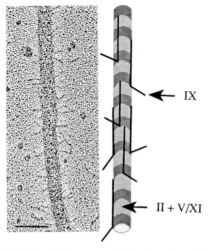

Fig. 1. Left: rotary shadowing of a collagen fibril from bovine vitreous humor. Bar = 100 nm. Right: a model to show the organization of type IX collagen on the surface of a fibril containing type II and type V/XI collagen.

fibrils of a constant diameter by a mechanism that is still poorly understood for this family of molecules in any tissue (29–31).

Interestingly, an appreciable amount of the soluble type II collagen in pig (32) or bovine vitreous (33), is present as a procollagen form in which the non-collagenous amino terminus is still present (called pN-collagen). In later work, two spliced forms of the type II pN-collagen molecule were identified in which exon 2 is either included or excluded from the processed transcript (34, 35). However, most of the soluble collagen in the bovine vitreous is type IX collagen (22, 33), for which the $\alpha 2(IX)$ chain is always in the proteoglycan form with a single chondroitin sulfate chain of Mr 15–60 K (22). The significance of these observations made on such a small percentage of the total collagen present in the vitreous is not clear. The soluble collagen may simply represent molecules that failed to be incorporated into fibrils during fibrillogenesis. However, very little is known concerning the origin and formation of the collagen fibrils present in the mammalian vitreous. In the chicken eye, by *in situ* hybridization, type II collagen is synthesized throughout the neural retina whereas type IX is synthesized specifically at the ciliary body (*36, 37*). It may be that the vitreous collagen fibrils initially form without type IX collagen and that only later is type IX collagen incorporated at the fibril surface. However, not all of the type IX molecules become incorporated into fibrils and some may remain in solution as single molecules.

Non-collagenous proteins are also associated with the collagen fibrils of the vitreous, and we have recently identified and cloned a novel protein called vitrin, that possesses two von Willebrand A domains as potential collagen-binding domains. Vitrin was discovered as it remains associated with the collagen fibrils during centrifugation of the vitreous but is partially released from the fibrils by high salt concentrations (*12*). Interestingly, vitrin appears quite unrelated to other von Willebrand containing proteins (*38*) except for a protein found in the cochlea of the ear called 5B2 (*39*), for which partial cDNA sequences are presently available in GenBank for the chicken, mouse, and human forms of the molecules and the full-length protein sequence for the human and mouse forms of the molecule (*40*).

2. Hyaluronan

Numerous studies have shown that the major glycosaminoglycan of the mammalian vitreous is hyaluronan (*19, 41–43*). More recent studies involving rotary shadowing of human vitreous suggest that lateral associations of hyaluronan molecules occur, the molecules apparently forming networks (Fig. 2) (*6*). Self-association of hyaluronan molecules has been described in several previous studies (*44–47*) but it is not established that hyaluronan can bind directly to the collagen fibrils. Such a potential collagen/hyaluronan interaction is likely to be weak as all or most of the hyaluronan can be simply removed by washing the collagen fibrils after centrifugation of the vitreous

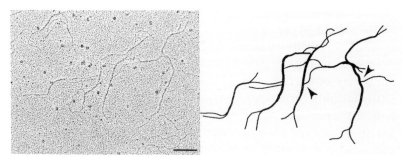

Fig. 2. Left: rotary shadowing of hyaluronan from human vitreous. Bar = 100 nm. Right: drawing to show the interactions of hyaluronan molecules to form a network.

into a pellet. One study has suggested that the single chondroitin sulfate chain of type IX collagen at the NC3 domain of the $\alpha2(IX)$ chain may interact with hyaluronan (33). Previously, chondroitin 6-sulfate was shown to interact specifically with hyaluronan in an agglutination assay using a mixture of beads derivatized with either hyaluronan or chondroitin sulfate (48).

3. Proteoglycans

Recent studies have shown the presence of a versican-like molecule in the bovine vitreous (49) which, together with link protein and hyaluronan, is likely to form aggregates within the vitreous. Interestingly, aggrecan does not appear to be present within the vitreous (49) and earlier studies based on immuno-detection would suggest that both decorin and biglycan are also absent from the vitreous (33).

4. Glycoproteins

Recently, appreciable amounts of glycoproteins potentially related to the mucins were isolated from the vitreous but their function remains unknown (P. Bishop, personal communication). These results probably relate to earlier studies which indicated the presence of unusual glycoproteins in the vitreous

(*42, 43*). Some of these proteins could apparently be labeled with tritiated fucose and were shown to turn over quite rapidly (*50–53*).

5. Other Proteins

The major protein of the vitreous is serum albumin which constitutes slightly less than half of the total protein as judged by sodium dodecyl sulfate-polyacrylamide gel electrophoresis (SDS-PAGE) followed by Coomassie Blue staining (*43*). Numerous other bands are also present on these gels but these proteins remain largely unidentified. In one study, immunological evidence was presented that the protein known as COMP is present in the vitreous (*54*). Such analyses are potentially very relevant to recent results showing that the chondrocytes of patients with pseudoachondroplasia contain intracellular accumulations of both COMP and type IX collagen (*55*). Mutations in COMP are known to result in either pseudoachondroplasia or the related condition known as multiple epiphyseal dysplasia (*56, 57*). Interestingly, a mutation in the $\alpha2(IX)$ collagen chain has also been described leading to multiple epiphyseal dysplasia (*58*). However, eye involvement in these disorders of cartilage has, to date, not been described.

Recently, we performed a systematic analysis by N-terminal amino acid sequencing of the proteins present in the bovine vitreous after SDS-PAGE and Western blotting. Most of the N-termini were blocked but exceptions include amyloid precursor protein (*59*), interphotoreceptor retinoid binding protein (ref. *60*; Z-X. Ren, and R. Mayne, unpublished results), and a novel protein called vitrin (*12*).

6. Microfibrils

In addition to the collagen fibrils, a second beaded fibril can be identified throughout the vitreous gel and separated from the collagen fibrils by isopycnic centrifugation (*61*). These microfibrils were previously observed in the chicken vitreous (*62*) and subsequently identified as being derived from the zonular fibrils of the lens (*5, 63, 64*). However, it would appear that some of these fibrils are widely dispersed throughout the vitreous. Interestingly, other studies have identified microfibrils of type VI collagen in the human and bovine vitreous (*65*).

II. STABILITY OF THE VITREOUS WITH AGE

During the aging process in homo sapiens the vitreous begins to liquify in selected areas by a process known as syneresis. This process can often progress until the vitreous detaches from the retina. However, if the vitreous remains locally attached to the retina, sufficient force may sometimes be generated to cause a retinal detachment. At present, very little is known about the mechanism of syneresis although electron microscopic analysis suggests that a localized collapse of the collagen gel is involved (*13*). Experimental treatment of the vitreous with enzymes that digest hyaluronan such as testicular hyaluronidase, streptococcal hyaluronidase or chondroitinase ABC all result in the collapse of the vitreous gel with aggregation of the collagen fibrils (*4*). Other studies have recently shown that even the normal vitreous has hyaluronidase activity and that turnover of hyaluronan may involve this enzyme (*66*). However, in this study, the vitreous in pathological conditions was not investigated for hyaluronidase activity. It is possible that syneresis arises from a failure of the hyaluronan network to keep the collagen fibrils separated from each other. Other studies show that the vitreous also contains matrix metalloproteinase-2 (MMP-2) (*67*), which cannot degrade soluble type II collagen, but will degrade type V/XI collagen and type IX collagen fragments (*68*). MMP-2 was also found to disrupt the vitreous gel *in vitro* releasing soluble protein but not glycosaminoglycan from the gel (*68*). Interestingly, the collagen fibrils of a collapsed vitreous are often laterally associated with each other suggesting a more ordered structure for the normal vitreous than is immediately apparent (*20*) (see Fig. 3). Such aggregates of collagen fibrils cannot be easily separated by high salt or by chaotrophic agents

Fig. 3.   Electron micrograph of a collapsed vitreous obtained after treatment with streptococcal hyaluronidase (*4*). Note the large number of parallel-aligned collagen fibrils which remain separated from each other. Bar = 100 nm.

and may relate to the lateral aggregates of collagen fibrils commonly observed in the older human vitreous (*69*).

III.   REORGANIZATION OF VITREOUS BY CELLS

Normally, except for hyalocytes, cells are prevented from entering the vitreous cavity by the inner limiting membrane of the retina. However, injury to the eye, diabetes mellitus and a variety of inflammatory and idiopathic conditions can all result in cells entering and proliferating in the vitreous cavity (*13*). It is also likely that cells enter the vitreous cavity from the macula where the inner limiting membrane is very thin (70). Such cells can assemble an extracellular matrix to form an epiretinal membrane and can also exert tractional forces on the vitreous cavity potentially causing both macular hole formation (17) and retinal detachment (13).

In appearance, the inner limiting membrane is a typical basement membrane, containing known basement membrane components such as laminin (71), fibronectin (71), type IV collagen (72, 73), and entactin/nidogen (74). However, more recent studies suggest that the inner limiting membrane may be more specialized. Analyses with monoclonal antibodies specific for each of the six chains of type IV collagen (75) showed that the inner limiting membrane contains only the $\alpha3$(IV), $\alpha4$(IV), $\alpha5$(IV), and $\alpha6$(IV) collagen chains and not the $\alpha1$(IV) and $\alpha2$(IV) chains (ref. 76; R. Lu and R. Mayne, unpublished data) in a similar manner to the glomerular basement membrane (75). In addition, by immunofluorescent staining using chain-specific antibodies the laminin chains of the inner limiting membrane appear to be $\alpha2$, $\beta2$, and $\gamma1$ (76, 77). In other studies, the $\beta1$ chain of integrin could be detected at the inner limiting membrane (78), as well as prominent immunofluorescent staining for dystroglycan (79). Both of these cell adhesion molecules are therefore likely to be involved in the attachment of the inner limiting membrane to the Müller cells.

There are other studies which suggest the presence of a unique glycoconjugate in the human inner limiting membrane (80) as well as a potentially novel chondroitin sulfate proteoglycan (81). Interestingly, degradation of this chondroitin sulfate proteoglycan with chondroitinase ABC results in the separation of the vitreous body from the retina (81), and therefore potentially provides an enzymatic method to facilitate a vitrectomy.

Further progress in determining the structure of the inner limiting membrane is most likely to occur as the result of peeling this membrane from the surface of the retina by vitreoretinal surgeons as part of the procedure for repair of macular holes (*82-84*). We have found that it is possible to obtain sufficient quantities of this membrane from surgery to perform a variety of analyses (R. Lu, R. Mayne, W.B. Priester, R. Morris, and C.D. Witherspoon, unpublished results).

The origin of the cells that give rise to idiopathic epiretinal membranes is not well established and morphological criteria were variously reported to be retinal pigmented epithelium, fibrous astrocytes, fibrocytes, myofibroblasts, and macrophages (*85, 86*). Use of antibodies to glial fibrilliary acid protein (GFAP) clearly showed that many of the cells present in epiretinal membranes are glial in origin (*87*). Examination of idiopathic epiretinal membranes by immunofluorescent staining documented the presence of collagen type I, II, III, and IV (*88, 89*), together with type V collagen, laminin, and fibronectin (*88*). Similar results to these were obtained on analysis of i) epiretinal membranes that developed during proliferative vitreoretinopathy after retinal detachment surgery (*90*) and ii) epiretinal membranes of proliferative diabetic retinopathy (*91*). This suggests that a generalized fibrotic response gives rise to epiretinal membrane formation and that a variety of different cell types may contribute.

Further studies on the formation and proliferation of epiretinal membranes and their involvement in vitreous reorganization are likely to require the development of *in vitro* model systems. The ability to grow Müller cells (*92-94*) and retinal pigmented epithelial cells in culture (*95, 96*) has made it possible to develop model systems in which these cells contract either a native collagen gel or isolated vitreous (*97-102*). In one model system, which involves contraction of the bovine vitreous by fibroblasts *in vitro*, the integrin called $\alpha 2\beta 1$ was implicated (*103*). However, this study did not show that this integrin is important for attachment and contraction of the vitreous by either Müller cells or retinal pigmented epithelial cells. In another important study with therapeutic implications for proliferative diabetic retinopathy, it was recently shown that an antagonist to $\alpha v\beta 3$ and $\alpha v\beta 5$ integrins will inhibit neovascularization of the retina (*104*).

IV. GENETIC DISEASES INVOLVING VITREOUS COLLAPSE

Several different genetic diseases have been described in which the vitreous collapses, sometimes leading to a retinal detachment (*18*). In Stickler syndrome, this is always associated with joint involvement and several mutations in type II collagen have been described (*105-107*). Curiously, these mutations usually result in a premature stop codon during translation. Only one half of the normal amount of type II collagen is therefore likely to be formed as the mutant chains will be unable to enter into triple helix formation which requires an intact carboxyl terminus. It seems that the presence of only half the normal amount of type II collagen results in a poorly organized vitreous matrix structure which usually is located at the front of the eye (*108*). Numerous other mutations in type II collagen have been described which do not appear to affect the vitreous but affect the cartilages (*25, 109*). In Marshall syndrome, the phenotypic effects on the eye are similar but there is no joint involvement. Recently linkage and mutations have also been found in the $\alpha 1(XI)$ chain resulting in either a Stickler or a Marshall phenotype (*26, 27*). However, not all Stickler or Marshall families involve mutations in these two genes (*110, 111*). The discovery that the vitreous contains the $\alpha 2(V)$ chain and not the $\alpha 2(XI)$ chain (*9*) resulted in the identification of a subclass of patients who have mutations in $\alpha 2(XI)$ (*112*). These patients have many characteristics of Stickler patients including the facial features but without eye involvement (*113*). In another condition, known as erosive vitreoretinopathy (*114*), none of the different collagen types of the vitreous is involved and this disease together with Wagner syndrome are both linked to chromosome 5q13-14 (*115*). The nature of the protein involved is at present unknown although the genes for both link protein and versican are at this same locus (*116, 117*).

The single mutation in type IX collagen so far described for patients with multiple epiphyseal dysplasia does not apparently result in a failure of the vitreous structure (*58*). This mutation occurs in the COL3 domain of the $\alpha 2$(IX) chain which is close to the amino terminus of the molecule and may affect the formation of the non-collagenous NC4 domain (*3, 25*). This suggests that in cartilage the triple helix of COL3 assembles first before the assembly of the NC4 domain of the $\alpha 1$(IX) chain can proceed correctly. It is the potential interaction of NC4 with other matrix components that is critical for the successful function of type IX collagen in cartilage. In vitreous, the NC4 domain of type IX collagen is missing (*33*) and so mutations in COL3 may be more easily tolerated.

SUMMARY

The vitreous gel provides a relatively simple model system with only a few known components to investigate the structural interactions of connective tissue macromolecules. In addition, collapse of the vitreous gel has important clinical consequences as it can cause a retinal detachment. A summary was presented of the known components of the vitreous gel including the collagen fibrils, the network of hyaluronan molecules and the non-collagenous proteins of the vitreous. Present knowledge of the structural components of the inner limiting membrane was also provided together with a discussion of the cells and connective tissue macromolecules that form epiretinal membranes. Finally, a brief discussion was presented of current knowledge concerning genetic diseases of the vitreous.

Acknowledgments
The authors would like to thank Pauline M. Mayne for her assistance in the preparation of the manuscript. They would also like to thank Brad Priester, M.D. for his critical reading of this manuscript. Original work was supported by NIH EY09908.

REFERENCES

1. Lazarus, H.S. and Hageman, G.S. (1994) *Arch. Ophthalmol.* **112**, 1356-1362.
2. Balazs, E.A. (1982) In *Ocular Anatomy, Embryology and Teratology* (Jakobiec, F.A., ed.), pp. 425-440, Harper and Row, Philadelphia.
3. Brewton, R.G. and Mayne, R. (1994) In *Extracellullar Matrix Assembly and Structure* (Yurchenco, P.D. and Birk, D.E., eds.), pp. 129-170, Academic Press, San Diego.
4. Mayne, R., Brewton, R.G., and Ren, Z-X. (1997) In *Biochemistry of the Eye* (Harding, J.J., ed.), pp.135-143, Chapman and Hall, London.
5. Ren, Z-X., Brewton, R.G., and Mayne, R. (1991) *J. Struct. Biol.* **106**, 57-63.
6. Brewton, R.G. and Mayne, R. (1992) *Exp. Cell Res.* **198**, 237-249.
7. Swann, D.A. and Sotman, S.S. (1980) *Biochem. J.* **185**, 545-554.
8. Ayad, S. and Weiss, J.B. (1984) *Biochem. J.* **218**, 835-840.
9. Mayne, R., Brewton, R.G., Mayne, P.M., and Baker, J.R. (1993) *J. Biol. Chem.* **268**, 9381-9386.
10. Vaughan, L., Mendler, M., Huber, S. *et al.* (1988) *J. Cell Biol.* **106**, 991-997.
11. Mayne, R., van der Rest, M., Bruckner, P., and Schmid, T.M. (1995) In *Extracellular Matrix: A Practical Approach* (Haralson, M.A. and Hassell, J.R., eds.), pp. 73-97, IRL Press, Oxford.
12. Ren, Z-X., Maier, A., and Mayne, R. (1997) *Mol. Biol. Cell* **8:S**, 64a.
13. Streeten, B.A.W. (1982) In *Pathology of Ocular Disease* (Garner, A. and Klintworth, G.K., eds.), pp. 1383-1419, Dekker, New York.
14. Aaberg, T.M. and Machemer, R. (1982) In *Pathology of Ocular Disease* (Garner, A. and Klintworth, G.K., eds.), pp. 1351-1381, Dekker, New York.
15. Wilkinson, C.P., Rice, T.A., Michels, R.G., and Hengst, T. (1996) *Michels Retinal Detachment*, Mosby-Yearbook, St. Louis.
16. Ryan, S.J. (1985) *Am. J. Ophthalmol.* **100**, 188-193.
17. Gass, D.M. (1994) *Am. J. Ophthalmol.* **119**, 752-759.
18. Snead, M.P. (1996) *Eye* **10**, 653-663.
19. Berman, E.R. (1991) In *Biochemistry of the Eye* (Blakemore, C., ed.), pp. 291-307, Plenum Press, New York.
20. Scott, J.E. (1992) *Eye* **6**, 553-555.
21. Nishimura, I., Muragaki, Y., and Olsen, B.R. (1989) *J. Biol. Chem.* **264**, 20033-20041.
22. Bishop, P., McLeod, D., and Ayad, S. (1992) *Biochem. Biophys. Res. Commun.* **185**, 392-397.
23. Linsenmayer, T.F., Gibney, E., Igoe, F. *et al.* (1993) *J. Cell Biol.* **121**, 1181-1189.
24. Keene, D.R., Oxford, J.T., and Morris, N.P. (1995) *J. Histochem. Cytochem.* **43**, 967-979.

25. Olsen, B.R. (1995) *Curr. Opin. Cell Biol.* **7**, 720–727.
26. Richards, A.J., Yates, J.R.W., Williams, R. *et al.* (1996) *Hum. Mol. Genet.* **5**, 1309–1343.
27. Sirko-Osadsa, D.A., Zlotogora, J., Tiller, G.E., Knowlton, R.G., and Warman, M.L. (1996) *Am. J. Hum. Genet.* **59**, A17.
28. Griffith, A.J., Sprunger, L.K., Sirko-Osadsa, D.A., Tiller, G.E., Meisler, M.H., and Warman, M.L. (1998) *Am. J. Hum. Genet.*, in press.
29. Andrikopoulos, K., Liu, X., Keene, R., Jaenisch, R., and Ramirez, F. (1995) *Nature Genet.* **9**, 31–36.
30. Marchant, J.K., Hahn, R.A., Linsenmayer, T.F., and Birk, D.E. (1996) *J. Cell Biol.* **135**, 1415–1426.
31. Li, Y., Lacerda,A., Warman, M.L. et al. (1995) *Cell* **80**, 423–430.
32. Yang, C., Notbohm, H., Açil, Y., Heifeng, R., Bierbaum, S., and Müller, P.K. (1995) *Biochem. J.* **306**, 871–875.
33. Bishop, P.N., Crossman, M.V., McLeod, D., and Ayad, S. (1994) *Biochem. J.* **299**, 497–505.
34. Bishop, P.N., Reardon, A.J., McLeod, D., and Ayad, S. (1994) *Biochem. Biophys. Res. Commun.* **203**, 289–295.
35. Sandell, L.J., Morris, N.P., Robbins, J.R., and Goldring, M.B. (1991) *J. Cell Biol.* **114**, 1307–1319.
36. Newsome, D.A., Linsenmayer, T.F., and Trelstad, R.L. (1976) *J. Cell Biol.* **71**, 59–67.
37. Linsenmayer, T.L., Gibney, E., Gordon, M.K., Marchant, J.K., Hayashi, M., and Fitch, J.M. (1990) *Invest. Ophthalmol. Vis. Sci.* **31**, 1271–1276.
38. Colombatti, A. and Doliana, R. (1996) *Molecular Biology Intelligence Unit: The Superfamily with von Willebrand Factor VA Domains*, R.G. Landes Company, Austin.
39. Robertson, N.G., Khetarpal, U., Gutiérrez-Espeleta, G.A., Bieber, F.R., and Morton, C.C. (1994) *Genomics* **23**, 42–50.
40. Robertson, N.G., Skvorak, A.B., Yin, Y. *et al.* (1997) *Genomics* **46**, 345–354.
41. Meyer, K. and Palmer, J.W. (1934) *J. Biol. Chem.* **107**, 629–634.
42. Swann, D.A. (1980) *Int. Rev. Exp. Pathol.* **22**, 1–63.
43. Balazs, E.A. and Denlinger, J.L. (1984) In *The Eye* (Davson, H., ed.), vol. IA, pp. 533–589, Academic Press, New York.
44. Turner, R.E., Lin, P., and Cowman, M.K. (1988) *Arch. Biochem. Biophys.* **265**, 484–495.
45. Yanaki, T. and Yamaguchi, T. (1990) *Biopolymers* **30**, 415–425.
46. Scott, J.E., Cummings, C., Brass, A., and Chen, Y. (1991) *Biochem. J.* **274**, 699–705.
47. Mikelsaar, R-H. and Scott, J.E. (1994) *Glycoconjugate J.* **11**, 65–71.
48. Turley, E.A. and Roth, S. (1980) *Nature* **283**, 268–271.
49. Bishop, P., Reardon, A., McLeod, D., Heinegård, D., and Sheehan, J. (1996)

Invest. Ophthalmol. Vis. Sci. **37**, S787.

50. Rhodes, R.H., Mandelbaum, S.H., Minckle, D.S., and Cleary, P.E. (1982) *Exp. Eye Res.* **34**, 921–931.

51. Haddad, A., de Almeida, J.C., Laicine, E.M., Fife, R.S., and Pelletier, G. (1990) *Exp. Eye Res.* **50**, 555–561.

52. Haddad, A., Laicine, E.M., de Almeida, J.C., and Costa, M.S.A. (1990) *Exp. Eye Res.* **51**, 139–143.

53. Andre, J.C., Haddad, A., Fife, R.S., and Pelletier, G. (1992) *Exp. Eye Res.* **55**, 65–71.

54. Nguyen, B.Q. and Fife, R.S. (1986) *Exp. Eye Res.* **43**, 375–382.

55. Maddox, B.K., Keene, D.R., Sakai, L.Y. *et al.* (1997) *J. Biol. Chem.* **272**, 30993–30997.

56. Briggs, M.D., Hoffman, S.M.G., King, L.M. *et al.* (1995) *Nature Genet.* **10**, 330–336.

57. Hecht, J.T., Nelson, L.D., Crowder, E. *et al.* (1995) *Nature Genet.* **10**, 325–329.

58. Muragaki, Y., Mariman, E.C.M., van Beersum, S.E.C. *et al.* (1996) *Nature Genet.* **12**, 103–105.

59. Ren, Z-X., Baker, J.R., and Mayne, R. (1997) *Invest. Ophthalmol. Vis. Sci.* **38**, S306.

60. Wiggert, B., Lee, L., Rodrigues, M., Hess, H., Redmond, T.M., and Chader, G.J. (1986) *Invest. Ophthalmol. Vis. Sci.* **27**, 1041–1049.

61. Davison, P.F. and Seery, C.M. (1993) *Curr. Eye Res.* **12**, 107–114.

62. Wright, D.W. and Mayne, R. (1988) *J. Ultrastruct. Mol. Struct. Res.* **100**, 224–234.

63. Wallace, R.N., Streeten, B.W., and Hanna, R.B. (1991) *Curr. Eye Res.* **10**, 99–109.

64. Kielty, C.M., Davies, S.J., Phillips, J.E., Jones, C.J.P., Shuttleworth, C.A., and Charles, S.J. (1995) *J. Med. Genet.* **32**, 1–6.

65. Bishop, P., Ayad, S., Reardon, A., McLeod, D., Sheehan, J., and Kielty, C. (1996) *Graefe's Arch. Clin. Exp. Ophthalmol.* **234**, 710–713.

66. Schwartz, D.M., Shuster, S., Jumper, M.D., Chang, A., and Stern, R. (1996) *Curr. Eye Res.* **15**, 1156–1162.

67. Brown, D., Hamdi, H., Bahri, S., and Kenney, M.C. (1994) *Curr. Eye Res.* **13**, 639–647.

68. Brown, D.J., Bishop, P., Hamdi, H., and Kenney, M.C. (1996) *Curr. Eye Res.* **15**, 439–445.

69. Sebag, J. and Balazs, E.A. (1989) *Invest. Ophthalmol. Vis. Sci.* **30**, 1867–1871.

70. Fine, B.S. and Yanoff, M. (1972) *Ocular Histology. A Text and Atlas*, Harper and Row, New York.

71. Kohno,T., Sargente, N., Ishibashi, T., Goodnight, R., and Ryan, S.J. (1987) *Invest. Ophthalmol. Vis. Sci.* **28**, 506–514.

72. Ishizaki, M., Westerhausen-Larson, A., Kino, J., Hayashi, T., and Kao, W-Y. (1993) *Invest. Ophthalmol. Vis. Sci.* **34**, 2680–2689.
73. Sarthy, V. (1993) *Invest. Ophthalmol. Vis. Sci.* **34**, 145–152.
74. Dong, L-J. and Chung, A.E. (1991) *Differentiation* **48**, 157–172.
75. Ninomiya, Y., Kagawa, M., Iyama, K-I. *et al.* (1995) *J. Cell Biol.* **130**, 1219–1229.
76. Ljubimov, A.V., Burgeson, R.E., Butkowski, R.J. *et al.* (1996) *J. Histochem. Cytochem.* **44**, 1469–1479.
77. Toti, P., de Felice, C., Malandrini, A., Megha, T., Cardone, C., and Villanova, M. (1997) *Neuromuscular Disorders* **7**, 21–25.
78. Brem, R.B., Robbins, S.G., Wilson, D.J. *et al.* (1994) *Invest. Ophthalmol. Vis. Sci.* **35**, 3466–3474.
79. Montanaro, F., Carbonetto, S., Campbell, K.P., and Lindenbaum, M. (1995) *J. Neurosci. Res.* **42**, 528–538.
80. Russell, S.R., Shepard, J.D., and Hageman, G.S. (1991) *Invest. Ophthalmol. Vis. Sci.* **32**, 1986–1995.
81. Hageman, G.S. and Russell, S.R. (1992) *Exp. Eye Res.* **55** (Suppl. 1), 216.
82. Yoon, H-S., Brooks, H.L., Capone, A.Jr., L'Hernault, N.L., and Grossniklaus, H.E. (1996) *Am. J. Ophthalmol.* **122**, 67–75.
83. Williams, D.F. (1997) *Invest. Ophthalmol. Vis. Sci.* **38**, S750.
84. Morris, R., Witherspoon, C.D., Kuhn, F., and Priester, B. (1997) *Vitreoretinal Surg. Technol.* **8**, 4:1–2.
85. Kampik, A., Kenyon, K.R., Michels, R.G., Green, W.R., and de la Cruz, Z.C. (1981) *Arch. Ophthalmol.* **99**, 1445–1454.
86. Smiddy, W.E., Maguire, A.M., Green, W.R. *et al.* (1989) *Ophthalmology* **96**, 811–821.
87. Hiscott, P.S., Grierson, I., Trombetta, C.J., Rahi, A.H.S., Marshall, J., and McLeod, D. (1984) *Br. J. Ophthalmol.* **68**, 698–707.
88. Scheiffarth, O.F., Kampik, A., Günther, H., and von der Mark, K. (1988) *Graefe's Arch. Clin. Exp. Ophthalmol.* **226**, 357–361.
89. Okada, M., Ogino, N., Matsumura, M., Honda, Y., and Nagai, Y. (1995) *Ophthalmic Res.* **27**, 118–128.
90. Jerdan, J.A., Pepose, J.S., Michels, R.G. *et al.* (1989) *Ophthalmology* **96**, 801–810.
91. Jerdan, J.A., Michels, R.G., and Glaser, B.M. (1986) *Arch. Ophthalmol.* **104**, 286–290.
92. Wakakura, M. and Foulds, W.S. (1988) *Invest. Ophthalmol. Vis. Sci.* **29**, 892–900.
93. Hicks, D. and Courtois, Y. (1990) *Exp. Eye Res.* **51**, 119–129.
94. Guidry, C. (1996) *Invest. Ophthalmol. Vis. Sci.* **37**, 740–752.
95. Li, W., Stramm, L.E., Aguirre, G.D., and Rockey, J.H. (1984) *Exp. Eye Res.* **38**, 291–304.
96. Campochiaro, P.A., Jerdan, J.A., and Glaser, B.M. (1986) *Invest. Ophthal-

mol. Vis. Sci. **27**, 1615–1621.

97. Forrester, J.V., Docherty, R., Kerr, C., and Lackie, J.M. (1986) *Invest. Ophthalmol. Vis. Sci.* **27**, 1085–1094.
98. Mazure, A. and Grierson, I. (1992) *Invest. Ophthalmol. Vis. Sci.* **33**, 3407–3416.
99. Guidry, C.M., McFarland, R.J., Morris, R., Witherspoon, C.D., and Hook, M. (1992) *Invest. Ophthalmol. Vis. Sci.* **33**, 2429–2435.
100. Hunt, R.C., Fox, A., Al Pakalnis, V. *et al.* (1993) *Invest. Ophthalmol. Vis. Sci.* **34**, 3179–3186.
101. Guidry, C. (1997) *Invest. Ophthalmol. Vis. Sci.* **38**, 456–468.
102. Hardwick, C., Feist, R., Morris, R. *et al.* (1997) *Invest. Ophthalmol. Vis. Sci.* **38**, 2053–2063.
103. Kupper, T.M. and Ferguson, T.A. (1993) *FASEB J.* **7**, 1401–1406.
104. Friedlander, M., Theesfeld, C.L., Sugita, M. *et al.* (1996) *Proc. Natl. Acad. Sci. USA* **93**, 9764–9769.
105. Ahmad, N.N., Ala-Kokko, L., Knowlton, R.G. *et al.* (1991) *Proc. Natl. Acad. Sci. USA* **88**, 6624–6627.
106. Ritvaniemi, P., Hyland, J., Ignatius, J., Kivirikko, K.I., Prockop, D.J., and Ala-Kokko, L. (1993) *Genomics* **17**, 218–221.
107. Ahmad, N.N., Dimascio, J., Knowlton, R.G., and Tasman, W.S. (1995) *Arch. Ophthalmol.* **113**, 1454–1457.
108. Snead, M.P., Payne, S.J., Barton, D.E. *et al.* (1994) *Eye* **8**, 609–614.
109. Prockop, D.J. and Kivirikko, K.I. (1995) *Annu. Rev. Biochem.* **64**, 403–434.
110. Fryer, A.E., Upadhyaya, M., Littler, M. *et al.* (1990) *J. Med. Genet.* **27**, 91–93.
111. Bonaventure, J., Philippe, C., Plessis, G. *et al.* (1992) *Hum. Genet.* **90**, 164–168.
112. Vikkula, M., Mariman, E.C.M., Lui, V.C.H. *et al.* (1995) *Cell* **80**, 431–437.
113. van Steensel, M.A.M., Buma, P., de Waal Malefijt, M.C., van den Hoogen, F.H.J., and Brunner, H.G. (1997) *Am. J. Med. Genet.* **70**, 315–323.
114. Brown, D.M., Kimura, A.E., Weingeist, T.A., and Stone, E.M. (1994) *Ophthalmology* **100**, 694–704.
115. Brown, D.M., Graemiger, R.A., Hergersberg, M. *et al.* (1995) *Arch. Ophthalmol.* **113**, 671–675.
116. Osborne-Lawrence, S.L., Sinclair, A.K., Hicks, R.C. *et al.* (1990) *Genomics* **8**, 562–567.
117. Iozzo, R.V., Naso, M.F., Cannizzaro, L.A., Wasmuth, J.J., and McPherson, J.D. (1992) *Genomics* **14**, 845–851.

Extracellular Matrix-Cellular Interaction: Molecules to Diseases (Y. Ninomiya et al., eds.), pp. 289–308, Japan Sci. Soc. Press, Tokyo/S. Karger, Basel (1998)

Chondrogenic Differentiation of Mouse Clonal Cell Line ATDC5; Differentiation-dependent Expression of Extracellular Matrix Genes

YUJI HIRAKI,[*1] CHISA SHUKUNAMI,[*1] AND KIYOTO ISHIZEKI[*2]

*Department of Biochemistry, Osaka University Faculty of Dentistry, Suita, Osaka 565-0871,[*1] and Department of Oral Anatomy, Iwate Medical University School of Dentistry, Morioka, Iwate 020-0023,[*2] Japan*

Long bone is formed by replacement of a cartilaginous bone rudiment (endochondral bone formation). Skeletal elements begin to be faintly discernible as the particular regions in which cells become more densely packed within mesenchyme. When the cells are in a condensation stage, they begin to produce large amounts of cartilage-specific molecules such as type II collagen and aggrecan (chondrogenic differentiation). During endochondral bone development, mesenchymal cells pass through at least three distinct differentiation stages: 1) prechondrogenic stage; 2) proliferating chondrocyte stage; and 3) hypertrophic and calcifying chondrocyte stage. Thus, as shown in Fig. 1, the cells undergo an orderly series of changes which involves the transitions of prechondrogenic cells to proliferating chondrocytes through cellular condensation (early-phase differentiation).

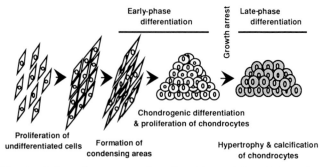

Fig. 1. Schematic representation of multistep differentiation of prechon-drogenic cells.

Chondrocytes at the proliferating stage then proceed to the hypertrophic and calcifying stage (late-phase differentiation). In permanent cartilage, *e.g.* normal articular cartilage on the joint surface, chondrocytes do not undergo the late-phase conversion of phenotype. The late-phase differentiation is characterized by a several-fold increase in cell volume and a marked increase in alkaline phosphatase (ALPase) activity. Calcified chondrocytes allow invasion of blood vessels into cartilage, leading to formation of bony tissue.

This biphasic transition of the cellular phenotype is accompanied by a change in collagen gene activation: prechondrogenic mesenchymal cells express type I collagen mRNA. The early-phase differentiation is characterized by inductive expression of type II and IX collagen genes as well as the aggrecan gene (*1, 2*). The late-phase differentiation is characterized by the onset of expression of type X collagen gene (*2, 3*). Cellular hypertrophy is also accompanied by reduction of aggrecan and type II collagen expression (*2, 3*).

Cultured limb mesenchymal cells exhibit an expression profile of the genes that closely resembles that in cytodifferentiation of chondrocytes *in vivo* (*4*). However, their usefulness has been limited by requirements for an extremely high seeding density of cells in culture and unavailability of a homogeneous cell population (5). Thus, analysis of multistep chondrogenic differentiation awaits the establishment of a clonal cell line that mimics cellular behaviors seen *in vivo*. Cell lines such as RCJ3.1 and CFK2 exemplify the utility of a chondrogenic clonal cell line for the study of skeletal development (6, 7). These cell lines, however, were isolated from fetal calvariae which do not develop via the endochondral pathway but via the intramembranous pathway of bone formation.

The clonal cell line, ATDC5, was isolated from the feeder-independent teratocarcinoma stem cell line AT805 on the basis of chondrogenic potentials in the presence of insulin (8). The high frequency of chondrogenic differentiation of the cells enables monitoring of the phenotypic transitions in terms of molecular and biochemical markers (9). ATDC5 cells reproduce the orderly transition of the differentiation stages of chondrocytes, including calcification (10). β-Glycerophosphate has been a culture supplement used for the stimulation of mineralization *in vitro*, even though it sometimes gives artifactual mineral deposition. However, for induction of calcification in the ATDC5 culture, no supplementation of β-glycerophosphate was required (10). Thus, the cell line is the first example of a chondrogenic cell line which provides an excellent model for the late-phase differentiation of chondrocytes in the absence of β-glycerophosphate supplementation, and offers an opportunity to analyze the molecular mechanism underlying multistep differentiation of chondrocytes during endochondral bone formation.

I. GROWTH PATTERN AND EARLY-PHASE CHONDROGENIC DIFFERENTIATION

ATDC5 cells are maintained in DMEM/Ham's F-12 or αMEM supplemented with 10 μg/ml transferrin, 3×10^{-8} M sodium selenite and 5% fetal bovine serum (FBS). Cells express transcripts for the type I collagen gene, and they rapidly proliferate with a short doubling time of 16 hr. The cells exhibit an elongated fibroblastic morphology at the subconfluent stage in either culture medium. On day 3 of culture, the cells stop growing as a result of contact inhibition at confluence to form a monolayer. Under these conditions, the saturation cell density was approximately 1.41×10^5 cells/cm². No Alcian blue-positive cartilage nodule appears in these cultures (Fig. 2). There is neither accumulation of cartilage matrix nor expression of cartilage-specific transcripts.

Alternatively, the cells were grown in differentiation medium containing $10 \mu g/ml$ insulin. Insulin did not significantly affect cell doubling time or the morphology of the cells in the subconfluent culture. Upon attainment of confluency, the cells similarly ceased to grow by contact inhibition. But, in the presence of insulin, the culture re-entered the growth state having a longer cell doubling time within the three days after confluency to eventually form numerous cartilage nodules by day 21. A transient condensation of cells with an elongated spindle-like

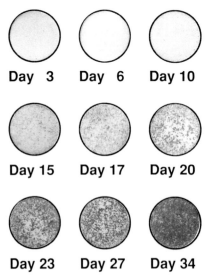

Day 3   Day 6   Day 10

Day 15   Day 17   Day 20

Day 23   Day 27   Day 34

Fig. 2.   Formation of cartilage nodules (Alcian blue staining).

morphology preceded the formation of nodules, as seen in mesenchymal condensation in limb bud chondrogenesis. The nodular structures formed were composed of proliferating cells with a round morphology characteristic of chondrocytes. Logarithmic growth of cells in the post-confluent phase occurred for about 4 days, from day 6 to day 10. The doubling time of the cells during this period was approximately 48 hr consistent with that reported previously (*8*). Similar to insulin, insulin-like growth factor-I (IGF-I; 300 ng/ml) induced cellular condensation and the subsequent formation of cartilage nodules.

As shown in Fig. 2, Alcian blue-positive cartilage nodules were first detectable between day 6 and day 10, and increased in size. Then the apparent growth rate of the culture decreased, probably because a portion of the differentiated cells began to mature. This post-confluent growth of cells, however, continued for 2 weeks until day 21, and then ceased in DMEM/Ham's F-12 medium. On day 21, the total number of cells reached a maximum at 9.1×10^5 cells/cm². Cells surrounding cartilage nodules retained their fibroblastic morphology and presumably did not grow during expansion of the nodules. Assuming that these cells were quiescent, 85% of the total cell population at the end of the growth stage of nodules was estimated to be differentiated chondrocytes.

Even after cessation of post-confluent growth, the Alcian blue-positive area appeared to expand by accumulation of cartilage matrix due to maturation of cells (Fig. 2). The timing when differentiated cells stop growing in cartilage-nodules appears to depend on maturation of cartilage matrix organization. In αMEM, ATDC5 cells began chondrogenic differentiation with a similar kinetics seen in DMEM/Ham's F-12 medium. However, growth of differentiated cells stopped earlier, around day 14 in αMEM. αMEM contains ascorbic acid which is known to stimulate maturation of collagen fibrils. Supplementation of ascorbic acid (50 μg/ml) in DMEM/Ham's F-12 medium similarily resulted in the earlier cessation of cell growth in ATDC5 cell culture.

Chondrogenic differentiation of ATDC5 cells was further characterized by an expression profile of cartilage-characteristic extracellular matrix genes such as aggrecan, type II collagen and type IX collagen. Transcripts for these genes were undetectable in undifferentiated ATDC5 cells. Undifferentiated ATDC5 cells on day 2 clearly expressed type I collagen mRNA which remained detectable throughout the experimental period. In contrast, transcripts for type II and type IX collagen genes were undetectable on day 2. They were readily detectable on day 14 and became maximal around day 21. Analysis of cartilage nodules by transmission electron microscopy (TEM) indicated that differentiated ATDC5 cells were surrounded by thin type II collagen fibrils (approximately 20 nm in diameter) as found in primary cultured chondrocytes (3), but no thick fibrils characteristic of type I collagen were found. Expression of aggrecan gene was also induced as the cells became differentiated.

In cultures of prechondrogenic mesenchymal cells from chick limb buds, there is a first phase in which all cells behave as fibroblasts, exhibiting contact inhibition. These cells rapidly proliferate at this stage, but there is a considerable increase in doubling time of the cells during chondrogenesis (11). Prechondrogenic mesenchymal cells express type I collagen mRNA, but no type II collagen mRNA. As the cells differentiate into chondrocytes, they initiate the synthesis of type II collagen and cease type I collagen synthesis, while they still contain type I collagen transcripts (1). A marked increase in type II collagen mRNA level occurs coincidentally with the condensation stage in chondrogenesis. Thereafter, a continuous and progressive increase in the accumulation of type II collagen mRNA occurs (1). With respect to their chondrogenic potential, morphological appearance, growth behavior, and collagen gene expression, ATDC5 cells appear to retain the characteristics of prechondrogenic stem cells.

Cellular condensation is an important prerequisite for initiation of chondrogenesis. In the presence of insulin, the cell density of ATDC5 culture increased by 50% or less during the 3 post-confluent days. This observation was compatible with an increase in packing cell density during chick limb-bud chondrogenesis *in vivo*. At this stage, cells lose contact inhibition and grow beyond confluence to produce two or three layers of cells. It has been reported that fibroblast growth factor (FGF) signaling plays a role in support of the proliferation of limb-bud mesenchyme and limb pattern formation (*12–15*). Since no mitotic activity is required for the onset of condensation in chick limb chondrogenesis *in vitro*, cellular condensation is mainly mediated by a change of cell motility and cell-cell interactions (*16, 17*). In fact, FGF-2 markedly stimulated the proliferation of confluent undifferentiated ATDC5 cells, but failed to support formation of cellular condensation. Thus, the mitogenic action of insulin does not seem to be of primary importance for induction of condensation. Insulin probably affects cell morphology and motility during differentiation of ATDC5 cells *in vitro*.

IGF-I has been implicated in early limb development and chondrogenesis (*18*). In cultures of ATDC5 cells, IGF-I also induces cellular condensation and formation of cartilage nodules. Therefore, it is possible that insulin acts through an IGF-I signaling pathway. However, IGF-I was less effective than insulin for chondrogenic induction of the cells. Bassas *et al.* reported the presence of insulin immunoreactivity and bioactivity in early chick embryos (*19*). They also found insulin receptors as well as IGF-I receptors on plasma membranes isolated from limb buds (*19*). Telford and co-workers demonstrated the presence of transcripts for insulin receptors and IGF-I receptors in early mouse embryos by reverse transcription-polymerase chain reaction (RT-PCR), and suggested a possibility that these embryos respond to exogenous (*e.g.*, maternal) sources of insulin (*20*). Undifferentiated ATDC5 cells express the transcripts for both insulin receptors and IGF-I receptors. Therefore, it is possible that insulin induces differentiation of ATDC5 cells through an insulin-specific signaling pathway, although the physiological relevance of insulin action to chondrogenic induction remains to be elucidated.

II. LATE-PHASE DIFFERENTIATION OF CHONDROCYTES

In the presence of insulin, hypertrophic cells are discernible in the central regions of the nodular structures under a phase-contrast microscope by day 24. Figure 3 shows the time-course of Alizarin red staining of ATDC5 cell culture. The cells became progressively hypertrophic and no Alizarin red positive nodules were seen in the culture by day 28. On that day, calcified regions began to appear in the cartilage nodules as tiny spots, and they propagated throughout the cartilage nodules with time. No mineral deposition was seen in the flat cell monolayer surrounding the cartilage nodules. On day 35, the central regions of these nodules look dark under a phase-contrast microscope in

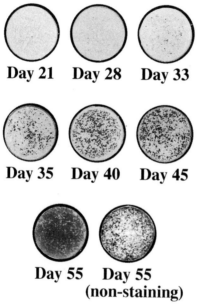

Fig. 3. Time-course of calcification in the culture of differentiated ATDC5 cells (Alizarin red staining).

ATDC5 cell culture. The hypertrophic cells occupied the cartilage nodules as the major cell-population. Alizarin red densely stained the light-impenetrable regions of the nodules, indicating mineralization of hypertrophic chondrocytes. By day 55, the culture was filled with calcified cartilage nodules. Heavily mineralized matrix was readily recognized without staining as white precipitates covering the culture plates (Fig. 3; non-staining).

The ultrastructure of ATDC5 cells was examined throughout the course of differentiation by transmission EM. Cells prior to chondrogenic differentiation retained an ultrastructure characteristic of undifferentiated immature cells: the cells were preserved with large oval euchromatin-rich nuclei containing prominent nucleoli, the profile of rough endoplasmic reticulum, and several mitochondria. The territorial matrix was composed of amorphous fine materials. On day 24, differentiated cells inside the cartilage nodules revealed an ultrastructure characteristic of early hypertrophic chondrocytes seen *in vivo* (*3, 21, 22*). Matrix vesicles are found in the various zones of the epiphyseal growth plate (*21*). Their association with mineral deposition is typically initiated in the zone of hypertrophy *in vivo*. In ATDC5 culture, dense aggregates of fine needle-like crystals were also found in association with matrix vesicles that were in the territorial matrix. Occurrence of matrix vesicles was also noted in the extracellular matrix of differentiated cells. By day 35, many of the matrix vesicles were heavily mineralized. Mineralized matrix vesicles in the extracellular matrix were fused with one another and formed calcospherites at the early stage of mineralization.

Ballock and Reddi studied induction of the hypertrophic phenotype of primary cultured chondrocytes in a chemically-defined medium (*23*). They suggested that insulin and thyroxine were important regulators for the survival and hypertrophy of chondrocytes, although they showed no indication of calcification in the culture. Once initial calcification took place in the culture of ATDC5 cells, it propagated similarly in the absence of serum as long as insulin was present. In the absence of insulin, hypertrophic morphology could not be maintained.

Ascorbic acid facilitated the appearance of hypertrophic cells in the cartilage nodules, as well as the elevation of ALPase activity and expression of type X collagen mRNA (Fig. 4). In its absence, calcification was rarely observed. The supplementation of DME/F12 medium with ascorbic acid induced a phenotypic transition of proliferating chondrocytes to hypertrophic cells. Iscove's modified Dulbecco's medium supplemented with ascorbic acid also worked well for the expression of the hypertrophic phenotype. In ATDC5 cell culture, we usually employ αMEM, which contains 50 μg/ml ascorbic acid in the standard formula, for induction of the hypertrophic phenotype in the cells. Thus, the cells exhibited the entire spectrum of cellular stages from undifferentiated to calcified in αMEM containing 5% FBS, insulin, transferrin, and selenite. Proliferation of differentiated cells ceased by day 15, one week earlier than that in DME/F12 medium. The final cell density at the end of the proliferating

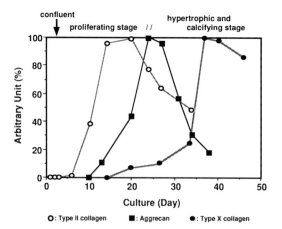

Fig. 4. Time-course of marker genes associated with chondrogenic differentiation. Total RNA was isolated on the indicated day of culture and analyzed by slot blot. Each mRNA level was quantified by scanning densitometry of the slot blot.

stage in αMEM was about 4.1×10^5 cells/cm², as compared to approximately 9.1×10^5 cells/cm² in DME/F12 medium. This resulted in the formation of smaller cartilage nodules where mineral deposition occurred. Thus, ATDC5 cells were usually cultured in DEM/F12 medium for the initial three weeks, and then in αMEM for the remaining time of culture.

It has been difficult to generate stable cell lines that express the phenotype of chondrocytes, since the phenotype is not stably maintained. Multiple passages of the cells resulted in a rapid loss of the differentiated phenotype of chondrocytes. However, constitutive expression of the *myc* oncogene kept avian chondrocytes in the proliferating stage without loss of type II collagen synthesis (*24*). Expression of the *myc* oncogene also allowed for the isolation of immortalized mouse limb bud cells with chondrogenic potentials (*25*). Mallein-Gerin and Olsen (*26*) reported generation of MC615 cells having the phenotype of proliferating chondrocytes by retroviral expression of the large T antigen of simian virus 40 (SV40) in mouse rib chondrocytes. The cells expressed type II, IX, and XI collagen as well as aggrecan and link protein. However, there was no sign of cellular hypertrophy or type X collagen expression. Immortalization of chondrocytes prevented these cells from phenotypic conversion into hypertrophic and calcifying chondrocytes. When chondrocytes were immortalized by a temperature-sensitive SV40 large T antigen, expression of type X collagen was upregulated upon growth arrest by shifting the temperature to a nonpermissive temperature (*27*). Therefore, withdrawal from the growing cell cycle may be an important prerequisite for the late-phase differentiation of chondrocytes.

In ATDC5 cell culture, formation of hypertrophic cells was also preceded by the growth arrest of proliferating chondrocytes. The phenotypic conversion into hypertrophic cells proceeded in association with the induction of type X collagen gene expression and reduction of the level of type II collagen mRNA (Fig. 4). This observation was compatible with the previous findings during differentiation of chondrocytes *in vitro* and *in vivo* (*2, 28,*

29). Ascorbic acid has been suggested to be an important supplement of culture medium for induction of type X collagen expression (*3, 30, 31*). In ATDC5 cells, the supplementation of culture medium with ascorbic acid resulted in the early onset of type X collagen gene expression. However, even in the absence of ascorbic acid, the expression of type X collagen mRNA was evident on day 21 in cells which had been cultured in DME/F12 medium. Thus, type X collagen expression *per se* does not require the presence of ascorbic acid in culture, although it facilitated a higher level of type X collagen expression and maintenance of hypertrophic phenotype.

β-Glycerophosphate (usually 4–20 mM) has been an important culture supplement for the stimulation of extracellular calcification of chondrocytes as well as osteoblasts (*3, 30, 32*), although it sometimes results in artifactual mineralization in culture. It should be noted that calcification was initiated and propagated successfully in the absence of β-glycerophosphate in the culture of ATDC5 cells. However, calcification was markedly facilitated in the atmosphere of a lower CO_2 concentration (3% CO_2 in air). Since a slight reduction in the efficiency of chondrogenesis and the subsequent growth of cartilage nodules were observed under 3% CO_2 in air, the cells were maintained in 5% CO_2 for the initial three weeks. The culture was then shifted to 3% CO_2 in air for facilitation of the mineralizing process.

III. ACQUISITION OF PTH/PTHrP RESPONSIVENESS AND NEGATIVE REGULATION OF DIFFERENTIATION BY PTH/PTHrP SIGNALING

When limb mesenchymal cells move away from the influence of the apical ectodermal ridge, their cAMP content participates in triggering chondrogenic differentiation (*33*). Elevation of cAMP content of the subridge mesenchymal cells precludes the necessity of cells passing through a condensation phase prior to overt cartilage formation (*33*). Thus, intracellular cAMP is hypothesized to be a key second messenger for regulation of limb chondrogenesis (*34, 35*). Parathyroid hormone (PTH) shares a common receptor with PTH-related peptide (PTHrP), which was originally identified as a causative agent of humoral hypercalcemia of malignancy (*36–38*). Growth plate chondrocytes express PTH/PTHrP receptors which utilize cAMP as a second messenger (*39*). Unlike PTH, the expression of PTHrP has been demonstrated in a variety of fetal and postnatal tissues in normal animals (*40, 41*). The tissue distribution of PTHrP and its receptor suggests that PTHrP plays a role as a paracrine regulator of cartilage growth and differentiation (*42–44*).

Binding studies and Scatchard analysis by $[^{125}I]PTH(1–34)$ indicated the presence of a single class of binding sites on the differentiated ATDC5 cells that exhibited an apparent dissociation constant (Kd) of 3.9 nM and a binding capacity (B_{max}) of 3.2×10^5 sites/cell. The specific binding component was identified as a diffuse 80 kDa band by chemical cross-linking of the labeled ligand. Taking advantage of the inductive chondrogenesis of ATDC5 cells, the differentiation-dependent expression of PTH/PTHrP receptors was demonstrated by monitoring the specific binding of $[^{125}I]PTH(1–34)$ (*9*). No specific binding was detectable on day 4, at which time no cartilage nodules were formed. However, only two days later the specific binding of $[^{125}I]PTH(1–34)$ was clearly detectable. In association with the formation of cartilage nodules, the specific binding increased markedly and reached a maximum of 11% of the added radiolabeled ligand on day 21. Thus, the expression of cell-surface binding sites for PTHrP is closely associated with chondrogenic differentiation of the cells.

The expression of a PTH/PTHrP mRNA in ATDC5 cells was studied by Northern blotting (*9*). There was no detectable hybridization to RNA extracted from undifferentiated cultures of ATDC5 cells on day 2. In contrast, on day 7 when the formation of cartilage nodules was observed, the 2.3–2.5 kb PTH/PTHrP-receptor transcript was clearly detectable. The time-course of changes in the PTH/PTHrP receptor mRNA level was determined by slot blot analysis, and compared to that of type II collagen mRNA. Densitometric analysis clearly indicated that the inductive expression of the PTH/PTHrP receptor gene coincided with expression of the type II collagen gene in a logarithmic manner. Therefore, expression of the PTH/PTHrP receptor gene is also associated with chondrogenic differentiation.

Moreover, secretion of PTHrP, an endogenous ligand for the PTH/PTHrP receptor, from ATDC5 cells was demonstrated: the medium conditioned from day 2 to day 5 contained a detectable level of PTHrP (4.4 pM). Concentration of PTHrP in the conditioned medium decreased as cell differentiation progressed, and was below the detection limit (<0.1 pM) on day 52 (9). To elucidate the functional role of PTH/PTHrP signaling in chondrogenesis, we studied the effects of PTH analogs on chondrogenic differentiation of ATDC5 cells. The cells were incubated with 10^{-8} M PTH(1–34) or PTHrP(1–141), which activated adenylate cyclase to a maximal level. Both peptides completely inhibited cellular condensation and the subsequent formation of cartilage nodules in culture (9). In contrast, treatment with PTH(7–34) did not affect differentiation. Thus, when PTH/PTHrP receptors were activated continuously during the early-phase differentiation of the cells, these agents inhibited the formation of cartilage nodules. Phenotypic transition of the cells from the proliferating to the hypertrophic and calcifying stage was also inhibited by activation of PTH/PTHrP receptors (Fig. 5) (10).

Upon interaction with the ligands, PTH/PTHrP receptors activate both protein kinase A (PKA) and the protein kinase C (PKC) signal pathways (45, 46). Because the N-terminal amino acids of PTH fragments are required for adenylate cyclase activation (47), PTH(7–34) lacking the N-terminal seven amino acids binds to PTH/PTHrP receptors without activation of adenylate cyclase (48). Jouishomme (49) reported that the PKC activation domain lies within the 28–34 region of the PTH molecule. PTH(7–34) activates PKC with a relative potency of

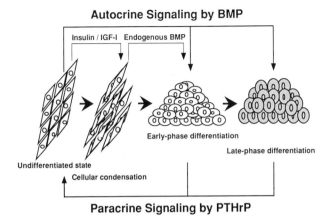

Fig. 5. Regulation of differentiation by counteracting signalings from BMP and PTHrP.

51% of that of PTH(1–34), suggesting that activation of the PKA pathway by the PTH/PTHrP receptor interferes with the early chondrogenesis through the cAMP/PKA signaling pathway.

Jüppner and his colleagues identified a single heterozygous nucleotide exchange in the PTH/PTHrP receptor gene of a patient with Jansen-type metaphyseal chondrodysplasia (*50, 51*). The nucleotide exchange causes a missense mutation in the PTH/PTHrP receptor. When the mutant receptor was expressed in COS cells, ligand-independent constitutive cAMP accumulation occurred in the cells. Thus, interference of chondrogenesis by constitutive activation of the PTH/PTHrP receptors may account for the abnormal formation of endochondral bone in this rare form of short-limbed dwarfism. In addition, homozygous mice carrying the PTHrP null mutation exhibited widespread abnormalities of endochondral bone development. Despite the wide distribution of PTHrP and its receptor transcripts in the body, there were no morphological abnormalities in other tissues (*52, 53*). Animals homozygous for the null mutant of the PTH/PTHrP receptor displayed essentially the same abnormalities (*54*). These results suggest the importance of PTHrP signaling in the control of multistep differentiation of chondrocytes during endochondral bone formation.

There is also growing evidence to suggest that bone morphogenetic proteins (BMPs) stimulate differentiation of cells of the chondroblastic lineage as well as those of the osteoblastic lineage (*25, 55, 56*). Transcripts for BMP genes have been localized in mesenchymal condensations and the perichondrium in mouse embryos (*57, 58*). Studies on *brachypodism* and *short ear* mice implied that BMPs mediate cell-cell interactions during condensation and subsequent chondrogenesis (*59, 60*). An amplified anabolic action of BMP has been documented during the phenotypic conversion of chondrocytes to hypertrophic cells (*23, 25, 39*). BMP may play a role in the process as an internal mediator for chondrogenic differentiation. We demonstrated that ATDC5 cells express BMP-4 and its receptor transcripts, while exogenous BMP was not required for chondrogenic differentiation or mineralization in the ATDC5 cell culture (*9, 10*). When added as a culture supplement, recombinant BMP-2 facilitated induction of early-phase differentiation and mineralization of ATDC5 cells. Intriguingly, precondensing ATDC5 cells responded to exogenous BMP-2 to undergo chondrogenic differentiation and skipped the condensation stage (*61*). Thus, BMP-dependent serine/threonine kinase signaling participates in the chondrogenesis of condensing ATDC5 cells. Since BMPs modulate PTHrP responsiveness of chondrocytes (*39*), chondrogenic differentiation must be coordinately regulated by BMP-dependent signaling and PTHrP-dependent signaling (Fig. 5). ATDC5 cells therefore provide a valuable opportunity for molecular analysis of multistep chondrogenesis.

SUMMARY

During the process of endochondral bone formation, proliferating chondrocytes give rise to hypertrophic cells, which then deposit a mineralized matrix to form calcified cartilage prior to replacement by bone. We established chondrogenic differentiation of the mouse embryonal carcinoma-derived clonal cell line ATDC5 as a model of chondrogenesis in the early stages of endochondral bone development. ATDC5 cells retain the properties of chondroprogenitor cells, and rapidly proliferate. Insulin induced chondrogenic differentiation of the cells in a postconfluent phase through a cellular condensation process, resulting in the formation of cartilage nodules, as evidenced by the expression of type II collagen and aggrecan genes. Moreover, under the appropriate culture conditions, the differentiated ATDC5 cells became hypertrophic at the center of cartilage nodules when the cells ceased to grow. Formation of hypertrophic chondrocytes took place in association with type X collagen gene expression. After five weeks of culture, mineralization of the culture could be discerned as Alizarin red-positive spots, which spread throughout the nodules. Thus, we demonstrated that ATDC5 cells keep track of the multistep differentiation process encompassing the stages from mesenchymal condensation to calcification *in vitro*. ATDC5 cells provide an excellent model to study the molecular mechanism underlying cartilage differentiation during endochondral bone formation.

REFERENCES

1. Kosher, R.A., Kulyk, W.M., and Gay. S.W. (1986) *J. Cell Biol.* **102**, 1151–1156.
2. Castagnola, P., Dozin, B., Moro, G., and Cancedda, R. (1988) *J. Cell Biol.* **106**, 461–467.
3. Gerstenfeld, L.C. and Lands, W.J. (1991) *J. Cell Biol.* **104**, 1435–1441.
4. Solursh, M. (1983) In *Cartilage* (Hall, B.K., ed.), vol. 2, pp. 121–141, Academic Press, New York.
5. Kulyk, W.M., Rodgers, B., Green, K., and Kosher, R.A. (1989) *Dev. Biol.* **135**, 424–430.
6. Grigoriadis, A.E., Heersche, J.N.M., and Aubin, J.E. (1990) *Dev. Biol.* **142**, 313–318.
7. Bernier, S.M. and Goltzman, D. (1993) *J. Bone Min. Res.* **8**, 475–484.
8. Atsumi, T., Miwa, Y., Kimata, K., and Ikawa, Y. (1990) *Cell Diff. Dev.* **30** 109–116.

9. Shukunami, C., Shigeno, C., Atsumi, T., Ishizeki, K., Suzuki, F., and Hiraki, Y. (1996) *J. Cell Biol.* **133**, 457–468.
10. Shukunami, C., Ishizeki, K., Atsumi, T., Ohta, Y., Suzuki, F., and Hiraki, Y. (1997) *J. Bone Min. Res.* **12**, 1174–1188.
11. George, M., Chepenik, K.P., and Schneiderman, M.H. (1983) *Differentiation* **24**, 245–249.
12. Niswander, L., Tickle, C., Vogel, A., Booth, I., and Martin, G.R. (1993) *Cell* **75**, 579–587.
13. Rousseau, F., Bonaventure, J., Legeai-Mallet, L. *et al.* (1994) *Nature* **371**, 252–254.
14. Shiang, R., Thompson, L.M., Zhu, Y.-Z. *et al.* (1994) *Cell* **78**, 335–342.
15. Cohn, M.J., Izpisúa-Belmonte, J.C., Abud, H., Heath, J.K., and Tickle, C. (1995) *Cell* **80**, 739–746.
16. Kulyk, W.M., Upholt, W.B., and Kosher, R.A. (1989) *Development* **106**, 449–455.
17. Oberlender, A. and Tuan, R.S. (1994) *Development* **120**, 177–187.
18. Geduspan, J.S. and Solursh, M. (1993) *Dev. Biol.* **156**, 500–508.
19. Bassas, L., Lesniak, M.A., Serrano, J., Roth, J., and de Pablo, F. (1988) *Diabetes* **37**, 637–644.
20. Telford, N.A., Hogan, A., Franz, C.R., and Schultz, G.A. (1990) *Mol. Reprod. Dev.* **27**, 81–92.
21. Arsenault, A.L. and Hunziker, E.B. (1988) *Calcif. Tiss. Int.* **42**, 119–126.
22. Ishizeki, K., Kuroda, N., and Nawa, T. (1992) *Anat. Embryol.* **185**, 421–430.
23. Ballock, R.T. and Reddi, A.H. (1994) *J. Cell Biol.* **126**, 1311–1318.
24. Quarto, R., Dozin, B., Tacchetti, C. *et al.* (1992) *Dev. Biol.* **149**, 168–176.
25. Rosen, V., Nove, J., Song, J.J., Thies, R.S., Cox, K., and Wozney, J.M. (1994) *J. Bone Min. Res.* **9**, 1759–1768.
26. Mallein-Gerin, F. and Olsen, B.R. (1993) *Proc. Natl. Acad. Sci. USA* **90**, 3289–3293.
27. Lefebvre, V., Garofalo, S., and de Crombrugghe, B. (1995) *J. Cell Biol.* **128**, 239–245.
28. Iyama, K., Ninomiya, Y., Olsen, B.R., Linsenmayer, T.F., Trelstad, R.L., and Hayashi, M. (1991) *Anat. Rec.* **229**, 462–472.
29. Iyama, K., Kitaoka, M., Monda, M., Ninomiya, Y., and Hayashi, M. (1994) *Histochem. J.* **26**, 844–849.
30. Leboy, P.S., Vaias, L., Uschmann, B., Golub, E., Adams, S.L., and Pacifici, M. (1989) *J. Biol. Chem.* **264**, 17281–17286.
31. Dozin, B., Quarto, R., Campanile, G., and Cancedda, R. (1992) *Eur. J. Cell Biol.* **58**, 390–394.
32. Tacchetti, C., Quarto, R., Campanile, G., and Cancedda, R. (1989) *Dev. Biol.* **132**, 442–447.
33. Kosher, R.A. and Savage, M.P. (1980) *J. Embryol. Exp. Morphol.* **56**, 91–105.

34. Révillion-Carette, F., Desbiens, X., Meunier, L., and Bart, A. (1986) *Differentiation* **33**, 121-129.
35. Rodgers, B.J., Kulyk, W.M., and Kosher, R.A. (1989) *Cell Diff. Dev.* **28**, 179 -187.
36. Jüppner, H., Abou-Samra, A.-B., Uneno, S., Gu, W.-X., Potts, J.T. Jr., and Segre, G.V. (1988) *J. Biol. Chem.* **263**, 8557-8560.
37. Shigeno, C., Yamamoto, I., Kitamura, N. *et al.* (1988) *J. Biol. Chem.* **263**, 18369-18377.
38. Jüppner, H., Abou-Samra, A.-B., Freeman, M. *et al.* (1991) *Science* **254**, 1024-1026.
39. Hiraki, Y., Inoue, H., Shigeno, C. *et al.* (1991) *J. Bone Min. Res.* **6**, 1373-1385.
40. Campos, R.V., Asa, S.L., and Drucker, D.J. (1991) *Cancer Res.* **51**, 6351-6357.
41. Schermer, D.T., Chan, S.D.H., Bruce, R., Nissenson, R.A., Wood, W.I., and Strewler, G.J. (1991) *J. Bone Min. Res.* **6**, 149-155.
42. Lee, K., Deeds, J.D., and Segre, G.V. (1995) *Endocrinology* **136**, 453-463.
43. Urenã, P., Kong, X.-F., Abou-Samura, A.-B. *et al.* (1993) *Endocrinology* **133**, 617-623.
44. van de Stople, A., Karperien, M., Löwik, C.W.G.M. *et al.* (1993) *J. Cell Biol.* **120**, 235-243.
45. Abou Samra, A.B., Jüppner, H., Westerberg, D., Potts, J.T. Jr., and Segre, G.V. (1989) *Endocrinology* **124**, 1107-1113.
46. Iida Klein, A., Varlotta, V., and Hahn, T.J. (1989) *J. Bone Min. Res.* **4**, 767-774.
47. Fujimori, A., Cheng, S.L., Avioli, L.V., and Civitelli, R. (1992) *Endocrinology* **130**, 29-36.
48. Horiuchi, N., Holick, M.F., Potts, J.T. Jr., and Rosenblatt, M. (1983) *Science* **220**, 1053-1055.
49. Jouishomme, H., Whitfield, J.F., Chakravarthy, B. *et al.* (1992) *Endocrinology* **130**, 53-60.
50. Schipani, E., Kruse, K., and Jüppner, H. (1995) *Science* **268**, 98-100.
51. Schipani, E., Jensen, G., Parfitt, A.M., and Jüppner, H. (1995) *J. Bone Min. Res.* **10** (Suppl. 1), S173.
52. Karaplis, A.C., Luz, A., Glowacki, J. *et al.* (1994) *Gene Dev.* **8**, 277-289.
53. Amizuka, N., Warshawsky, H., Henderson, J.E., Goltzman, D., and Karaplis, A.C. (1994) *J. Cell Biol.* **126**, 1611-1623.
54. Lanske, B., Karaplis, A., Luz, A., Mulligan, R., and Kronenberg, H. (1994) *J. Bone Min. Res.* **9** (Suppl. 1), S121.
55. Chen, P., Carrington, J.L., Hammonds, R.G., and Reddi, A.H. (1991) *Exp. Cell Res.* **195**, 509-515.
56. Katagiri, T., Yamaguchi, A., Komaki, M. *et al.* (1994) *J. Cell Biol.* **127**, 1755-1766.

57. Lyons, K.M., Pelton, R.W., and Hogan, B.L.M. (1990) *Development* **109**, 833–844.
58. Jones, C.M., Lyons, K.M., and Hogan, B.L.M. (1991) *Development* **111**, 531–542.
59. Kingsley, D.M., Bland, A.E., Grubber, J.M. *et al*. (1992) *Cell* **71**, 399–410.
60. Storm, E.E., Huynh, T.V., Copeland, N.G., Jenkins, N.A., Kingsley , D.M., and Lee, S.-J. (1994) *Nature* **368**, 639–643.
61. Shukunami, C., Ohta, Y., Sakuda, M., and Hiraki, Y. (1998) *Exp. Cell Res.*, in press.

Extracellular Matrix-Cellular Interaction: Molecules to Diseases (Y. Ninomiya et al., eds.), pp. 309–322,
Japan Sci. Soc. Press, Tokyo/S. Karger, Basel (1998)

Type IIA Procollagen May Function to Present Bone Morphogenetic Proteins during Endochondral Bone Formation

LINDA J. SANDELL,[*1] YONG ZHU,[*1]
DOUGLAS R. KEENE,[*2] AND ANUSH
OGANESIAN[*3]

*Department of Orthopaedic Surgery, Washington University
School of Medicine, St. Louis, Missouri 63110,[*1] Shriner's
Hospital for Crippled Children, Portland, Oregon 97201-
3095,[*2] and Department of Pathology, University of
Washington, Seattle, Washington 98195,[*3] U.S.A.*

I. INTRODUCTION

Type II procollagen is expressed as two splice forms: one form, type IIB is synthesized by chondrocytes and is the major extracellular matrix component of cartilage. A second form, type IIA, is present in chondrogenic mesenchyme and perichondrium. Type IIA procollagen contains an additional 69 amino acid cysteine-rich domain in the NH_2-propeptide. From studies examining the mRNA expression pattern of type IIA procollagen, we hypothesized that this additional protein domain plays a role in the chondrogenesis. Studies have shown that type IIA procollagen mRNA precedes type IIB procollagen mRNA expression during formation of the endochondral skeleton. For example, type IIA procollagen mRNA is present in the somites, notochord, neuroepithelia, and pre-chondrogenic mesenchyme of mouse and human embryos, and in pre-cartilaginous condensations and perichondrium during development of avian long bones. In tissues that undergo chondrogenesis, the mRNA splice form switches from type IIA to type IIB procollagen upon differentiation into chondroblasts.

1.   Expression of Type IIA Procollagen

Type IIA procollagen is expressed prior to terminal differentiation in a variety of tissues in addition to pre-cartilage mesenchyme. It is present transiently in embryonic tissues such as developing skin, kidney, aorta (*1–4*), epithelium, and some basement membranes (*5, 6*) where its function is completely unknown. In all of these tissues, including cartilage, bone morphogenetic proteins (BMP) (also called growth and differentiation factors, GDF) play a role in the induction of cellular differentiation. For example, BMP-7 is important for kidney development (*7*) and transforming growth factor-β (TGFβ) is important for skin development.

2.   Function of the Alternatively Spliced Domain: NH₂-propeptide

The fibril-forming collagens, types I, II, III, V, and XI, are synthesized as procollagens with amino (NH_2)- and carboxyl (COOH)- terminal extension peptides. Cleavage of the NH_2- and COOH-terminal extension peptides generally occurs extracellularly by specific neutral proteases presumably prior to collagen fibrillogenesis. In collagen types I and II, the removal of the propeptides is concomitant with fibrillogenesis. However, in types III and V collagen, there is evidence for retention of the NH_2-propeptide in the mature collagen molecule. It has been suggested that the NH_2-propeptide may play a role in fibrillogenesis and feedback control of collagen biosynthesis. Studies on embryogenesis of skin show that the NH_2-propeptides of type I and type III collagens participate in fibrillogenesis and may regulate fibril diameter. A definitive role of the collagen NH_2-propeptide has not been elucidated.

Analysis of the coding sequence of type IIA procollagen exon 2 shows the presence of 10 invariant cysteine residues that are conserved across collagens, the alpha 1 chain of type I, types III and V collagens, and thrombospondins 1 and 2. Nine of the 10 residues are also found in residues 1,815–1,886 of human von Willebrand factor. In the human amino propeptide, residues 38, 41, 47, 82, 84, and 92 corresponding to glycine, tyrosine, tryptophan, proline, glycine, and proline, respectively, are conserved across collagens, thrombospondins and von Willebrand factor. Recently two new proteins have been isolated that contain multiple linear copies of a domain with similar conservation, the *Drosophila* protein, sog and the *Xenopus* protein, chordin. These proteins function in dorsal-ventral patterning by binding members of the TGFβ superfamily to control the distribution of the growth factor. Sog acts as an antagonist of the TGFβ superfamily member, decapentaplegic. Chordin functions by antagonizing the ventral BMP signaling pathway by binding BMP4 to inhibit its attachment to receptor and induction of subsequent dorsalizing activities. Piccolo and colleagues have shown that chordin inhibits the activity of BMP4 by binding and removing it from the site of receptor (*22*).

Preceding any studies on the functional role of type IIA procollagen it is necessary to establish whether type IIA procollagen is deposited and remains in the extracellular matrix. We have previously shown that type IIA procollagen can be detected in the media of primary cultures of young human costal chondrocytes (*8*). However, isolation of type IIA procollagen and production of antisera for its immunolocalization have been hampered by the inability to extract sufficient amounts of type IIA procollagen from tissue and the lack of a cell line or primary cell culture capable of producing adequate amounts of this protein. To overcome these problems, exon 2 encoding the unique portion of the NH_2-propeptide was produced as a recombinant fusion protein and expressed in bacterial cells. The recombinant protein was used to immunize rabbits and the resulting polyclonal antisera employed to localize type IIA procollagen in selected human embryonic tissues.

II. LOCALIZATION AND FUNCTION OF TYPE IIA PROCOL-LAGEN

1. Cloning and Expression of the Type II Collagen Exon 2 cDNA

Polymerase chain reaction (PCR) was used to amplify a 207 bp region of the human type IIA collagen gene encoding amino acids 29 to 97. A DNA band corresponding to the calculated molecular mass of the fragment was isolated from the gel, purified and subcloned into the pGEX-2T expression vector in order to generate a fusion to GST to allow the purification of the fusion protein by affinity to glutathione sepharose. The expressed exon 2 fusion protein was approximately 37 kDa by sodium dodecyl sulphate-polyacrylamide gel electrophoresis (SDS-PAGE). The fusion protein was released from the cells following sonication and a large proportion was found in the soluble fraction. The yield of the fusion proteins from 500 ml cell culture was 4 mg. Liberation of the exon 2 propeptide from its fusion protein was achieved enzymatically via incubation with thrombin. Rabbits were injected with the purified protein as described. Subsequent Western blot analysis shows that the antiserum reacted with GST fusion protein from the *Escherichia coli* lysate and the isolated exon 2 protein fragment, whereas no reactivity was detected by preimmune serum.

2. Specificity of Type IIA Procollagen NH_2-propeptide Antiserum

In Fig. 1A, the band shown in lane 2 is collagenase-sensitive, reacts with type II triple helical antiserum and is not observed when samples were unreduced. In addition, the band was positive for immunoreaction with antiserum to COOH-pro-

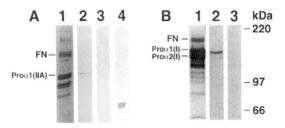

Fig. 1. Preparation and characterization of type IIA procollagen NH₂-propeptide antiserum. PCR primers used for amplification of exon 2 were designed to amplify a 207 base pair product from nucleotide 87 (coding for the N-terminal Gln residue) through nucleotide 294 of the exon 2 sequence. The forward 27-mer primer was 5′-AGTGGATCCCAGGAG-GCTGGCAGCTGT-3′, containing *Bam*HI site and three additional nucleotides at the 5′-end. The reverse 29-mer primer was 5′-AGTGAAT-TCTGAACTGGCAGTGGCGAGGT-3′, containing additional nucleo-tides to protect the restriction site, an *Eco*RI site, and a stop codon at its 5′-end. The PCR product was purified from agarose gel slices, digested with *Bam*HI and *Eco*RI and cloned into a pGEX 2T vector (Pharmacia, Biotech, Piscataway, NJ). New Zealand White rabbits were immunized with purified exon 2-GST fusion protein. Briefly, 400 μg of fusion protein in 1 ml isotonic saline was emulsified with 1 ml of Freund's complete adjuvant and injected subcutaneously in multiple sites at the back of each animal. Specificity of type IIA procollagen antiserum. [³H]-proline la-beled medium was run on SDS-PAGE and transferred onto PVDF membrane. Specific lanes of the gel were subjected to autoradiography. (A) Proteins from cultures of human 54-day gestation embryonic ribs were precipitated, applied to a 6% SDS-PAGE, and run for 4 hr under reducing conditions. Lane 1, autoradiograph showing the position of proα(IIA) and fibronectin. Strips of the membrane were incubated with type IIA NH₂-propeptide polyclonal antiserum (lane 2) or preimmune serum (lane 3). Lane 4, immunoblot of isolated α1(II) pepsinized collagen chain reacted with type II antiserum, showing migration of the mature type II collagen. (B) Proteins from human skin fibroblast culture medium were precipitated, applied to a 6% SDS-PAGE, and run for 2 hr under reducing conditions. Lane 1, autoradiograph showing the proα1(I) and proα2(I) chains of type I collagen as well as fibronectin. Lanes 2 and 3, immuno-blots of culture medium proteins. The strips of the membrane were in-cubated with anti-procollagen α1(I) monoclonal antibody (lane 2) and procollagen type IIA NH₂-propeptide polyclonal antiserum (lane 3). Antibodies were used at 1 : 500 dilution. Molecular weight markers are shown at right.

peptide. These results strongly suggest that it is pNC procollagen. To compare the migration of the type IIA procollagen with mature collagen, type II collagen alpha chains were run in lane 4 and immunoreacted with the rat anti-collagen triple helical domain antiserum.

Protein sequence comparisons indicate that human type IIA NH_2-propeptide is most closely related to human $\alpha 1(I)$ NH_2-propeptide (9). In order to determine whether there is cross-reactivity between the $\alpha 1(I)$ and $\alpha 1(II)$ procollagens, radiolabeled proteins from human skin fibroblast culture medium were separated by electrophoresis and subjected to Western blot analysis using antisera to both type I amino propeptide and type IIA amino propeptide. Type IIA antiserum does not show any cross-reactivity with type I procollagen at a dilution of 1:500 (Fig. 1B, lane 3). In addition, no cross reactivity with recombinant $\alpha 1(I)$ propeptide (kindly provided by Dr. Paul Bornstein) was observed (data not shown).

3. Immunolocalization of Type IIA Procollagen

The distribution of the NH_2-propeptide was investigated by immunohistochemistry. Localization of type IIA NH_2-propeptide has been determined in many embryonic tissues (10). Here we show the distribution of type IIA procollagen NH_2-propeptide in young cartilage from a 50 day human embryo (Fig. 2). These cells no longer make type IIA procollagen, but synthesize type IIB procollagen. The type IIA procollagen remains in the tissue lying in the interterritorial matrix, having been displaced there by the newly synthesized collagen. Prior to cartilage differentiation, type IIA procollagen was synthesized by the chondroprogenitor cells (data not shown).

4. Electron Microscopic Immunolocalization of Type IIA Procollagen Fibrils

To determine the molecular localization of the NH_2-propeptide, immunohistochemistry of type IIA procollagen in embryonic chondrogenic tissue was performed and visualized using

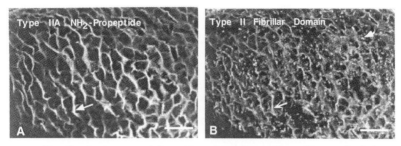

Fig. 2. Immunoreactivity with embryonic cartilage. Human fetal tissue (50 and 57 days gestation) was provided by the Central Laboratory for Human Embryology at the University of Washington. For immunofluorescence two polyclonal antibodies were used. Rabbit antisera against human type IIA procollagen NH_2-propeptide (*10*), and rat antiserum against bovine type II triple helical collagen (*23*). Images were collected on a Bio-Rad MRC600 scanning laser confocal microscope mounted on a Nikon Optiphot base. Data was collected using a Nikon 20X/0.50 n.a. dry objective. The Bio-Rad A1-A2 filters were used with an argon laser light source producing excitation at 514 nm and collecting emissions at 520–560 nm (green, FITC) or >600 nm (red, Cy-3). Optical sections were about 2 μm in thickness with the 20X objective. Full frame (768×512) 8-bit images were collected for analysis. Type IIA procollagen NH_2-propeptide was localized in extracellular matrix (arrow) of cartilage of 50 days human fetal cartilage (panel A), while type II fibrillar domain was detected in extracellular matrix (arrow) and in cytoplasm (arrowhead) (panel B).

electron microscopy. Antiserum to the NH_2-propeptide was used to localize the procollagen in tissue (Fig. 3). The results demonstrate localization of antibody-bound gold particles on the surface of collagen fibrils present in cartilage/perichondrial tissue. To further clarify the position of the NH_2-propeptide within the fibrils, individual fibrils were released from tissue matrix by shearing in ammonium bicarbonate buffer using a tissue homogenizer, incubated with type IIA specific antibody, then further incubated with 10-nm gold secondary antibody conjugate. After incubation in type IIA antibody, protrusions from the fibril surface can be seen in the type IIA-containing fibril. The identity of the protrusions as primary antibody is

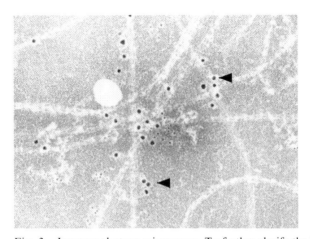

Fig. 3. Immuno-electron microscopy. To further clarify the localization of type IIA procollagen NH_2-propeptide within the fibrils, cartilage containing perichondrium was sheared in 0.2 M ammonium bicarbonate, pH 7.6 using an Omni International 2000 homogenizer. The homogenate was washed three times with resuspension in phosphate buffered saline (PBS) and centrifugation at $600 \times g$ for 5 min. The resulting homogenate was either directly deposited onto carbon coated grids and stained with 3% phosphotungstic acid, pH 7.0, was labeled only with primary antibody (1: 5 in PBS) prior to staining, or it was labeled with primary antibody followed by secondary antibody 10 nm gold conjugate (1:3 in bovine serum albumin (BSA)) prior to staining. Type IIA procollagen NH_2-propeptide was detected on the surface of the fibrils (arrow).

confirmed by secondary antibody-gold conjugate (black dots). A determination of periodicity following gold conjugate is complicated by the additional length of the complex (primary antibody-secondary antibody-gold particles) and by some secondary antibodies carrying more than one gold particulate. Therefore, the estimate of gold particle spacing was made from the primary antiserum photomicrographs. Taken together, these results indicate that the pN-collagen is present within the type II collagen fibril and found at periodic locations, some of which correspond to the 67 nm repeat of the collagen molecule.

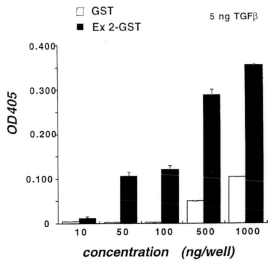

Fig. 4. Solid phase binding assay of TGF-β1 to GST-exon 2 fusion protein. 96-Well plates were coated with 5 ng TGF-β1, blocked with 3% BSA and incubated with various concentrations of Exon 2-GST fusion protein or GST alone for 2 hr at 37°C. Secondary antibody conjugated with alkaline phosphatase was added and color reaction developed using p-nitro-phenylphosphatase as substrate. Absorbance was determined at 405 nm using an enzyme-linked immunosorbent assay (ELISA) Microtiter Reader.

5. Binding of Type IIA Procollagen NH$_2$-propeptide to TGFβ

Lastly, we asked whether type IIA NH$_2$-propeptide could bind to members of the TGFβ superfamily. Saturable binding was shown between recombinant GST-exon 2 protein and recombinant TGFβ (Fig. 4). We now have preliminary evidence that suggests that the function of the type IIA NH$_2$-propeptide is to bind members of the TGFβ superfamily of morphogenetic proteins.

III. SIGNIFICANCE

A great deal of information is available on the localization of mRNA encoding the type IIA alternate splice form of type II procollagen, however, no data has existed correlating the mRNA with the tissue distribution of the protein. For a number of years we have known that the protein can be secreted into the culture media of chondrogenic costal cells (8). Recently, the type IIA and type IIB pN procollagen splice forms have been isolated and identified in bovine vitreous (11). We produced and character- ized antisera that react specifically with the cysteine rich, NH_2- propeptide domain of type IIA procollagen and localize this type IIA procollagen NH_2-propeptide in regions of embryonic tissue that contain type IIA procollagen mRNA. Furthermore, we identified type IIA procollagen chains in cultured embryonic cartilage. These results taken together with the isolation of type IIA pN collagen from bovine vitreous strongly suggest that in the extracellular matrix a population of the type IIA NH_2- propeptides remain bound to the triple-helical portion of the secreted collagen molecule.

The molecular form of type IIA procollagen NH_2-propep- tide *in vivo* can now be directly determined by these studies. We show directly that the NH_2-propeptide antiserum reacts with the collagen fibril in the extracellular matrix at regular intervals consistent with distribution of the NH_2-termini. Other data using antiserum to the COOH-propeptide demonstrates that the collagen fibril does not contain the COOH-propeptide. In con- trast, in tissue culture of embryonic ribs, pNC collagen was present and no pN collagen was detected. This result is not unexpected because the enzymes necessary to cleave the propep- tides are not accessible. Consistent with this interpretation, the fibroblast culture medium only contained pNC forms of type I collagen. Support for the developmental regulation of type II collagen splice forms in chondrogenesis is the finding of type IIA to type IIB collagen ratios in fetal and adult vitreous. In these interesting studies of Bishop *et al.* (*11*), type IIA and type IIB pN collagens were isolated from bovine vitreous. In young tissue, the ratio of type IIA pN to type IIB pN was 5 : 1 whereas in the older tissue, the ratio was closer to 1.

There are other proteins that contain the type IIA-procollagen NH_2-propeptide domain, which begins to suggest diverse functional roles for this domain. Three procollagen chains, $\alpha1(I)$ (*12*), $\alpha1(III)$ (*13*), and $\alpha2(V)$ (*14*) possess similar cysteine-rich NH_2-propeptides, including the same placement of the cysteine residues as well as many others (compared by Ryan and Sandell) (*15*). A similar domain is found in thrombospondins I and II (*16*) where the cysteines and other amino acids are conserved and von Willbrand factor where 9 of 10 cysteine residues are conserved. The function of the collagen propeptides is predicted to be a feedback inhibitor of collagen synthesis (*17*) and a regulator of fibril diameter (*18*). In thrombospondins, this domain is thought to be involved in regulation of angiogenesis (*19*). It has been shown recently that two additional proteins, sog (short of gastrulation), in *Drosophila* (*20*) and chordin, in *Xenopus* (*21*), have multiple homologous domains conserving all of the cysteines and many other amino acids. The NH_2-propeptide-like domains in sog and chordin function to regulate the availability of members of the TGFβ superfamily, decapentaplegic and bone morphogenetic protein 4, respectively, by direct binding (*22*), and thus participate in establishing the dorsalventral body pattern. Now that we have shown type IIA NH_2-propeptide in developing skeletal tissue, future studies can be directed at determining whether such a binding function is possible for type IIA procollagen.

In summary, the presence of type IIA procollagen in cells prior to terminal differentiation suggests that the unique domain in the type IIA procollagen, the NH_2-propeptide, is involved in developmental events. The finding that type IIA procollagen is deposited into the matrix of human embryos, retains the unique NH_2-propeptide domain, and binds to TGFβ, a cartilage induction factor lends support to the hypothesis that type II collagen, *i.e.*, type IIA procollagen, functions in the early embryo and, potentially, is the portion of the type II collagen molecule that plays a role in the induction of chondrogenesis in receptive cells. These results taken together with the similar function of chordin and BMP4 (and sog : decapentaplegic) strongly suggests a molecular paradigm that connects the process of dorsal-ventral patterning in the embryo with patterning of the skeleton.

SUMMARY

Type II procollagen is expressed as two splice forms: one form, type IIB is synthesized by chondrocytes and is the major extracellular matrix component of cartilage. The other form, type IIA, contains an additional 69 amino acid cysteine-rich domain in the NH_2-propeptide and is synthesized by chondrogenic mesenchyme and perichondrium. We have hypothesized that the additional protein domain plays a role in chondrogenesis. The present study was designed to determine the localization of the type IIA NH_2-propeptide in the extracellular matrix and to determine its function during chondrogenesis. Immunofluorescence histochemistry employing antibodies to two domains of the type IIA procollagen molecule was used to localize the NH_2-propeptide and fibrillar domain of the type IIA procollagen molecule during chondrogenesis in a developing human long bone (stage XXI). Tissue fluorescence was visualized by confocal microscopy. Type IIA procollagen fibrils were identified in the extracellular matrix by immunoelectron microscopy. Immunoelectron microscopy revealed type IIA procollagen fibrils labeled with antibodies to NH_2-propeptide at an approximately 70 nm interval, directly indicating that the NH_2-propeptide remains attached to the collagen molecule in the extracellular matrix. As differentiation proceeds, the cells switch synthesis from type IIA to type IIB procollagen, and the newly synthesized type IIB collagen displaces the type IIA procollagen into the interterritorial matrix. To initiate studies on the function of type IIA procollagen, binding was tested between recombinant NH_2-

propeptide and various growth factors known to be involved in chondrogenesis. Solid phase binding assay indicated that type IIA procollagen NH_2-propeptide bound to $TGF\beta$, a cartilage induction factor. Taken together, these results suggest that the NH_2-propeptide of type IIA procollagen could participate in the extracellular matrix distribution of bone morphogenetic proteins in chondrogenic tissue.

Acknowledgments

This study was funded by National Institutes of Health grant R0136994, the Department of Veterans Affairs, Seattle, WA, and the Department of Orthopaedic Surgery, Washington University School of Medicine. Also, thanks to Gloria Hoch for her help with the preparation of this manuscript.

REFERENCES

1. Cheah, K.S.E., Lau, E.T., Au, P.K.C., and Tam, P.P.L. (1991) *Development* **111**, 945–953.
2. Sandberg, M.M., Hirvonen, H.E., Elima, K.J.M., and Vuorio, E.I. (1993) *Biochem. J.* **294**, 595–602.
3. Swiderski, R.E., Daniels, K.J., Jensen, K.L., and Solursh, M. (1994) *Dev. Dynamics* **200**, 294–304.
4. Lui, V.C.H., Ng, L.J., Nicholls, J., Tam, P.P.L., and Cheah, K.S.E. (1995) *Dev. Dynamics* **203**, 198–211.
5. Kosher, R.A. and Solursh, M. (1989) *Dev. Biol.* **131**, 558–566.
6. Mallein-Gerin, F. and Garrone, R. (1990) *Biol. Cell* **69**, 9–16.
7. Karsenty, G., Luo, G., Hofmann, C., and Bradley, A. (1996) *Ann. N.Y. Acad. Sci.* **785**, 98–107.
8. Sandell, L.J., Morris, N., Robbins, J.R., and Goldring, M.R. (1991) *J. Cell Biol.* **114**, 1307–1319.
9. Ryan, M.C., Sieraski, M., and Sandell, L.J. (1990) *Genomics* **8**, 41–48.
10. Oganesian, A., Zhu, Y., and Sandell, L. (1997) *J. Histochem. Cytochem.* **45**, 1469–1480.
11. Bishop, P.N., Reardon, A.J., McLeod, D., and Ayad, S. (1994) *Biochem. Biophys. Res. Commun.* **203**, 62–85.
12. Chu, M.L., De Wet, W., Bernard, M. *et al.* (1984) *Nature* **310**, 337–340.
13. Liau, G., Yamada, Y., and de Crombrugghe, B. (1985) *J. Biol. Chem.* **260**, 531–536.

14. Woodbury, D., Benson-Chanda, V., and Ramirez, F. (1989) *J. Biol. Chem.* **264**, 2735-2738.
15. Ryan, M.C. and Sandell, L.J. (1990) *J. Biol. Chem.* **265**, 10334-10339.
16. Bornstein, P. and Sage, E.H. (1994) *Methods Enzymol.* **245**, 62-85.
17. Fouser, L., Sage, E.H., Clark, J., and Bornstein, P. (1991) *Proc. Natl. Acad. Sci. USA* **88**, 10158-10162.
18. Fleischmajer, R., Olson, B.R., Timpl, R., Perlish, J.S., and Lovelace, O. (1990) *Proc. Natl. Acad. Sci. USA* **580**, 161-175.
19. Good, D.J., Polverini, P.J., Rastinejad, F. *et al.* (1990) *Proc. Natl. Acad. Sci. USA* **87**, 6624-6628.
20. Francois, V., Solloway, M., O'Neill, J.W., Emery, J., and Bier, E. (1994) *Genes Dev.* **8**, 2602-2616.
21. Sasai,Y., Lu, B., Steinbeisser, H., Geissert, D., Gont, L.K., and De Robertis, E.M. (1994) *Cell* **79**, 779-790.
22. Piccolo, S., Sasai, Y., Lu, B., and De Robertis, E. (1996) *Cell* **86**, 589-598.
23. Cremer, M.A. and Kang, A.H. (1988) *Int. Rev. Immunol.* **4**, 65-81.

*Extracellular Matrix-Cellular Interaction: Molecules to Diseases (Y. Ninomiya et al., eds.), pp. 323–337,
Japan Sci. Soc. Press, Tokyo/ S. Karger, Basel (1998)*

Cartilage Collagens and Degenerative Diseases of the Joints and Spine

TOMOATSU KIMURA*

*Department of Orthopaedic Surgery, Osaka Rosai Hospital,
Sakai, Osaka 591-0035, Japan*

Osteoarthritis (OA) is a common joint disorder and is almost universal in the elderly. It is a multifactorial disease characterized by the degeneration of articular cartilage and shows widely varying progression and severity. OA also shows variation in its clinical presentation, partly depending on the joints involved in each individual. Various pathophysiological processes have been implicated in the development of OA and the importance of mechanical as well as metabolic factors has been documented (*1*). However, it is now evident that OA is not the result of simple wear and tear of the articular cartilage during aging. In fact, there are many elderly people who do not show any evidence of degenerative joint disease. Recent investigations are beginning to provide a picture of disease prevalence and etiology. Some of the findings suggest that OA may result primarily from subtle pathological changes in articular cartilage that occur at a much younger age.

*Present address: Department of Orthopaedic Surgery, Osaka University Medical
School, Suita, Osaka 565-0871, Japan

Degenerative disease of the spine is another leading cause of musculoskeletal disability in elderly people. Intervertebral disc degeneration and spondylosis are examples of such diseases and again seem to be multifactorial disorders. The structure of the intervertebral discs differs significantly from that of the articular cartilage. However, these two cartilaginous structures in different parts of the skeleton share some biochemical components as well as viscoelastic characteristics when subjected to loading and deformation. In addition, degeneration of these two structures sometimes occurs concomitantly in humans (2). This chapter will focus on the general concepts of cartilage degeneration and current knowledge about risk factors, with particular emphasis on the role of cartilage matrix collagens.

I. ARTICULAR CARTILAGE DEGENERATION AND OSTEOAR-THRITIS

OA is defined as a functional disorder of joints accompanied by degeneration and loss of articular cartilage. The cartilage shows fibrillation, erosion, and fissure formation indicating disruption of the matrix. Biochemically, these histological changes are associated with increased water content, loss of proteoglycans, and minimal changes in collagen content. The chondrocytes, partly responding to the physical and biochemical changes of the matrix, show increased metabolic activity (3) and occasionally undergo proliferation to form cell clusters. Changes in bony structure also occur, which include osteophyte formation, subchondral bone sclerosis, and cyst formation.

Over the years, several terms have been used to designate the degeneration of cartilage. As yet, there is no agreed definition of OA as well as no universally accepted diagnostic criteria or classification system. The Kellgren and Lawrence criteria are among the most widely used epidemiologic criteria for OA and are based on radiologic changes (4). The American College of Rheumatology criteria involves a mixture of clinical and radiographic features (5, 6). As seen in Table I, which is a

TABLE I
Classification by OA Subsets[a]

Primary (idiopathic)
Localized:
Hands and feet (*e.g.*, Herberden's nodes, talonavicular)
Knee
Hip
Spine
Others (*e.g.*, shoulder, wrist, ankle)
Generalized:
3 or more joint areas listed above
Secondary
Post-traumatic
Congenital or developmental diseases
Localized
Hip diseases (*e.g.*, shallow acetabulum, Legg-Calvé-Perthes disease)
Mechanical and local factors (*e.g.*, obesity, hypermobility, valgus/varus deformity)
Generalized
Osteo-chondrodysplasias
Metabolic diseases (*e.g.*, mucopolysaccharidoses, ochronosis)
Calcium deposition disease
Other bone and joint disorders (*e.g.*, avascular necrosis, rheumatoid arthritis)
Other miscellaneous diseases (*e.g.*, endocrine diseases, neuropathic arthropathy, caisson disease, Kashin-Beck disease)

[a] Modified from Altman *et al.* (*5, 6*). It should be noted that some of the primary (idiopathic) OA may turn out to be a milder phenotype of a subset classified in secondary OA.

modification of one such criteria, OA is a heterogeneous condition and shows huge variation in expression. It is clear that in addition to "primary" changes, various other "secondary" factors may be responsible for the onset of OA. It is likely that the common risk factors for the condition include age, obesity, abnormal biomechanics, joint trauma, and genetic predisposition (*1*). However, OA of different joints (including the spine) may involve a different balance of risk factors. For example, OA of the knee is predominantly found in women and is occasionally linked to obesity, whereas OA of the hip has little association with obesity and shows an equal sex incidence in Western countries (7).

II. GENETIC FACTORS AND OSTEOARTHRITIS

Patients with OA in one joint often have the disease in many joints. This condition is termed generalized osteoarthritis (GOA) and involves multiple joints such as those of the hands, knees, and spine (8). Study of these patients indicates that most of the involvement occurs in upper limb non-weight-bearing joints. GOA sometimes shows familial clustering, being more common in older women and may be inherited in a polygenic pattern. This tendency is more obvious in patients with Heberden's and Bouchard's nodes (nodal disease), and close relatives of patients with such nodal disease tend to develop the nodal type of GOA (GOA with nodal disease). On the other hand, families of patients with non-nodal disease tend to have non-nodal GOA with a lower rate of hereditary transmission. Among patients with only one or two joints affected by OA, however, there is no proved evidence of increase of OA in their relatives.

It is now becoming more likely that genetic factors, in addition to environmental factors, play a part in the onset of GOA. There has been active investigation of the genetic factors which cause GOA. Attention has been focused on the possible role of genetic mutations in components of the articular cartilage matrix and this will be discussed later in this chapter. Various other hereditary conditions also affect joints and cause degeneration of the articular cartilage. For example, chondrocalcinosis (calcium crystal deposition disease), hemochromatosis (iron storage disease), ochronosis (alkaptonuria), mucopolysaccharidoses, and other metabolic diseases are known to be related to the onset and progression of OA. Although the pathogenesis of these joint lesions is not completely clear, these conditions affect joint metabolism directly or indirectly and eventually cause deterioration of the cartilage matrix, leading to cartilage degen-

eration.

Another potential genetic connection can be found in hip OA. Primary concentric hip OA may be partly related to unknown genetic factors. Even some secondary hip OA due to acetabular dysplasia may also have a hereditary background. Acetabular dysplasia is characterized by a shallow acetabulum and is not uncommon among Asian populations. It accounts for a substantial proportion of "secondary" hip OA in women. However, its association with certain human leucocyte antigen (HLA) haplotypes suggests the role of a genetic factor in secondary OA as well. Certainly, more evidence is necessary to understand the genetic background of such mono-articular or pauci-articular OA.

III. COLLAGENS AND HERITABLE CARTILAGE DISORDERS

Recent investigation of the genetic defects which cause GOA and/or familial OA has been focused on analysis of the cartilage matrix components. Articular cartilage consists of a highly hydrated extracellular matrix that includes a mixture of collagen fibrils and proteoglycan aggregates, providing the cartilage with its unique physiological properties (9). The collagen fibrils provide the cartilage with its shape and high tensile strength, while the proteoglycan aggregates create a high osmotic pressure and help give cartilage its resilience by resisting compressive loads. The collagen fibrils in articular cartilage are heterotypic fibrils composed of types II, IX, and XI collagens. These collagens, by forming a fibrillar network and interacting with other matrix components, play a critical role in maintenance of the biomechanical properties of cartilage (9). Since the collagen fibril network resists the osmotic pressure of charged proteoglycans, a molecular defect that causes a failure of this fibrillar network would result in loss of the unique physiological properties of cartilage. Softening of the cartilage matrix, for example, may occur and lead to cartilage degeneration under repeated stress or with the aging process.

Since type II collagen is the major collagenous component of hyaline cartilage, it is reasonable to hypothesize that mutations in the gene for type II collagen (*COL2A1*) can cause systemic diseases affecting cartilage. In fact, a deletion mutation of the *COL2A1* gene was first identified in 1989 in a certain form of chondrodysplasia, spondyloepiphyseal dysplasia (SED) congenita, which is a developmental disease mainly affecting growth cartilage (*10*). Since then, various mutations have been identified in various forms of chondrodysplasia (Table II). Mutations introducing a premature termination codon into the *COL2A1* gene were identified as a cause of Stickler syndrome, which features OA combined with ophthalmopathy (*11*). Mutations of the *COL2A1* gene were also identified in chondrodysplasia with a more severe phenotype, such as achondrogenesis type II, hypochondrogenesis, and Kniest dysplasia (*12, 13*). In addition, alleles of the *COL2A1* gene were shown to be linked to familial OA. Ala-Kokko *et al.* demonstrated a single base mutation that converted the codon for arginine at position 519 in the triple helical region to the codon for cysteine in a family with primary generalized OA associated with mild chondrodysplasia (*14*). Several other families have been reported to show the same Arg519Cys mutation in *COL2A1*. Affected members of such families usually begin to develop joint pain and stiffness in the second and third decades and from then on develop typical radiographic evidence of OA in multiple joints. Clinically, these family members show subtle to mild signs of chondrodysplasia and seem to belong to the late-onset SED subset. Different mutations of *COL2A1* have also been reported to be responsible for late-onset SED (*15*). Thus, the data currently available about mutations suggest that the type II collagen gene may harbor mutations that cause a broad spectrum of rare and less rare diseases, and late-onset SED may link chondrodysplasia to more common diseases such as familial OA.

Mutations of other cartilage collagens (types IX, X, and XI) have also been identified in various forms of chondrodysplasia (Table II). A mutation of the $\alpha2$(IX) collagen gene causes a

TABLE II
Molecular Defects in the Chondrodysplasias

Gene	Disease
Structural protein of cartilage	
COL2A1	Achondrogenesis type II, hypochondrogenesis, SED congenita, SEMD type Strudwick, Kniest dysplasia, Stickler syndrome type 1, late-onset SED, familial OA
(*Col9a1*)[a]	OA, intervertebral disc degeneration, chondrodysplasia
COL9A2	Multiple epiphyseal dysplasia (EDM2)
COL10A1	Metaphyseal dysplasia type Schmid
(*Col10a1*)	Spondylometaphyseal dysplasia
COL11A1	Stickler syndrome type 2
(*Col11a1*)	Chondrodysplasia (cho)
COL11A2	Stickler syndrome, OSMED
COMP	Pseudoachondroplasia (PSACH) Multiple epiphyseal dysplasia (EDM1)
(*Aggrecan*)	Chondrodysplasia (cmd)
Growth factor receptor	
FGFR3	Achondroplasia (ACH), hypochondroplasia (HCH), thanatophoric dysplasia (TD type I & II)
PTH/PTHRP receptor	Metaphyseal dysplasia type Jansen
Growth factor	
CDMP-1	Acromesomelic dysplasia type Hunter-Thompson
(*Gdf5*)	Brachypodism (bp)
(*Bmp5*)	Short ear (se)
Transcription factor	
SOX9	Campomelic dysplasia
CBFA1	Cleidocranial dysplasia (CCD)
Others	
DTDST	Diastrophic dysplasia, achondrogenesis type IB, atelosteogenesis type II
ARSE	Chondrodysplasia punctata XR
Cathepsin K	Pycnodysostosis

[a] Molecular defects observed in mice are shown in parentheses. SED, spondyloepiphyseal dysplasia; SEMD, spondyloepimetaphyseal dysplasia; OSMED, otospondylomegaepiphyseal dysplasia.

subset of multiple epiphyseal dysplasia (*16*). In addition, mutations at the C-terminal peptide-coding domain of type X collagen are responsible for Schmid metaphyseal dysplasia (*17,*

18). Mutations of the $\alpha 1(XI)$ and $\alpha 2(XI)$ collagen genes have been identified in different subsets of Stickler syndrome (*19, 20*). Through these studies of humans as well as through the generation of genetically engineered experimental animals, we are starting to understand the phenotypic consequences of mutations of the collagen genes. Although the mechanism by which collagen mutations lead to connective tissue fragility has not been fully clarified, some of the pathological process can be explained based on the studies of type I collagen mutations. Most mutations of collagen cause the synthesis of structurally abnormal but partially functioning α chains and these mutant α chains may prevent zipper-like folding of the collagen triple helix (*12*). Unfolded or incompletely folded molecules accumulate in the cells and are degraded because they are unstable and secreted slowly, resulting in loss of the protein that has been produced. Haploinsufficiency is also a cause of reduced collagen synthesis, as observed in a subset of Stickler syndrome (*11*). Other types of mutations may have little or no effect on the folding of collagen molecules. However, changes of molecular morphology could potentially cause disturbances in the assembly of fibrils. This would probably affect the morphogenesis of cartilaginous primordias or cause deterioration of cartilage matrix integrity resulting in loss of durability. Of course, the collagen mutations leading to various forms of chondrodysplasia would be far more complex and require investigation in the future. Nevertheless, there remains a possibility that subtle changes in the morphology and stability of the collagen molecules could alter articular cartilage matrix architecture. Such articular cartilage may become vulnerable to repetitive and long-term mechanical stress during aging.

IV. HEREDITARY PREDISPOSITION TO INTERVERTEBRAL DISC DEGENERATION

Spondylosis involves chronic degeneration of the intervertebral discs and consequent osteophytosis of the related vertebral bodies. It is one of the major problems in the aging population since it causes discomfort, back pain, and sometimes clinical syndromes of radiculopathy and myelopathy. The onset and progression of disc degeneration/spondylosis often seem to be associated with aging. However, in addition to age-related changes, environmental factors, such as occupations involving heavy lifting, trauma, and vibration, have been suggested as possible risk factors for disc degeneration (21, 22). Similarly, accumulation of major and minor mechanical insults during sporting activity also leads to accelerated disc degeneration (23). Smoking is probably another possible factor for structural changes in the spine (24).

Despite the importance of various environmental factors including mechanical overload, several lines of evidence suggest that genetic factors underlie the onset of structural and degenerative changes in the spine. Spondylolisthesis, Schmorl nodes, and scoliosis have been suggested to show familial clustering (25–27). Radiologic studies of twins by King (28, 29) demonstrated similarities in osteophyte formation and anterior degeneration of the thoracic discs, suggesting the presence of genetic elements that may determine spinal degeneration. Another analysis of cervical spondylosis in twins demonstrated concordance of degenerative changes and again suggested the presence of some genetic influence (30). A recent magnetic resonance imaging study of identical twins has further indicated that genetic influences are important determinants of disc degeneration (31, 32). Lumbar disc herniation in juveniles also shows a hereditary predisposition and familial clustering (33, 34). Such familial disc herniation is presumed to result from premature degeneration of the intervertebral discs. In addition, the herniation in a certain young population could potentially arise from acceleration of the process seen in adults, to which at least one genetic trait may contribute (35). These studies have raised the possibility that genetic factors, together with environmental factors, influence the onset or progression of various degenerative diseases of the spine.

What, then, are the possible genetic factors influencing the onset or progression of disc degeneration? The intervertebral disc is a cartilaginous tissue composed of an outer layer of fibrocartilage (the annulus fibrosus) and central myxoid tissue (the nucleus pulposus). Collagen and proteoglycan account for about 80% of the dry weight of the disc, but collagen predominates in the annulus fibrosus and proteoglycan in the nucleus pulposus. Type II collagen is predominant in the nucleus pulposus, whereas type I collagen is concentrated in the outer annulus fibrosus, forming bundles and lamellar sheets that provide great mechanical strength. Other collagens (such as types III, V, VI, IX, and XI) are also found in the intervertebral discs (36). The structural characteristics of the intervertebral discs are largely dependent on these collagens, and other extracellular matrix components. Therefore, it is quite reasonable to assume that a genetic defect in one of these matrix components or factors regulating matrix metabolism could lead to disc degeneration.

In humans, the genetic factors responsible for spondylosis and intervertebral disc degeneration are still unknown; however, analysis of genetically engineered animals can give us some clues. We have previously generated transgenic mice bearing various $\alpha 1$(IX) collagen mutations (37, 38). The phenotype of the transgenic mice varied among mice bearing different $\alpha 1$(IX) gene mutations and even within the individual lines. Nevertheless, several mouse lines showed a mild epiphyseal chondrodysplasia phenotype with dwarfism (38). Among other lines of transgenic mice, we previously found that two hemizygous lines expressing truncated $\alpha 1$(IX) chains were initially quite normal and had no signs of chondrodysplasia. However, these mice developed OA of the knee joints with aging (37, 39). We also examined the spines of these OA-predisposed mice in detail, because patients with primary OA sometimes have severe spondylosis (2) and the spine is one of the sites affected in GOA (Table I). Radiological studies of the transgenic mice initially revealed little evidence of osteophyte formation in the cervical and lumbar spine; however, the mice started to show increasing thoraco-lumbar kyphosis with aging. In addition, cervical disc space narrowing accompanied by mild osteophyte formation became evident at 12–18 months of age (*40*). Such changes were not prominent in the normal mice. Histological examination indicated that, at 12–14 months of age, most of the intervertebral discs of the transgenic as well as control mice exhibited shrinkage of the nucleus pulposus and focal loss of the lamellar structure in the annulus fibrosus. In the transgenic mice, however, disc degeneration was more advanced (Fig. 1). Partial

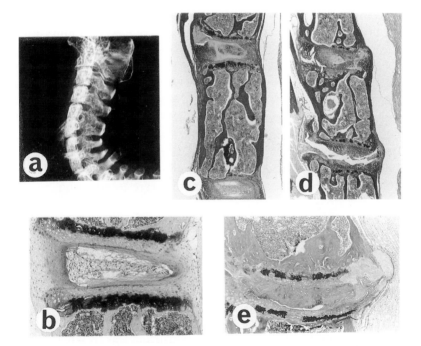

Fig. 1. Cervical spine of the transgenic mice. (a) Radiograph of 18-month-old mouse showing mild disc space narrowing and osteophyte formation. (b, c, and d) Midsagittal sections from 6, 12, and 20-month-old mice, respectively. (e) Midsagittal sections from 20-month-old mouse showing herniation of disc material with impingement on the spinal cord. (Modified from Kimura *et al.* (*40*), with permission).

disruption of the annulus fibrosus and end plate was occasionally observed in these mice. The nucleus pulposus was replaced by consolidated fibrous tissue and the nuclear-annular demarcation was obscure. Later in life, disc degeneration became quite advanced in the transgenic mice, and they sometimes developed anterior and posterior herniation of disc material. Although degenerative changes of the discs were also observed in normal mice as a part of the aging process, the degeneration was more advanced in transgenic mice than in control mice.

These results obtained using transgenic mice indicate that genetic changes, including those of the type IX collagen genes, may influence the structural integrity of the articular cartilage and intervertebral discs. The findings also suggest that mutations affecting type IX collagen may be responsible for relatively common diseases such as familial OA and intervertebral disc degeneration. At present, mutations of the $\alpha 1(IX)$ collagen gene like those we generated in transgenic mice have not been identified in humans. It is probably rare for degenerative changes of the human joints and spine to have a purely genetic etiology. However, it is quite possible that allelic variants of type IX collagen gene may account for phenotypic heterogeneity and influence various diseases ranging from rare to common.

V. MUTATIONS OF COLLAGEN: RISK FACTOR FOR CARTILAGE DEGENERATION?

Is human OA or disc degeneration at least partly caused by mutations of the collagen genes or some other genes? At present we do not have a conclusive answer, although studies by several groups suggest that approximately 2% of familial OA is due to type II collagen mutations (41). It is still unclear to what extent abnormalities of other cartilage collagens or other matrix molecules are responsible for the onset of cartilage degeneration. It may be important, however, to remember that collagen is a large triple-helical molecule that consists of repeated Gly-X-Y amino acid triplets, and this repetitive sequence is necessary for proper zipper-like folding of the collagen triple-helix. Therefore, mutation in any part of this repetitive sequence could potentially affect proper folding of the molecule. In other words, collagen is a molecule that is quite sensitive to mutations. In fact, it is not uncommon to find polymorphism within the collagen genes. Most of the changes found within exons are silent mutations that do not cause amino acid substitution. However, we occasionally find amino acid substitutions in apparently normal individuals. These "variations" in the amino acid sequences of collagen chains, depending on the chemical nature of the substitution, could potentially cause very minute changes in the morphology and stability of the triple-helix. Hence, we predict that allelic variants of cartilage collagen genes may be one of the risk factors for cartilage degeneration.

Analysis of such molecular variations of the large collagen genes is not an easy task, but new technologies, such as the DNA chip-base assay which can be assembled containing hundreds of thousands of oligonucleotides with predetermined sequences and can be used for detection of polymorphisms, are now becoming a reality (42). We believe that future identification of many variations in the amino acid sequences of the collagen genes will help to clarify the role of genetic factors in degeneration of the articular cartilage and intervertebral discs.

SUMMARY

Degenerative diseases of joints and the spine are thought to develop as a result of multiple influences such as aging, mechanical stress, and metabolic factors. Increasing evidence suggests, at least under some conditions, that genetic factors are an important predisposing factor that cause accelerated degeneration of the articular cartilage and discs in response to environmental and extrinsic factors such as mechanical loading. Possible molecular defects are now beginning to be elucidated in humans and in animal models. Certain inherited forms of OA are caused by mutations in genes expressed in cartilage, such as type II collagen gene. In humans, it is still not clear to what extent the molecular defects of matrix components or other molecules act as predisposing factors for degeneration. However, new technologies of molecular biology are being developed and will allow us to define the exact molecular cause of such common diseases.

REFERENCES

1. Peyron, J.G. and Altman, R.D. (1992) In *Osteoarthritis—Diagnosis and Medical/Surgical Management* (Moskowitz, R.W. *et al.*, eds.), pp. 15-37, WB Saunders Co., Philadelphia.
2. Kellgren, J.H. and Lawrence, J.S. (1958) *Ann. Rheum. Dis.* **17**, 388–397.
3. Mankin, H.J. and Brandt, K.D. (1992) In *Osteoarthritis—Diagnosis and Medical/Surgical Management* (Moskowitz, R.W. *et al.*, eds.), pp. 109–154, WB Saunders Co., Philadelphia.
4. Kellgren, J.H. and Lawrence, J.S. (1957) *Ann. Rheum. Dis.* **16**, 494–502.
5. Altman, R., Asch, E., Bloch, D. *et al.* (1986) *Arthritis Rheum.* **29**, 1039–1049.
6. Altman, R.D. (1995) *J. Rheumatol.* **22** (Suppl. 43), 42–43.
7. Felson, D.T. (1988) *Epidemiol. Rev.* **10**, 1–28.
8. Kellgren, J.H. and Moore, R. (1952) *Br. Med. J.* **1**, 181–187.
9. Mow, V.C., Ratcliffe, A., and Poole, A.R. (1992) *Biomaterials* **13**, 67–97.
10. Lee, B., Vissing, H., Ramirez, F., and Rogers, D. (1989) *Science* **244**, 978–980.
11. Ahmad, N.N., Ala-Kokko, L., Knowlton, R.G. *et al.* (1991) *Proc. Natl. Acad. Sci. USA* **88**, 6624–6627.
12. Prockop, D.J. and Kivirikko, K.I. (1995) *Annu. Rev. Biochem.* **64**, 403–434.
13. Rimoin, D.L. (1996) *Am. J. Mol. Genet.* **63**, 106–110.
14. Ala-Kokko, L., Baldwin, C.T., Moskowitz, R.W., and Prockop, D.J. (1990) *Proc. Natl. Acad. Sci. USA* **87**, 6565–6568.
15. Reginato, A.J., Passano, G.M., Neumann, G. *et al.* (1994) *Arthritis Rheum.* **37**, 1078–1086.
16. Muragaki, Y., Mariman, E.C.M., van Beersum, S.E.C. *et al.* (1996) *Nature Genet.* **12**, 103–105.
17. Warman, M.L., Abbott, M., Apte, S.S. *et al.* (1993) *Nature Genet.* **5**, 79–82.
18. McIntosh, I., Abbott, M.H., and Francomano, C.A. (1993) *Hum. Mut.* **5**, 121–125.
19. Vikkula, M., Mariman, E.C.M., Lui, V.C.H. *et al.* (1995) *Cell* **80**, 431–437.
20. Richards, A.J., Yates, J.R., Williams, R. *et al.* (1996) *Hum. Mol. Genet.* **5**, 1339–1343.
21. Frymoyer, J.W. (1988) *N. Engl. J. Med.* **318**, 291–299.

22. Anderson, G.B.J. (1991) In *The Adult Spine: Principles and Practice* (Frymoyer, J.W., ed.), pp. 107–146, Raven Press, New York.
23. Sward, L., Hellstrom, M., Jacobsson, B., Nyman, R., and Peterson, L. (1991) *Spine* **16**, 437–443.
24. Ernst, E. (1993) *Br. J. Rheumatol.* **32**, 239–242.
25. Wiltse, L.L. (1962) *J. Bone Joint Surg.* **44**-A, 539–560.
26. Wynne-Davies, R. and Scott, J.H.S. (1979) *J. Bone Joint Surg.* **61**-B, 301–305.
27. Cowell, H.R., Hall, J.N., and MacEwen, G.D. (1972) *Clin. Orthop.* **86**, 121–131.
28. King, J.B. (1968) *Clin. Radiol.* **19**, 315–317.
29. King, J.B. (1971) *Clin. Radiol.* **22**, 375–378.
30. Palmer, P.E., Stadalnick, R., and Arnon, S. (1984) *Skel. Radiol.* **11**, 178–182.
31. Battie, M.C., Haynor, D.R., Fisher, L.D., Gill, K., Gibbons, L.E., and Videman, T. (1995) *J. Bone Joint Surg.* **77**-A, 1662–1670.
32. Battie, M.C., Videman, T., Gibbons, L.E., Fisher, L.D., Manninen, H., and Gill, K. (1995) *Spine* **20**, 2601–2612.
33. Matsui, H., Terahata, N., Tsuji, H., Hirano, N., and Naruse, Y. (1992) *Spine* **17**, 1323–1328.
34. Nelson, C.L., Janecki ,C.J., Gildenberg, P.L., and Gerard, S. (1972) *Clin. Orthop.* **88**, 142-150.
35. Varlotta, G.P., Brown, M.D., Kelsey, J.L., and Golden, A.L. (1991) *J. Bone Joint Surg.* **73**-A, 124–128.
36. Roberts, S., Menage, J., Duance, V., Wotton, S., and Ayad, S. (1991) *Spine* **16**, 1030–1038.
37. Nakata, K., Ono, K., Miyazaki, J. *et al.* (1993) *Proc. Natl. Acad. Sci. USA* **90**, 2870–2874.
38. Tsumaki, N., Nakata, K., Adachi, E., Olsen, B.R., Ono, K., and Kimura, T. (1994) *Orthop. Trans.* **18**, 382.
39. Kimura, T., Nakata, K., Tsumaki, N., and Ono, K. (1994) In *Cell Mechanics and Cellular Engineering* (Mow, V.C. *et al.*, eds.), pp. 445–456, Springer-Verlag, New York.
40. Kimura, T., Nakata, K., Tsumaki, N. *et al.* (1996) *Int. Orthop.* **20**, 177–181.
41. Ritvaniemi, P., Körkkö, J., Bonaventure, J. *et al.* (1995) *Arthritis Rheum.* **38**, 999–1004.
42. Hacia, J.G., Brody, L.C., Chee, M.S., Fodor, S.P.A. and Collins, F.S. (1996) *Nature Genet.* **14**, 441–447.

Extracellular Matrix-Cellular Interaction: Molecules to Diseases (Y. Ninomiya et al., eds.), pp. 339–351,
Japan Sci. Soc. Press, Tokyo / S. Karger, Basel (1998)

Cartilage Hypertrophy: Isolation and Characterization of Upregulated Genes

THOMAS F. LINSENMAYER AND MARIA NURMINSKAYA

Department of Anatomy and Cellular Biology, Tufts University Medical School, Boston, Massachusetts 02111, U.S.A.

Growth and remodeling of skeletal elements can be advantageously studied in the epiphyseal growth zone. Here, a continuum of changes in cartilaginous matrices occurs as young growth cartilage matures and becomes hypertrophic, ultimately to be replaced by a marrow cavity and/or bony tissue (*1*). During cartilage hypertrophy, a number of changes in the matrix components are known to occur, including: (a) the acquisition of synthesis of collagen type X (*2–5*), (b) the loss of synthetic capacity for collagen types II and IX (*6, 7*) with concomitant covalent crosslinking of type II (*8*), (c) a decrease in proteoglycan synthesis with the complete loss of a small proteoglycan, PG-Lb (*9, 10*), and (d) an increase in activity of metalloproteinases (*11*) and alkaline phosphatase and calcification of cartilage (*12–14*). Presumably, this cascade, which we call the "hypertrophic program", renders the matrix susceptible to removal during marrow cavity formation and/or to participation in endochondral bone formation.

When a chondrocyte enters the hypertrophic program, how

many additional genes are upregulated and how many are down-regulated? Which of these changes are inherent within chondro-cytes themselves, and which require interactions with additional factors within the embryo?

To begin to answer these questions we have identified and characterized genes that are upregulated during cartilage hyper-trophy (*15*). For this we have employed a modification of the polymerase chain reaction (PCR)-mediated subtractive hybridi-zation method of Wang and Brown (*16*). The method was used to eliminate the mRNAs "common" to all chondrocytes, leaving those specific for hypertrophic cells. Using this modified proce-dure we have isolated 50 individual cDNA fragments for the upregulated genes; 18 have been tentatively identified, and two are being further studied.

I. EXPERIMENTAL APPROACHES

Since the procedures employed are critical to this study, we will present them in some detail.

Cultured chondrocytes at different stages of maturation were used as sources of RNAs. Hypertrophic chondrocytes were from the hypertrophic zone (*1, 2, 17*) of the tibia. These were "aged" in culture for 2–4 weeks and passaged several times before use, ensuring that the cells had progressed to the hypertro-phic stage (*3, 18*). Nonhypertrophic chondrocytes were from the caudal 1/3 of the sternum and were cultured for only 3–4 days before use.

Poly A+ RNA was isolated and cDNAs were obtained by reverse transcription using oligo (dT) as a primer. In prepara-tion for PCR amplification, the double-stranded cDNAs from hypertrophic and nonhypertrophic cells were digested with Sau3AI, and then the cDNAs were ligated with a specific double-stranded adaptor (8 nmoles). Each adaptor (dGACACT-CTCGAGACATCACCGTCC plus its complement for the tracer cDNA amplification and dAGCACTCTCCAGCCTCT-CACCGCA and its complement for the driver cDNA) was designed to have a 4-base, 3′ protruding end complementary to a Sau3AI-generated terminus of the cDNA. For production of a double-stranded adaptor, equimolar amounts of the complementary synthetic oligonucleotides were annealed at 45°C for 10 min. These then were ligated to the Sau3AI-digested cDNA fragments by overnight incubation with T4 ligase at 14°C. The unligated adaptors were removed from the reaction mixtures by electrophoresis in a 1.4% low-melting-agarose gel, and the smear containing the ligated cDNAs was cut out and used directly for PCR amplification.

The adaptor-ligated cDNA fragments for both the tracer (hypertrophic) and driver (nonhypertrophic) cDNAs were amplified by PCR. Each PCR reaction ($100 \mu l$) contained the adaptor-ligated cDNA in low-melting agarose ($2–3 \mu l$), AmpliTaq buffer (PerkinElmer Cetus) with 2.5 mM $MgCL_2$, 5U of Taq polymerase (PerkinElmer Cetus), and 1 μg of the appropriate primer: dGACACTCTCGAGACATCACCGTCC for the tracer cDNA amplification or dAGCACTCTCCAGCCTCTCACCGCA for the driver cDNA. Amplification was performed for 30 cycles of 94°C for 1 min, 56°C for 1 min, and 72°C for 2 min, with a 25 sec autoextension per cycle.

The amplified cDNAs ($30–50 \mu g$ of each) were labeled with Photoprobe LongArm Biotin, according to the directions of the manufacturer (Vector Laboratories).

For the subtractive hybridization, 35 μg of the biotinylated, non-hypertrophic driver cDNA was mixed with 1.5 μg of the nonbiotinylated, hypertrophic tracer cDNA. The mixture was precipitated with ethanol, redissolved in 10 μl of H_2O and denatured at 100°C. Ten μl of $2 \times$ hybridization buffer (1.5 M NaCl, 50 mM Hepes, pH 7.5, 10 mM EDTA, 0.2% SDS) was added and hybridization was performed at 68°C overnight. Following hybridization, the mixture was diluted to 0.5 M NaCl and the biotinylated, hybridized molecules were removed by streptavidin binding followed by phenol extraction (19). This binding/extraction procedure was performed a total of 4 times. The last phenol extraction was followed by a chloroform-ether extraction. A second subtractive hybridization (4 hr) of the tracer (hypertrophic) cDNA was performed with another aliquot of biotinylated nonhypertrophic driver cDNA (15 μg), and the hybridized molecules were removed with streptavidin, as just described.

The nonhybridized (upregulated hypertrophic tracer) cDNA fragments were amplified by PCR using the dGACACT-CTCGAGACATCACCGTCC primer, resulting in subtracted cDNA, called "2H-cDNA".

In a parallel set of subtractions, non-hypertrophic driver cDNA was treated in the same way with biotinylated hypertrophic cDNA and PCR amplified with dAGCACTCTCCAGCC-TCTCACCGCA primer, resulting in subtracted sternal cDNA.

The cDNA fragments obtained after amplification were ligated with the pCRII vector for transformation into competent *Escherichia coli* cells (TA cloning kit, Invitrogen). Colony hybridization of the randomly picked colonies was performed as described (*20*).

II. UPREGULATED GENES

1. Identification of Upregulated mRNAs by Subtractive Hybridization

The strategy employed to identify upregulated genes in hypertrophic, tibial chondrocytes is diagrammed in Fig. 1. It is a modified version of the PCR-based gene expression screen strategy of Wang and Brown (*16*). In the original method the same adaptor was ligated to both the cDNA population used for subtraction of the common gene products (the "driver" cDNA) and the cDNA population containing the desired, upregulated gene products (the "tracer" cDNA). In the present study, we used different adaptors for amplification of the driver cDNA (from nonhypertrophic chondrocytes) and the tracer cDNA (from hypertrophic chondrocytes). We found that this modification eliminates the amplification of driver cDNA fragments

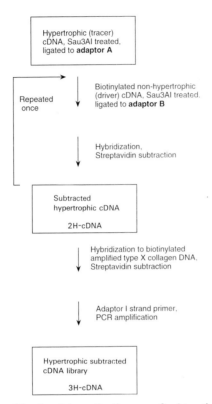

Fig. 1. Schematic diagram of subtractive hybridization.

remaining in the hybridization mixture after streptavidin extraction of the biotinylated driver.

The hypertrophic chondrocyte cDNA fragments enriched by 2 cycles of subtractive hybridization with cDNA fragments from nonhypertrophic sternal cells (the 2H-cDNA in Fig. 1) were reamplified by PCR, yielding products in the 0.2–2.0 kb range. These were cloned into the plasmid pCRII yielding a 2H-cDNA library. Twenty randomly chosen colonies were assayed for the sizes of their inserts by PCR, using the insert-ligated adaptor primer. The sizes of the cloned inserts ranged from 0.2 to 0.8 kb.

It has been reported previously (*16*) that after subtractive hybridization, the enrichment of one predominant sequence can occur. This also happened here, since two thirds of our clones were for collagen type X cDNA. To enrich for the less abundant mRNAs, the sequences for type X collagen were subtracted from the 2H-cDNA by hybridization with a biotinylated 845 bp fragment of collagen type X cDNA (Fig. 1). The resulting enriched hypertrophic cDNA fragments (3H-cDNA) were cloned into the plasmid pCRII. Although a few colonies (less than 10%) still showed positive hybridization for collagen type X, the subtraction was sufficient for further analysis of the less abundant upregulated mRNAs.

2. Analysis of the Upregulated Genes

For further identification and analyses of the upregulated cDNA fragments, we used direct sequencing as the fastest and potentially the most informative method. The first set of cDNA fragments that were sequenced consisted of 38 clones from the 2H-cDNA library which did not hybridize positively to type X collagen cDNA. The second set consisted of 23 clones from the 3H-cDNA library. Thus, in total, we isolated and sequenced 50 individual cDNA sequences for genes upregulated in hypertrophic chondrocytes.

By sequence analysis, 18 inserts (about 30% of the total number) have identities with entries in GenBank. These (summarized in Table I) included proteins thought to be involved in tissue differentiation, translation factors, and the enzymes transglutaminase and adseverin, and the carbohydrate binding protein galectin.

To examine the extent to which these mRNAs are differentially expressed in hypertrophic cartilage, several of them were analyzed by Northern blots of RNAs from hypertrophic and nonhypertrophic cells. Depending on the gene, the range of

TABLE I

Summary of the Identified Upregulated Genes in Chondrocyte Hypertrophy (the Blastn Search Program was used)

Clone number	Frequency	Gene with closest identity	Source	% of identity
1		Plasma transglutaminase (positions 620–855 in full-length cDNA)	Human	71
2		Plasma transglutaminase (positions 1240–1400 in full length cDNA)	Human	80
3	2	Galectin-3	Rabbit	71
4		Homolog to membrane receptor protein	Human	84
5		51C protein	Human	78
6		Vgr-1 protein	Mouse	80
7		Adseverin	Bovine	78
8		Glycyl tRNA synthetase	Human	79
9		Glycogen phosphorylase	Human	72
10	113	Collagen type X	Chicken	96
11		Alternative splicing factor SF2p33	Human	82
12		Eukaryotic initiation factor 4 g	Human	79
13	2	Elongation factor EF-1	Chicken	97
14	3	Transactivation factor p17	Mouse	83
15		Nucleolar protein No38	Chicken	96
16		Ribosomal protein S2 (sop)	Drosophila	87
17		Ribosomal protein S7	Human	83
18	3	Ribosomal protein S18	Rat	79

upregulation was from several fold to a complete qualitative difference.

3. Characterization of Individual cDNAs for Transglutaminase and Galectin

A) Transglutaminase

Two clones had high identity of the nucleotide sequences to the different regions of human plasma transglutaminase, also called coagulation factor XIIIa. Northern blot analysis produced a single band of 3.9 kb, which is similar to the plasma form in humans (*21*). We extended the length by 5′- and 3′-RACE and obtained a 1.6 kb clone encoding approximately the NH$_2$-terminal half of the transglutaminase. The overall identity of the amino acid sequence of the cloned chicken transglutaminase with human coagulation factor XIIIa is 75% (compared to 56% identity with the human keratinocyte enzyme and 60% with the human tissue transglutaminase), suggesting that it is this form of the molecule that is upregulated in the hypertrophic cells. More recently, we (Nurminskaya and Linsenmayer, in preparation) have identified the tissue form of the molecule as also being present, as had been previously reported (*22*). In our hands, however, the tissue form is distributed uniformly throughout nonhypertrophic and hypertrophic cartilage.

B) Galectin

The cDNA insert of one clone showed 71% identity to rabbit galectin-3. To extend the insert in the clone, we employed 5′- and 3′-RACE. The sequence identity with galectin-3 of rabbit and mouse is 64–65%. The C-terminal end contains a globular domain with 78% identity to the carbohydrate-binding domain of rabbit galectin-3. In addition, the derived amino acid sequence has a potential 27-residue N-terminal signal peptide with a hydrophobic core. As the mammalian galectins do not contain a signal peptide and are thought to be secreted by a non-classical secretory pathway (*23, 24*), it is possible that the cloned chicken chondrocyte protein is not the avian homologue of the rabbit galectin-3 but represents a galectin-related protein. Alternatively, this difference may reflect an evolutionary divergence of mammalian and avian species.

III. DISCUSSION OF UPREGULATED GENES

In this study we employed a modified version of the previously described gene expression strategy of Wang and Brown (*16*). In our modification, we used different pairs of PCR adaptor primers for the tracer (hypertrophic) and driver (nonhypertrophic) cDNAs. This allowed for more specific amplification for the desired tracer cDNA by preventing amplification of the driver cDNA fragments that might have been incompletely removed during the streptavidin subtraction.

Approximately 70% of the clones in the subtracted hypertrophic chondrocyte cDNA library contained the same fragment of the type X collagen cDNA. In the original study by Wang and Brown (16) this was also noted, with 90% of their clones containing one predominant insert.

Of the approximately 50 genes we have observed as being upregulated *in vitro*, some have already been reported as being upregulated in hypertrophic cartilage *in vivo*, such as type X collagen, transglutaminase and galectin. Others are reported here for the first time as being upregulated in the hypertrophic chondrocytes. These include certain translational factors and ribosomal proteins, and an intracellular modifier of the actin cytoskeleton, adseverin.

The presence of transglutaminase activity and protein has been reported in both articular and growth plate cartilages. In cell cultures of rabbit articular chondrocytes, Demignot *et al.* (25) observed two enzymatic activities for transglutaminase that differed in their cellular localization and their responses to retinoic acid. In the epiphyseal growth plate of the rat, Aeschlimann *et al.* observed immunohistochemically-detectable transglutaminase activity using antibodies against the tissue form of the molecule (26). The immunoreactivity was present intracellularly in the proliferative and maturation zones, but it became externalized in the lower hypertrophic zone. The highest activity was within the hypertrophic chondrocytes adjacent to the mineral deposits of the calcified cartilage (22). It was proposed that a major matrix substrate for this enzyme was osteonectin (22).

In none of these studies was the form of transglutaminase definitively identified by cDNA or protein sequencing, and at least four forms of the enzyme exist. Sequence analysis of the cDNA that we have isolated suggests that it is the chicken homolog of the "a" subunit of plasma transglutaminase, factor XIIIa. In chicken, the only form of transglutaminase previously characterized is from erythrocytes, a form that we think is likely to be tissue form (27).

That the form of transglutaminase we detected in hypertrophic cartilage is factor XIIIa-like appears to conflict with the previously published immunohistochemical identification of the cartilage molecule as tissue transglutaminase (26). The antibodies employed in that study were against the tissue form of the molecule and were reported as having no crossreactivity with factor XIIIa. Since, however, that study was done on rat, the differences might simply reflect a species difference. Alternatively, hypertrophic cartilage may contain more than one form of the enzyme.

The known substrates for factor XIIIa include extracellular (28, 29) and cytoskeletal proteins (30). This, coupled with the immunohistochemical results (22) showing the growth plate transglutaminase to be intracellular in nonhypertrophic cartilage and extracellular in the hypertrophic zone, raises the possibility that the enzyme performs multiple roles.

The other cDNA for an upregulated gene we characterized is for a protein of the galectin family. The galectins are a class of lectins that crosslink carbohydrate chains on cell surfaces and/or in the extracellular matrix (31). The chicken galectin we isolated, by the criteria described, is the chicken homolog of mammalian galectin-3. There are, however, some differences from the mammalian molecule, most notably the presence of a putative signal peptide. Putative interacting substrates include fibronectin (32), which in cartilage has been previously shown to be localized specifically within the hypertrophic matrix (33). A recent study employing *in situ* hybridization on a number of mouse tissues has also shown the molecule to be upregulated in hypertrophic cartilage (34).

Approximately half the remaining upregulated genes we identified are associated with translation.

Many of the upregulated cDNAs isolated in this study have not been identified. Some may play essential yet unknown roles in chondrocyte differentiation, but further investigation will be required to identify these genes and determine their functions in the hypertrophic program.

SUMMARY

To isolate genes upregulated during cartilage hypertrophy we employed a modification of the PCR-mediated subtractive hybridization method of Wang and Brown (*16*). Cultures of hypertrophic tibial chondrocytes and non-hypertrophic sternal cells were used for RNA isolation. Among 50 individual cDNA fragments isolated as upregulated genes, 18 were tentatively identified by their similarities to entries in the GenBank database, whereas the other 32 showed no similarity. The identified genes included translational and transcriptional regulatory factors, ribosomal proteins, the enzymes transglutaminase and glycogen phosphorylase, type X collagen, and the carbohydrate-binding protein galectin. Two of these, transglutaminase and galectin, were cloned and further characterized.

Acknowledgment
This work was supported by NIH grant HD 23681.

REFERENCES

1. Stocum, D.L., Davis, R.M., Leger, M., and Conrad, H.E. (1979) *J. Embryol. Exp. Morphol.* **54**, 155–170.
2. Schmid, T.M. and Conrad, H.E. (1982) *J. Biol. Chem.* **257**, 12444–12450.
3. Schmid, T.M. and Linsenmayer, T.F. (1983) *J. Biol. Chem.* **258**, 9504–9509.
4. Gibson, G.J., Beaumont, B.W., and Flint, M.H. (1984) *J. Cell Biol.* **99**, 208–216.
5. Capasso, O., Quarto, N., Descalzi-Cancedda, F., and Cancedda, R. (1984) *EMBO J.* **3**, 823–827.
6. Linsenmayer, T.F., Chen, Q., Gibney, E. *et al.* (1991) *Development* **111**, 191–196.

7. Oshima, O., Leboy, P.S., McDonald, S.A., Tuan, R.S., and Shapiro, I.M. (1989) *Calcif. Tiss. Int.* **45**, 182–192.

8. Chen, Q., Fitch, J.M., Gibney, E., and Linsenmayer, T.F. (1993) *Dev. Dynam.* **196**, 47–53.

9. Shinomura, T., Kimata, K., Oike, Y., Maeda, N., Yano, S., and Suzuki, S. (1984) *Dev. Biol.* **103**, 211–220.

10. Shinomura, T. and Kimata, K. (1992) *J. Biol. Chem.* **267**, 1265–1270.

11. Sakiyama, H., Inaba, N., Toyoguchi, T. *et al.* (1994) *Cell Tiss. Res.* **277**, 239–245.

12. Leboy, P.S., Vaias, L., Uschmann, B., Golub, E., Adams, S.L., and Pacifici, M. (1989) *J. Biol. Chem.* **264**, 17281–17286.

13. Stein, G.S., Lian, J.B., and Owen, T.A. (1990) *FASEB J.* **4**, 3111–3123.

14. Wuthier, R.E. and Register, T.C. (1995) In *The Chemistry and Biology of Mineralized Tissue* (Butler, W.T., ed.), pp. 113–124, Ebsco Media Inc., Birmingham.

15. Nurminskaya, M. and Linsenmayer, T.F. (1996) *Dev. Dynam.* **206**, 260–271.

16. Wang, Z. and Brown, D.D. (1991) *Proc. Natl. Acad. Sci. USA* **88**, 11505–11509.

17. Kim, J.J. and Conrad, H.E. (1977) *J. Biol. Chem.* **252**, 8292–8299.

18. Schmid, T.M. and Conrad, H.E. (1982) *J. Biol. Chem.* **257**, 12451–12457.

19. Sive, H.L. and John, T.S. (1988) *Nucl. Acids Res.* **16**, 10937.

20. Sambrook, J., Fritsch, E.F., and Maniatis, T. (1989) *Molecular Cloning: A Laboratory Manual*, Cold Spring Harbor Laboratory Press, Cold Spring Harbor, New York.

21. Greenberg, C.S., Birckbichler, P.J., and Rice, R.H. (1991) *FASEB J.* **5**, 3071–3077.

22. Aeschlimann, D., Kaupp, O., and Paulsson, M. (1995) *J. Cell Biol.* **129**, 881–892.

23. Cooper, D.N.W. and Barondes, S.H. (1990) *J. Cell Biol.* **110**, 1681–1691.

24. Lindstedt, R., Apodaca, G., Barondes, S.H., Mostov, K.E., and Leffler, H. (1993) *J. Biol. Chem.* **268**, 11750–11757.

25. Demignot, S., Borge, L., and Adolphe, M. (1995) *Biochim. Biophys. Acta: Mol. Cell Res.* **1266**, 163–170.

26. Aeschlimann, D., Wetterwald, A., Fleisch, H., and Pausson, M. (1993) *J. Cell Biol.* **120**, 1461–1470.

27. Weraarchakul-Boonmark, N., Jeong, J.-M., Murthy, S.N.P., Engel, J.D., and Lorand, L. (1992) *Proc. Natl. Acad. Sci. USA* **89**, 9804–9808.

28. Prince, C.W., Dickie, D., and Krumdieck, C.L. (1991) *Biochem. Biophys. Res. Commun.* **177**, 1205–1210.

29. Mosher, D.F. and Proctor, R.A. (1980) *Science* **209**, 927–929.

30. Zhu, Y., Tassi, L., Lane, W., and Mendelsohn, M.E. (1994) *J. Biol. Chem.* **269**, 22379–22384.

31. Barondes, S.H., Cooper, D.N.W., Gitt, M.A., and Leffler, H. (1994) *J. Biol.*

Chem. **269**, 20807–20810.

32. Sato, S. and Hughes, R.C. (1992) *J. Biol. Chem.* **267**, 6983–6990.

33. Chen, Q. and Linsenmayer, T.F. (1993) In *Fourth International Conference on Limb Development and Regeneration (Part B)* (Fallon, J.F. *et al.*, eds.), pp. 495–504, John Wiley & Sons Inc., New York.

34. Fowlis, D., Colnot, C., Ripoche, M.-A., and Poirier, F. (1995) *Dev. Dynam.* **203**, 241–251.

*Extracellular Matrix-Cellular Interaction: Molecules to Diseases (Y. Ninomiya et al., eds.), pp. 353–366,
Japan Sci. Soc. Press, Tokyo/S. Karger, Basel (1998)*

Collagen-induced Arthritis

LINDA K. MYERS,*1 KUNIAKI TERATO,*2
DAVID BRAND,*2 EDWARD F.
ROSLONIEC,*2 JEROME M. SEYER,*2
MICHAEL A. CREMER,*2 JOHN M.
STUART,*2 AND ANDREW H. KANG*2

*Department of Pediatrics*1 *and Department of Internal
Medicine,*2 *University of Tennessee, Memphis, Research
Service of the Veterans Affairs Medical Center, Memphis,
Tennessee 38163, U.S.A.*

The implication of immune response to collagen in the pathogenesis of arthritis is certainly not a new concept. Investigations by Steffen and Timpl in the 1960's demonstrated antibodies to collagen in patients with rheumatoid arthritis (RA) (*1*). This finding and other studies prompted Steffen to propose the hypothesis that autoimmunity to collagen is involved in the pathogenesis of RA. Additional studies of immune responses to collagens in RA by several other groups of investigators have been reported since then suggesting that a majority of patients with RA developed antibodies and/or T cell sensitivity to type II collagens (CII), the major constituent protein of articular cartilage (*2–5*). During the course of investigations of the role of collagen autoimmunity in arthritis, we discovered that immunization of certain strains of rats led to the development of inflammatory polyarthritis resembling RA (*6*). Subsequently, mice and nonhuman primates have also been found to be susceptible to type II collagen-induced arthritis (CIA) (*7–9*). We will attempt to briefly review and summarize the current status of the CIA model and its implications for the pathogenesis of RA.

I. DESCRIPTION OF CIA

CIA can be induced in susceptible strains of animals by the subcutaneous injection of purified CII emulsified in a suitable adjuvant (*10*). Susceptibility to CIA is associated with the expression of specific class II molecules of the major histocompatibility complex (MHC) (*11*). Using B10 congenic strains of mice, Wooley and his co-workers have shown that CIA susceptibility is linked to the class II molecules, I-A, and specifically the I-A^q and I-A^r alleles. Following immunization, clinically evident arthritis occurs usually from the 28th to the 35th day in approximately 90% of susceptible mice. Arthritis onset is usually rapid with swelling, heat and erythema, often symmetrically involving the lower and upper extremities. In the majority of cases, arthritis is persistent and chronic, and progresses to complete destruction of involved joints. Histologically, the lesion is a proliferative synovitis with infiltrating inflammatory cells. The earliest cell type identified in abundance in the synovium is the neutrophil which is gradually replaced by infiltrating monocytes and lymphocytes. Within 3 days of onset, erosions of cartilage and subchondral bone by pannus-like tissue are evident and healing follows slowly by fibrosis and ankylosis of the involved joints (*12*).

II. IMMUNOPATHOGENESIS OF CIA

1. Role of Humoral Response to CII

There is ample evidence that the destructive arthritis characteristic of CIA is initiated by the binding of complement fixing antibodies to epitopes exposed on CII molecules on the surface of articular cartilage (*13*). Deposition of IgG antibodies and the complement component C3 on the cartilage surface at the very onset of arthritis has been clearly demonstrated by immunofluorescence studies (*13*). Depletion of serum complements by cobra venom administration suppresses CIA until serum complement levels return towards normal levels (*14*). Genetically complement-deficient strains of mice are resistant to CIA (*15, 16*). Arthritis can be passively transferred to naive mice by polyclonal antibodies purified from the sera of CII-immunized arthritic DBA/1 mice (*17*). Moreover, we have shown that disease can also be induced using a selected combination of monoclonal antibodies (mAb) reactive with CII (*18*). Watson and Townes in our laboratory have observed that IgG2a, which is capable of fixing and activating complement in mice, is the predominant isotype/subclass present in arthritic animals (*19*). In addition, γ-interferon, a lymphokine produced by TH1 cells which upregulates IgG2a has been shown to enhance CIA (*20*). In contrast, IL-4, a lymphokine produced by TH2 cells which down-regulates IgG2a has been shown to suppress CIA (*21*). Therefore, it is likely that the induction of arthritis is not only dependent on the presence of antibodies to CII, but also on their epitope specificity, isotype, and subclass composition, which are regulated by T cells (*22*).

2. Role of T Cell Response to CII

Thus, while autoreactive antibodies initiate the inflammatory arthritis, it is also clear that T cells are critical to the induction of arthritis. As mentioned previously, susceptibility to CIA is strongly linked to the class II molecules of the MHC (*23*), and depletion of CD4+ T cells prevents arthritis. In addition, T cell-deficient nude mice are resistant to arthritis, and, as expected, produce little, if any, antibody on immunization with CII (*24*). Anti-T cell antibodies and anti-CD4 antibodies have also been shown to suppress CIA (*25–28*). T cells can also contribute to the severity of disease. Mice transgenic for a collagen-reactive T cell receptor develop more severe disease (*29*), while mice lacking important T cell receptor subsets develop milder arthritis (*30*).

Based on these observations, we have proposed that the

antigen-specific activation of T cells capable of promoting the production of antibodies of particular subclass and epitope specificities is a critical step in inducing CIA. We further suggest that stimulation of these arthritis-promoting T cells is dependent on the ability of the specific class II molecules to bind and present critical collagen antigenic peptide(s), and that only the presentation of these critical peptides will drive the production of arthritogenic antibodies.

As part of a systematic effort to identify the arthritogenic T and B cell epitopes, we analyzed the reactivity of the anti-CII antibodies isolated from arthritic DBA/1 mice for each of the CNBr peptides. We found that CB11 alone contained T cell and B cell epitopes sufficient to induce arthritis (31). Subsequent studies identified the major T cell and B cell epitopes located in CB11 (CII 124–402) (22, 31–35).

III. IDENTIFICATION OF AN IMMUNODOMINANT T CELL EPITOPE IN CB11 (CII 124–402)

The specific amino acid sequence of CB11 which can regulate autoreactive T cells was delineated by constructing a set of 93 consecutive overlapping 15-mer peptides, beginning at every third residue, and spanning the entire sequence of CB11 using the Mimotope Multipin Peptide Synthesis System (32). T cells from CII-immunized DBA/1 mice were cultured with these peptides and then assayed for proliferative response. Three peptides, 44, 45, and 46, generated the greatest responses (Fig. 1) and their sequences are shown in Table I. These data strongly suggest that the antigenic determinant contained within these 3 peptides is the CB11 immunodominant T cell determinant.

IV. TOLERANCE INDUCTION AND SUPPRESSION OF ARTHRITIS BY THE IMMUNODOMINANT T CELL EPITOPE IN CB11 (CII 124–401)

We have previously shown that native CII, when given as a

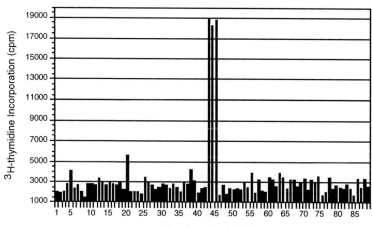

Fig. 1. Identification of major antigenic determinants in CB11 recognized by DBA/1 (I-A^q) mice using Mimotope synthetic peptides. T cells obtained from lymph nodes of CII-immunized mice were cultured with a series of overlapping peptides (Mimotope peptides, Chiron), 15 mers, beginning at every 3rd residue, representing the entire sequence of CB11. Cells were cultured with peptides for 72 hr, the last 18 hr in the presence of [^{3}H]-thymidine. Cells were harvested on fiberglass filters, and [^{3}H]-thymidine incorporation was measured. Reproduced with permission from ref. *37*. Copyright 1994. The American Association of Immunologists.

TABLE I

Identification of a Major Antigenic Determinant in CB11 Recognized by DBA/1 (I-A^q) Mice Using Mimotope Synthetic Peptides

Mimotope peptide[a]	CB11 Mimotope sequence					T cell stimulation
	255	260	265	270	275	
43 (CII 250–264)	GPKGQTGEBGIAGFK					−
44 (CII 253–267)	GQTGEBGIAGFKGEQ					+
45 (CII 256–270)	GEBGIAGFKGEQGPK					+
46 (CII 259–273)	GIAGFKGEQGPKGEB					+
47 (CII 262–276)	GFKGEQGPKGEBGPT					−

[a] Residues are numbered from the NH$_2$-terminus of CII. The letter B represents hydroxyproline. Reproduced with permission from ref. *32*. Copyright 1994. The American Association of Immunologists.

tolerogen intravenously prior to immunization, was capable of suppressing CIA (*35, 36*). This activity was also contained within CB11. Having identified CII 260–270 as an immunodominant T cell epitope, it was of interest to determine whether this epitope had functional significance in the induction and regulation of arthritis (*33*). Therefore, a synthetic peptide representing CII 260–270 was administered to mice as a neonatal tolerogen. Groups of mice tolerized with CII 260–270 had 31% arthritis, compared to 89% arthritis in untolerized controls ($p \leq 0.01$). Mean levels of antibodies to native CII were also significantly decreased ($p \leq 0.005$). A series of hexacosopeptide analogs of CII 245–270 containing substitutions based on type I collagen sequences or alanine substitutions for proline were synthesized as

TABLE II

Identification of Residues Critical for the Induction of Tolerance in Experimental Autoimmune Arthritis Utilizing CII Peptides with Amino Acid Substitutions

Peptides used to induce tolerance					Incidence of arthritis[a]
250	255	260	265	270	
P T G P L G P K G Q T G E L G I A G F K G E Q G P K					6/20 (30%)
I A G F K G E Q G P K					4/13 (31%)
A- -					2/7 (28%)
- S -					2/7 (28%)
- - - A -					6/14 (43%)
- - - - B -					3/7 (43%)
- - - - - A -					3/7 (43%)
- - - - - - - - N - - - - - - - - - - - - - - - - - -					3/10 (30%)
- - - - - - - - - S - - - - - - - - - - - - - - - - -					1/5 (20%)
- - - - - - - - - - B - - - - - - - - - - - - - - - -					9/14 (64%)
- - - - - - - - - - - - A- - - - - - - - - - - - - -					18/24 (75%)**
- - - - - - - - - - - - - B - - - - - - - - - - - - -					17/22 (77%)*
- - - - - - - - - - - - - - - N- - - - - - - - - - -					20/26 (77%)**
- - - - - - - - - - - - - - - - - - A- - - - - - - -					10/14 (71%)*
- D- - - - -					23/35 (66%)*
- T- - -					12/21 (57%)
- A-					8/16 (50%)

[a] In the absence of tolerogen, the incidence of arthritis was 16/18 (89%)**. B indicates hydroxyproline. *$p \leq 0.02$ using Fisher's Exact Test, **$p \leq 0.003$ using the same test.

shown in Table II. Substitutions based on type I collagen were used in these experiments because the amino acid sequence of type I collagen is very similar to CII, yet immunization of mice with type I collagen does not induce CIA. Each peptide was then administered to neonatal DBA/1 mice as a tolerogen to test its ability to suppress arthritis. Substitutions at positions 260, 261, 263, 264, and 266 significantly decreased the suppressive effect ($p \leq 0.003$ for 260, 261, and 263; $p \leq 0.02$ for 264 and 266, Table II) indicating that these amino acid residues are critical for suppression of arthritis.

V. DEVELOPMENT OF ANALOG PEPTIDE LIGAND CAPABLE OF SUPPRESSING CIA

1. Identification of CII Analog Peptides Incapable of Stimulating CII 245-270-specific T Cells

Although five critical residues of the immunodominant epitope were identified, it was unclear whether the substitutions interfered with binding to I-A^q, or alternatively, disrupted T cell receptor (TCR) recognition. Each of four analogs of CII 245-270 (Table III) was cultured with pooled spleen and lymph node cells from CII-immunized mice, and culture supernatants were

TABLE III
Amino Acid Sequence of Synthetic Peptide Analogs

Peptide[a]	246	250	254	258	262	266	270
CII 245-270	P T G P	L G P	K G Q T	G E L	G I A G F	K G E	Q G P K
CII 245-270 [s248, 249]	- - - A	B - -	- - - -	- - -	- - - - -	- - -	- - - -
CII 245-270 [s248, 251, 269]	- - - A	- - A	- - - -	- - -	- - - - -	- - -	- A -
CII 245-270 [s260, 261, 263]	- - - -	- - -	- - - -	- - -	A B - N -	- - -	- - -
CII 245-270 [s266, 267, 269]	- - - -	- - -	- - - -	- - -	- - - - -	D T -	A -
Type I 245-270	- S - A	B - -	- - N S	- - B	- A B - N	- - D	T - A -

[a] A "-" indicates identity at that position with wild type peptide CII 245-270. Amino acid residues are represented by a single letter code, with the letter "B" denoting hydroxyproline. Residues are numbered from the NH$_2$-terminus of CII. Reproduced with permission from ref. 37. Copyright 1994. The American Association of Immunologists.

tested for the presence of γ-interferon as an indicator of T cell stimulation. As shown in Fig. 2, substitution of alanine for proline at position 248, and hydroxyproline for leucine at position 249 had no effect on T cell stimulation compared to the response of T cells to the wild type peptide, CII 245–270. However, when the substitution at residue 248 was combined with an alanine for proline substitution at residues 251 and 269, the ability of the T cells to respond to this peptide was greatly reduced (25% of the wild-type peptide response). In contrast, the CII-primed T cells did not respond to two analog peptides, one containing substitutions at residues 260, 261, and 263, and the other at residues 266, 267, and 269 (Fig. 2) (*37*). All of these substitutions are based on type I collagen sequences and are non-conservative substitutions, with the exception of aspartic acid for glutamic acid at residue 266. These data indicated that the amino acid residue(s) at these positions are critical for

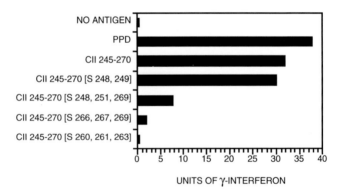

Fig. 2. Pooled spleen and lymph node cells from CII-immunized DBA/1 mice were cultured with either PPD, CII 245–270 synthetic peptide, or an analog of CII 245–270. Seventy-two hours later, the supernatants were tested for the presence of γ-interferon by ELISA. Two of the analogs, CII 245–270[s266, 267, 269] and CII 245–270[s260, 261, 263], stimulated less than 5 units of γ-interferon production by the T cells. Reproduced with permission from ref. *37*. Copyright 1993. The American Association of Immunologists.

I-A^q-restricted presentation of the CII 245–270 peptide to T cells.

2. Competitive Inhibition of I-A^q-restricted Antigen Presentation by CII Analog Peptides

The inability of the analog peptides to stimulate T cells lies in either the disruption of peptide binding to the I-A^q molecule or the inability of the TCR to recognize the peptide. In order to

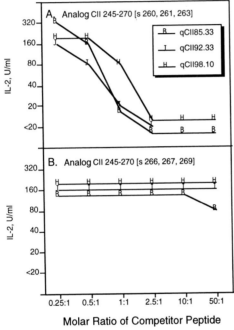

Fig. 3. Competitive inhibition of antigen presentation to CII 260–270 specific T-cell hybridomas. DBA/1 spleen cells were pulsed with various molar ratios of analog peptide CII 245–270[s260, 261, 263] (panel A) or CII 245–270[s266, 267, 269] (panel B) and wild type peptide CII 245–270, washed, and tested for their ability to present antigen to CII 245–270 specific T-cell hybridomas. IL-2 production by the T-cells was determined by the ability of culture supernatants to support the growth of IL-2-dependent HT-2 cells.

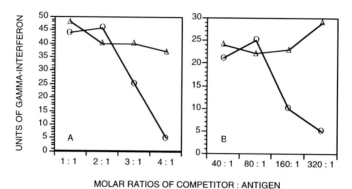

MOLAR RATIOS OF COMPETITOR : ANTIGEN

Fig. 4. Analog peptides of CII 245–270 were cultured in various molar ratios with the wild type peptide CII 245–270 (panel A), or native CII (panel B). Seventy-two hours later supernatants were collected and tested for the presence of γ-interferon by ELISA. As shown, CII 245–270[s260, 261, 263] (○) competitively inhibited T cell responses to both CII 245–270 and CII in a dose-dependent manner, while CII[s266, 267, 269] (△) had no effect. Reproduced with permission from ref. 37. Copyright 1994. The American Association of Immunologists.

determine whether the analog peptides could bind to the class II molecule, competitive antigen presentation assays were performed. Antigen presenting cells (APC) were pre-pulsed with various ratios of CII 245–270 and an analog peptide, washed, and tested for their ability to stimulate CII 245–270-specific T cell hybridomas. When APC were pulsed with CII 245–270 [s260, 261, 263] and CII 245–270 at molar ratios of 6.5 : 1 or greater, respectively, their ability to stimulate the T-cell hybridomas was greatly reduced (Fig. 3). In contrast, the CII 245–270 [s266, 267, 269] analog peptide did not compete with the wild-type peptide indicating that the residues altered in this analog are critical for the T cell recognition of the peptide.

When these same analog peptides were tested for their ability to inhibit the presentation of antigen to CII-primed, bulk T cells, similar results were observed (37). In these experiments, analog peptides were combined with either CII 245–270 or native CII at various molar ratios and added to spleen cells from CII-immunized DBA/1 mice. As was observed with the T-cell hybridomas, the addition of peptide CII 245–270[s260, 261, 263] to the T cell cultures significantly decreased responses to both CII 245–270 and CII in a dose dependent manner while CII 245–270[s266, 267, 269] had no significant effect (Fig. 4). The fact that higher molar ratios were required for inhibition of native CII than for inhibition of the CII 245–270 peptide is likely an indication that the CII molecule contains several class II binding epitopes.

3. Co-immunization of DBA/1 Mice with CII 245–270[s260, 261, 263] and Native CII

Since the analog peptide CII 245–270[260, 261, 263] inhibited the presentation of antigen *in vitro*, it was tested for its ability to inhibit the induction of arthritis *in vivo*. DBA/1 mice were immunized with either CII, CII plus CII 245–270, or CII plus CII 245–270[s260, 261, 263] and were observed for the development of arthritis. DBA/1 mice co-immunized with CII 245–270[s260, 261, 263] demonstrated a decrease in the inci-

TABLE IV

Suppression of Arthritis by Simultaneous Immunization with CII and an Analog Peptide

Peptide[a]	Molar ratio (CII : peptide)	Number of arthritic mice	Number of arthritic limbs
CII 245–270[s260, 261, 263]	1 : 0	10/12 (83%)	24/48 (50%)
	1 : 160	4/6 (67%)	10/24 (42%)
	1 : 320	6/12 (50%)	11/48 (23%)**
	1 : 480	0/10 (0%)*	0/40 (0%)**
CII 245–270	1 : 480	4/6 (67%)	8/24 (33%)

[a] DBA/1 mice were immunized intradermally with a single emulsion containing either CII 245–270[s260, 261, 263] and CII, or wild-type CII 245–270 and CII in complete Freund's adjuvant. *$p \leq 0.002$, **$p \leq 0.005$ using Fisher's Exact Test. Reproduced with permission from ref. *37*. Copyright 1994. The American Association of Immunologists.

dence of arthritis and number of arthritic limbs (Table IV). When native CII and CII 245-270[s260, 261, 263] were co-injected at a molar ratio of 1 : 480, arthritis did not develop (*37*). Simultaneous immunization with CII plus CII 245-270 did not alter the incidence of disease. Concordant with a decrease in the incidence and severity of arthritis, antibody production to native CII was also significantly decreased.

SUMMARY

In summary, we have identified $\alpha 1$(II)-CB11 (CII 124-402) as the region of CII which contains major T and B cell epitopes recognized by DBA/1 (I-A^q) mice. Within CB11, CII 260-267 has been identified as the immunodominant T cell epitope. Utilizing antigen-presentation inhibition assays, we have deduced specific residues important in T cell recognition. Based on these data, we developed a synthetic peptide analog which can suppress immune response to CII and prevent CIA when administered at the time of immunization with CII. Future investigations will be directed at understanding the mechanism of action of the analog and at optimizing the dosing schedule for the maximal therapeutic benefit. We believe that a similar approach to the therapy of RA may be warranted.

REFERENCES

1. Steffen, C. and Timpl, R. (1963) *Int. Arch. Allergy Appl. Immunol.* **22**, 333–340.
2. Steffen, C., Sanger, L., and Menzel, J. (1980) *Scand. J. Rheumatol.* **9**, 69–76.
3. Michaeli, D. and Fudenberg, H.H. (1974) *Clin. Immunol. Immunopathol.* **2**, 153–159.
4. Stuart, J.M., Postlethwaite, A.E., Townes, A.S., and Kang, A.H. (1980) *Am. J. Med.* **69**, 13–18.
5. Trentham, D.E., Dynesius, R.A., Rocklin, R.E., and David, J.R. (1978) *N. Engl. J. Med.* **299**, 327–332.
6. Trentham, D.E., Townes, A.S., and Kang, A.H. (1977) *J. Exp. Med.* **146**, 857–868.

7. Courtenay, J.S., Dallman, M.J., Dayan, A.D., Martin, A., and Mosedale, B. (1980) *Nature* **283**, 666–668.

8. Cathcart, E.S., Hayes, K.C., Gonnerman, W.A., Lazzari, A.A., and Franzblau, C. (1986) *Lab Invest.* **54**, 26–31.

9. Yoo, T.J., Kim, S.Y., Stuart, J.M. *et al.* (1988) *J. Exp. Med.* **168**, 777–782.

10. Rosloniec, E.F., Kang, A.H., Myers, L.K., and Cremer, M.A. (1997) In *Current Protocols in Immunology* (Coico, R. and Shevach, E., eds.), pp. 15.5.1–24, Wiley & Sons, New York, NY.

11. Wooley, P.H., Luthra, H.S., Stuart, J.M., and David, C.S. (1981) *J. Exp. Med.* **154**, 688–700.

12. Stuart, J.M., Townes, A.S., and Kang, A.H. (1985) *Ann. N.Y. Acad. Sci.* **460**, 355–362.

13. Stuart, J.M., Tomoda, K., Townes, A.S., and Kang, A.H. (1983) *Arthritis Rheum.* **26**, 1237–1244.

14. Morgan, K., Clague, R.B., Shaw, M.J., Firth, S.A., Twose, T.M., and Holt, P.J. (1981) *Arthritis Rheum.* **24**, 1356–1362.

15. Spinella, D.G., Jeffers, J.R., Reife, R.A., and Stuart, J.M. (1991) *Immunogenetics* **34**, 23–27.

16. Reife, R.A., Loutis, N., Watson, W.C., Hasty, K.A., and Stuart, J.M. (1991) *Arthritis Rheum.* **34**, 776–781.

17. Stuart, J.M. and Dixon, F.J. (1983) *J. Exp. Med.* **158**, 378–392.

18. Terato, K., Hasty, K.A., Reife, R.A., Cremer, M.A., Kang, A.H., and Stuart, J.M. (1992) *J. Immunol.* **148**, 2103–2108.

19. Watson, W.C. and Townes, A.S. (1985) *J. Exp. Med.* **162**, 1878–1891.

20. Snapper, C.M., Peschel, C., and Paul, W.E. (1988) *J. Immunol.* **140**, 2121–2127.

21. Mossman, T R. and Coffman, R.L. (1989) *Annu. Rev. Immunol.* **7**, 145–173.

22. Brand, D., Marion, T., Myers, L. *et al.* (1996) *J. Immunol.* **157**, 5178–5184.

23. Wooley, P.H., Dillon, A.M., Luthra, H.S., Stuart, J.M., and David, C.S. (1983) *Trans. Proc.* **15**, 180–186.

24. Klareskog, L., Holmdahl, R., Larsson, E., and Wigzell, H. (1983) *Clin. Exp. Immunol.* **51**, 117–125.

25. Ranges, G.E., Sriram, S., and Cooper, S.M. (1985) *J. Exp. Med.* **162**, 1105–1110.

26. Goldschmidt, T.J. and Holmdahl, R. (1991) *Eur. J. Immunol.* **21**, 1327–1330.

27. Brahn, E. and Trentham, D. (1984) *Cell Immunol.* **86**, 421–428.

28. Brahn, E. and Trentham, D.E. (1987) *Cell Immunol.* **109**, 139–147.

29. Mori, L., Loetscher, H., Kakimoto, K., Bluethmann, H., and Steinmetz, M. (1992) *J. Exp. Med.* **176**, 381–388.

30. Anderson, G.D., Banerjee, S., Luthra, H.S., and David, C.S. (1991) *J. Immunol.* **147**, 1189–1193.

31. Terato, K., Hasty, K.A., Cremer, M.A., Stuart, J.M., Townes, A.S., and

Kang, A.H. (1985) *J. Exp. Med.* **162**, 637–646.

32. Brand, D., Myers, L., Terato, K. *et al.* (1994) *J. Immunol.* **152**, 3088–3097.
33. Myers, L.K., Terato, K., Seyer, J.M., Stuart, J.M., and Kang, A.H. (1992) *J. Immunol.* **149**, 1439–1443.
34. Myers, L.K., Terato, K., Stuart, J.M., Seyer, J.M., David, C.S., and Kang, A. H. (1993) *J. Immunol.* **151**, 500–505.
35. Myers, L.K., Stuart, J.M., Seyer, J.M., and Kang, A.H. (1989) *J. Exp. Med.* **170**, 1999–2010.
36. Cremer, M.A., Hernandez, A.D., Stuart, J.M., Townes, A.S., and Kang, A.H. (1983) *J. Immunol.* **131**, 2995–3000.
37. Myers, L., Rosloniec, E., Seyer, J., Stuart, J., and Kang, A. (1993) *J. Immunol.* **150**, 4652–4658.

Extracellular Matrix-Cellular Interaction: Molecules to Diseases (Y. Ninomiya et al., eds.), pp. 367–370,
Japan Sci. Soc. Press, Tokyo/S. Karger, Basel (1998)

Perspectives

TOSHIRO OOYAMA

Department of Clinical Physiology, Toho University School of Medicine, Ota-ku, Tokyo 143-0015, Japan

The excessive action of collagenases leading to breakdown of both type II and type I collagens in specific extracellular matrix (ECM) is thought to be a major cause of rheumatoid arthritis (and osteoarthrosis). Degradation of type I collagen has been shown to take place through the action of collagenases at a helical cleavage site between Gly775 and Ile776 of the α1(I) chain and a homologous site in the α2(I) chain using the technique of targeted mutation in mice. Understanding the mechanism by which collagenase works may help in the design of new therapeutic approaches to inflammatory and degenerative joint diseases such as rheumatoid arthritis (RA) and osteoarthrosis (OA).

Is human OA or disc degeneration caused by mutations of the collagen genes? At present there is no conclusive answer to this question, although studies suggest that approximately 2% of familial OA is due to type II collagen mutations. The idea that the allelic variants of cartilage collagen genes may be one of the risk factors for cartilage degeneration should be tested in the near future using new DNA chip-base assay technology.

Fibrillins, one of the major constituents of microfibrils, appear to perform important functions in vascular and skeletal growth and in the maintenance of connective tissue integrity. Several functions of fibrillins have been distinguished by genetic analyses in humans and mice, however, the mechanisms behind these roles remain obscure. Structural information provided by analyses of the specific types of domains (calcium-binding EGF-like domains, 8 cys domain, hybrid domains, amino and carboxyl termini, pro/gly-rich regions) will provide the basis for understanding important molecular interactions (ligand interactions; binding to growth factors; assembly of microfibrils). Mutations in fibrillin I gene have already been identified in individuals with the Marfan syndrome and related phenotypes and further systematic investigations on fibrillins are required.

Aggrecan, a large chondroitin sulfate proteoglycan, forms huge aggregates by binding to both hyaluronan and link proteins. Extensive hydration of the chondroitin sulfate chains attached to the protein core results in a unique gel-like property and resistance to deformation which are characteristic of cartilage. Thus, aggrecan maintains the cartilage structure and functions. Deficiency in cartilage matrix including aggrecan is exemplified by the mutation-disease of a proteoglycan gene identified in mammals. Mice with proteoglycan (aggrecan) deficiency show a high incidence of spinal misalignment and movement problems which develop with age, primarily involving spastic paralysis of the hind limbs. This paralysis resembles spinal disc herniation or spondylo-myelopathy in older humans. The first test is yet to be done on human suspected of having this condition.

Elastin expression in arterial smooth muscle cells (SMCs) is thought to be inert but to be actively regulated by factors affecting the SMC growth state. For example, elastin expression was enhanced by inhibitors of SMC proliferation including minoxidil, heparin, and retinoic acid; heparin has been known to inhibit the formation of thickening of the intima after balloon injury in animals. In contrast, elastin expression was reduced by stimulators of SMC proliferation like EGF, K ion, angiotensin II, and phorbol ester. Thus, elastin gene expression and the SMC growth state are closely linked.

The same is true of collagen gene expression which is strongly linked to the proliferative state of arterial SMCs. B-myb, a member of the myb gene family, inhibits collagen and elastin gene expression in arterial SMCs without promoting progression of the cell cycle. These findings suggest that B-myb is an intracellular mediator of signals relating the SMC growth state to the level of matrix gene expression, and specifically driving the downregulation of elastin and collagen gene expression in proliferating SMCs. As K.E. Kepreos and G.E. Sonenshein have referred to in their chapter, B-myb might be an anti-fibrotic gene in vascular SMCs in the treatment of restenosis.

Collagen IV molecules assemble into a network of glomerular basement membrane though various types of interactions. The antigens that cause the nephritis have been identified as a short stretch of amino acid sequences of the NCI domains of the $\alpha3(IV)$ and $\alpha4(IV)$ chains. Measurement of autoantibody titer in the blood of nephritis patients will be possible in the near future, and the minimum requirements of an epitope for the nephritogenic antigen, the pathogenetic mechanism of nephritis, and possible therapeutic procedures will be determined. Information recently obtained in extensive studies on collagen-induced arthritis by L.K. Myers will be helpful in subsequent investigations.

Thus, knowledge of the structure and function of molecules and the regulation of gene expressions resulting from progress in the technologies of protein chemistry, cell biology, and genetic analysis of extracellular matrices is contributing greatly to our current understanding of diseases of the connective tissue. The function of this tissue, however, is diverse; based on the assembly of tissue and stage-specific extracellular matrices and on the regulation of and interplay between matrix-specific gene products, it seems that the application is limited to typical hereditary diseases like Marfan syndrome.

More information is needed not only about hereditary diseases but also the pathogenesis of age-related disease. Ageing is a determinant factor which influences the expression of matrix-specific gene products in degenerative disorders(OA) and fibrotic diseases including generalized atherosclerosis and restenosis of coronary atherosclerosis. The development of new technolgoy from an essentially different point of view will greatly aid our control.

Finally, I would like to thank Eisai Co., Ltd. for its long financial support to the Japanese meeting on Elastin Research and Symposium on Extracelluar Matrix-Cell Interaction.

Subject Index

371